Tropical agriculture

TROPICAL AGRICULTURE

The development of production fourth edition

Longman London and New York

Longman Group Limited
Longman House
Burnt Mill, Harlow, Essex, UK

*Published in the United States of America
by Longman Inc., New York*

First published 1982

British Library Cataloguing in Publication Data
Wrigley, Gordon
Tropical agriculture. — 4th ed.
1. Agriculture — Tropics
I. Title
630'. 913 SB111

Library of Congress Cataloging in Publication Data
Wrigley, Gordon.
Tropical agriculture.

Bibliography: p.
Includes index.
1. Agriculture—Tropics. I. Title.
S604.37.W74 1981 630'.913 81-8195
ISBN 0-582-46037-9 AACR 2

Printed in Singapore by Singapore National Printers (Pte) Ltd

CONTENTS

ACKNOWLEDGEMENTS

In the 20 years since I wrote the first edition of this book, the tropical world has seen many changes, mainly political, but fortunately some technical, which I trust are reflected in this new edition. Much of the important earlier work which remains the basis of tropical agriculture is still recorded here. The large increase in oil prices, inflation, and the serious reduction in the agricultural exports of many important tropical countries, have also meant a change in emphasis in this new edition.

In addition to the many who gave me their help and advice in preparing the previous edition, I have again been fortunate in having specialist assistance. The late Professor Hardy still continued to write even when his eyesight was failing, and offered me the free access to his many publication. The improvements to Chapter 1 owe a lot to 'Fred'. Anthony Johnston, Director of the Commonwealth Mycological Institute brought Table 4.5 into line with modern nomenclature and Dr Parker and his staff at the Commonwealth Institute of Entomology did the same for Table 4.6. Dr Howard Moore and Bernard Ryle improved the form and presentation of the Introduction and early Chapters Professor N. W. Simmonds cleared up a number of points in Chapter 3: and Dr G. A. Matthews checked the application section in Chapter 4.

John Purseglove whose four volumes on Tropical Crops I found invaluable, went meticulously through the whole draft, drawing my attention to certain errors and making important suggestions. He also brought the plant names and that I have used the past 40 years into the current nomenclature.

Once again Mrs Franklin put this edition together from illegible copy. The photographs are acknowledged individually, but I am particularly indebted to Rose Innes for his contribution to the section on natural grazing.

Despite every care in producing this book errors will inevitably appear and I will be grateful if these are drawn to my attention.

Linton, Cambridge, 1980.

INTRODUCTION

As a result of the world's population explosion, about twice as many people will need feeding in the year 2000 as were fed in 1960. This increase is now most rapid in the tropics, especially in Asia. The application of advances in medical science is largely responsible for this population growth and, by the same token, it is only a better understanding and a rapid application of advances in crop and animal production that can offer any hope of feeding these extra mouths. New industries cannot be developed with sufficient speed to employ any significant proportion of this rising population, so not only must food production be raised, but employment on the land must be increased.

Agriculture must not only provide food but in most tropical countries the cash on which the economy is based. The 'miracle' varieties of the 'Green Revolution' of the 1960s have made a contribution but not banished the prospect of famine in our time, and these once frequently used terms are now seldom heard. Sir Joseph Hutchinson (1970) suggested that the then current sufficiency of food in India had been more an evolution than a revolution, resulting from a steady increase in the use of nitrogen fertilizer and favourable monsoon rains, and he warned what might happen should the monsoons be poor or end early.

World food production has to meet two needs. First, a balance between the relatively constant demand per head and the variable yield per hectare resulting from seasonal variation; and second a steady increase over time as population increase continues (Hutchinson 1973). When a simple agricultural community produces enough for its needs, it makes no effort to produce more because there is nothing to do with it. Even if a market is available, a very small surplus often causes prices to slump, and the greater efforts of the growers are rewarded with a negligible return. Year-to-year variations in yields and prices can be evened out by storage, which is expensive; and local famines can be alleviated by mobility of stocks, but this is difficult because trade routes are often not established, and the people in need have not the money to buy.

The objective of agricultural development should be to provide a better life for the people, not just to ensure they do not die from starvation. Movement of people from the land and Western-style mechanization is often advocated as the solution for the development of tropical agriculture, but it is a poor development that moves people off the land to join the unemployed in the tropical urban slums. Mechanization in India created a few larger and more prosperous farmers at the expense of a lot of small farmers who joined the workless in the cities, and the same process is taking place on a larger scale in Brazil.

Agricultural development should be planned to make the fullest use of the limiting national resource. In tropical countries this is often foreign exchange, sometimes land, but rarely labour. For this reason where developments require large amounts of foreign exchange for machinery, fertilizers, chemicals, etc., it must be clear that such investment will make better use of labour and not reduce the labour requirements, the abundant resource. Where land is abundant, farmers, when encouraged by the right incentive to increase production, will just do more of what they are already doing. A small farmer must regard any of his own money spent on fertilizers, insecticides and better seed, as risk capital which in his society earns a high return. Consequently he expects that high return, perhaps 300 per cent, before he will risk his capital. A return of 10 per cent is of little interest, and when one reads research reports showing such returns, remember the statistics are not as significant as the economics. Small farmers require fertilizers and chemicals in small packets which are costly. Further, they require them in the local villages, which means high distribution costs. The local shopkeepers need a high margin on sales, particularly if selling on credit. Thus the cost of such materials is much higher than on farms in developed countries, from where they normally have to be transported. Often the true cost is camouflaged by government intervention either in the form of a subsidy or service, recouped by a high tax on the product when exported.

The technical knowledge needed to keep agricultural production ahead of population demand, at least until the end of this century, is probably available or can be readily adapted. The ways are obvious but sufficiently important to warrant repeating here, and explaining in greater detail later.

1. The prevention of soil erosion from the hills, in the crops, on the head waters of dams; and in pastures.
2. The fuller use of water resources, with emphasis on the conservation of rainfall - including mulching, and the development of irrigation under systems of complete water controls and economy in irrigation water use.
3. Growing acceptable crop varieties that make more efficient use of light, distribute the increased assimilates more to the desirable parts of the crops (e.g. cereals with a high grain : straw ratio), have a high protein content, and are tolerant of pests and diseases.
4. Timely planting to make the best use of the rainfall and soil nitrogen.

5. The maximum use of available organic manures and, where economic, purchased fertilizers.
6. Crop spacing, crop rotation, intercropping and relay cropping to make the maximum use of solar energy throughout the year and protect the soil from rainstorms and the hot sun.
7. Protecting crops against pests, diseases and competition from weeds.
8. The development of animal husbandry, including disease control, seeding of legumes in pastures, planting legume browse crops, improving the supply of drinking water, better utilization of grass and other local feeds not used by man, such as molasses; returning the manure to the crops where possible.
9. More efficient use of labour using improved machinery and tools, both hand and mechanical, particularly at periods of high labour demand.
10. A greater attention to the crop after harvest to reduce losses in threshing and storage.

As agricultural production increases, new job opportunities are created: transport drivers, mechanics, shopkeepers, implement makers and work in plants associated with agriculture such as fertilizer factories, seed cleaning and packaging plants, cotton spinning mills, coffee, tea and tobacco factories, meat and milk handling plants, breweries, fruit and vegetable canneries; and so the list grows.

In some cases, the changes needed to increase production cannot be introduced until certain services have been established: for example the growing of hybrid maize requires an organized seed industry, and the production of cash crops requires a buying network of stores, and roads to move the produce.

The continued development of agriculture depends upon the continuity of research effort. Fortunately, there are international research and training centres financed by a number of organizations and governments working in the tropics on food crops. The main ones are given below:-

Institute	**Main areas of activity**
International Rice Research Institute (IRRI), Los Banos, Philippines	Tropical rices, especially in Asia
Centro Internacional de Mejoramiento de Maiz y Trigo (CIMMYT) el Batan, Mexico	Maize and wheat; also other cereals (triticale, barley, sorghum)

Centro Internacional de Agricultura Tropical (CIAT), Palmira, Colombia	Cassava and beans; also maize and rice in collaboration with CIMMYT and IRRI. Also beef, pastures and farming systems. Latin American emphasis
International Institute of Tropical Agriculture (IITA), Ibadan, Nigeria	Grain legumes, roots and tubers; also maize and rice in collaboration with IRRI and CIMMYT. Also farming systems. African emphasis
Centro Internacional de Papa (CIP), Lima, Peru	Breeding potatoes for tropical latitudes
International Crops Research Institute for the Semi-Arid Tropics (ICRISAT), Hyderabad, India	Sorghum, millets, grain legumes (pigeon pea, chickpea). Dry land farming systems
International Laboratory for Research on Animal Diseases ILRAD. Nairobi, Kenya	Animal diseases
International Livestock Centre for Africa (ILCA), Addis Ababa, Ethiopia	Livestock production systems
International Centre for Agricultural Research in Dry Areas (ICARDA), Beirut [Egypt, Syria, etc.]	Barley, wheat, lentils and mixed farming systems. Sheep. Particular emphasis on Mediterranean areas rather than tropics

The first of these centres, the International Rice Research Institute established in 1960 has shown what is possible with the short rice varieties. CIMMYT was based on the Rockefeller Foundation centre founded in 1943.

Many governments, when examining their budgets, may question the high cost of research in a long-established industry, as Tanzania closed down Lyamungu as a coffee research centre, but the need for research is greater today than it ever was. For example a new disease or the disturbance of established markets can frequently be countered by research. Research expenditure can either be a most fruitful investment or an expensive extravagance, depending upon its direction and control and success. A practical advance which shows a 5 per cent increase in a major export crop will finance a research station for many years, but it is vital to abandon an investigation as soon as it becomes obviously unfruitful. Agricultural research is long term; for example in plant breeding, it can take 10-15 years from making a selected cross to distributing the new variety.

In order to achieve the growth in production, developing countries will need very large amounts of money for the necessary inputs as calculated by the Indicative World Plan (IWP) (FAO 1970, p. 180). Finance necessary to execute sound development schemes is available from the World Bank, private banks, and organizations such as the Commonwealth Development Corporation. Technical aid is also available from FAO, and national schemes such as the Overseas Development Administration in the UK and the American AID programme.

The more one sees of the traditional agricultural practices, apparently methodless and jumbled, the more one realizes how well adapted most of them are to continued production with the tools available, despite the many natural hazards (Allan 1965). Under increased pressure to feed a steadily growing population and a desire to produce more cash crops, these systems will often break down. This is seen clearly with nomadic cattle herding and shifting cultivation where the pressure of population gradually makes it more and more difficult to move to new areas. Without effective extension workers it will be difficult to advise farmers how to prevent this breakdown. However the speed at which certain appropriate new technologies have been adopted by the tropical farmer has been a surprise to many.

Whenever a change is made in one part of an existing agricultural system, the rest of the system should be examined and if necessary modified. For example, the new short-strawed rice varieties yield not much more than the best local variety unless fertilizers and insecticides are used, and irrigation water is controlled, hence the disappointment in many areas with these new varieties.

As a farmer increases his yield and sells his surplus, he is able to invest in animal or mechanical power for cultivation, irrigating or threshing which, if used wisely, further increases his production and takes more drudgery out of his life. However, any new technique must fit the grower's economic and social system, and must not put his whole way of life at risk; hence in some countries, the 'one-step-at-a-time' approach is being taken to development, as opposed to the 'package deal' of seed, fertilizer, insecticide, irrigation, etc., being introduced together.

To increase production a farmer needs an incentive. In general, the most effective incentive is money, yet all too often the grower gets too small a share of the market value of his produce and in some countries has to move it over the national boundary to obtain a fair return. If a government requires increased production they must of necessity set up a fair marketing system, preferably with guaranteed prices to encourage production to rise above subsistence level. Usury and restrictive systems of share cropping must be dealt with. To benefit from modern techniques, a fair system of land tenure, offering security to the competent farmer and the integration and redistribution of land holdings can often improve efficiency. Many of the changes will require capital, and the farmer may need credit facilities to tide him over until the benefits of improvements have been harvested. The reader will appreciate that marketing, credit and land tenure are very complex problems which, though vital to the theme, are outside the scope of this book.

The tropical farmer with relatively little education, and suffering from a complexity of diseases often aggravated by a low level of nutrition, cannot be expected to farm as efficiently, or to complete as long and hard a day's work as his counterpart in temperate countries. An improvement in health, while accentuating the population problem, goes a certain way in assisting in its solution.

In the early days of tropical planting, too many mistakes were made by transporting temperate farming methods to the tropics, with the consequent desolation from erosion, insolation, storms, etc. The pioneers of those days had no alternative, but to repeat such mistakes, as regrettably still occurs, is inexcusable. Tropical agriculture must rest on its own foundations and temperate techniques adapted and modified. Some systems used in Mediterranean and Middle-East countries are more appropriate than those from America or Western Europe. My approach has been to look at agriculture from within the tropics, few examples have been taken straight from temperate agriculture.

Specific agricultural operations and methods of cultivating individual crops have not been dealt with in this book as they vary considerably from region to region, and are effectively covered by local publications and the many excellent specialized publications in many languages now available on tropical crops and stock. The emphasis has not been on describing the ways in which crops and livestock are currently managed but more on where and how improvements might be made.

All the money that is spent on research and knowledge gained is wasted unless the results can be translated into practice, hence a well-organized education and extension service is essential. With the rising tide of literacy and more extensive ownership of transistor radios to receive local programmes, traditional methods of extension can be supplemented. The radio has a particular advantage in that it can deal both with seasonal topics and unforeseen situations. Films already play an important part in certain countries, and television may become more widely available to supplement extension service visual aids. The

FAO Fertilizer Programme showed how important it is to take the demonstration from the government farm to the farmer's own land where he can see and evaluate the benefits.

However, the extension service should never attempt to persuade the farming community to introduce changes into their system that are not soundly based and will show a worthwhile return to the farmer. In judging this, it is wrong to assume that the farmer does not value his leisure and is, therefore, likely to carry out proposed measures rather than do nothing; on the contrary, if a new technique such as the use of a herbicide means that a small farmer spends less of his days toiling with a machete, he may well regard this as a desirable improvement in his standard of living, and a worthwhile investment. Extension officers must remember that a grower's first consideration is the security he desires for himself and his family.

1 CROP ECOLOGY

'A world of green such as I had never seen before on earth, not even in my dreams' Charles Kingsley. 'At last - a Christmas in the West Indies.' This description of Trinidad is a generally-held view of the tropics, but to regard the tropics as an area covered by tropical rain forest is sadly incorrect. Indeed most of Africa is covered with grassland savanna. Similarly the soils of the tropics were long regarded as mainly red earths or laterites. The seasons of the tropics are described as the 'dry season', or the 'rainy or monsoon season', as rainfall is the most variable feature of tropical climate. 'Summer' and 'winter' though sometimes used are inappropriate as the day length changes little and the mean monthly and daily temperatures vary little. The growing season in the tropics is less restricted by temperature than in temperate climates. It is therefore important to examine the soil and climate in the tropics, to see how these ecological factors influence which crops are grown in the tropics and how they are grown.

SOIL

The soil cover of the earth is a very thin layer which provides support, food and water for plant growth. All soils consist of inorganic and organic matter, air and water. They are formed by the effect of climate, topography and living organisms on the parent rock over a period of time.

The original soil may remain on the surface, or may be subsequently covered by a deposit of volcanic ash as in much of Java, flood deposits from rivers as in the Central Plain of Thailand, or by wind-borne soil (loess). Soil is a material which is very much alive with millions of bacteria, fungi, algae and microscopic animals at work in every small part, breaking down plant and animal remains and carrying out many other complex chemical processes. In addition, there are the earthworms and termites which are important in moving the soil, aerating it and processing the dead plant material. Many tropical soils are acid, with small amounts of available nutrients.

For a fuller account of tropical soils, see Hardy (1970a), Mohr *et al.* (1972), Young (1976), Sanchez (1976), Bridges (1970), Buringh (1979).

Rocks

The rocks from which soils are formed can be put into three groups.

1. *Primary or igneous rock* which originally formed the earth's crust when it had cooled sufficiently to crystallize and harden into a mosaic of different mineral crystals. The slower the rate of cooling of the magma, the molten material of the earth, the larger the crystals formed. The deep seated rocks being cooled slowly formed large crystals; ejected volcanic material being cooled quickly formed small crystals, which may be too small to see without a microscope giving the rocks a uniform appearance. Many igneous rocks have both large and small crystals as a result of slowly cooling magma being ejected on to the earth's surface and cooling quickly (pyrochlasic rocks). Basalts are fine grained basic rocks of volcanic origin; granite, the most abundant of all acidic rocks, is coarse grained.
2. *Sedimentary rocks* were formed when the products of weathering of the igneous rocks, largely fine particles of rock, were transported by wind, or water, or in solution, and deposited.
3. *Metamorphic rocks* are igneous or sedimentary rocks, changed by heat and pressure caused by folding of the earth's crust or the intrusion of molten magma.

Rocks of all types are aggregates of minerals, hard inorganic materials with characteristic crystalline shapes and molecular constitutions, though their chemical composition varies due to the interchange of certain elements in these complex salts.

Only 10 minerals are really important in the formation of soils from igneous rocks, and of these 8 are silicates (Hardy 1970a).

Name	**Constituent elements**
Quartz	Si O
Feldspar (Orthoclase)	K Al Si O
Feldspar (Plagioclase)	Na Ca AlSi O
Mica (Muscovite)	K Al Si O
Mica (Biotite)	K Mg Fe AlSi O
Amphibole (Hornblende)	Ca Fe Mg Si O
Pyroxene (Augite)	Ca Fe Mg Si O
Olivine	Fe Mg Si O
Magnetite	Fe O
Apatite	Ca P O

Igneous rocks are formed from two or more of these minerals intercrystallized; for example granite, one of the most common crystalline rocks of West Africa, contains about 70 per cent feldspar, 20 per cent quartz, and 10 per cent other minerals such as mica or amphibole. The only essential element which the minerals forming primary rocks do not provide is nitrogen. Apatite is the sole source of phosphate in nature, while the primary sources of potash are weathered feldspars and micas. Other elements including trace elements such as sodium or manganese, are introduced into these rock structures by isomorphous substitution, where one ion replaces another of approximately the same size. As the introduced ion is not identical in size, it gives the crystal a structural weakness which speeds breakdown in weathering. The valency of the replacing cation may be different, and an extra cation may be taken up to restore electrical neutrality, again modifying the structure. For example the central silicon ions of feldspar, with a valency of four may be replaced by aluminium with a valency of three, or magnesium with a valency of two and the extra negative charges from the oxygen are satisfied by a mixture of potassium, sodium or calcium ions.

Quartz, the only mineral which makes no nutrient contribution to the soil, is extremely resistant to weathering, particularly chemical weathering, and is only broken down to smaller particles. For this reason quartz is the most abundant mineral in sedimentary rocks. Sand is practically all quartz.

Weathering of rocks

The weathering of rocks can be either by physical means, during which the material is broken down into smaller and smaller particles by heat, cold, wind and water; or chemical, where the small quantities of weak acids dissolved in rainwater cause hydrolysis, oxidation, carbonation and other chemical reactions with the minerals. As the constituent minerals react at different rates, the most easily weathered, such as olivine, break down first and are leached out, causing the rocks to disintegrate.

According to Van t'Hoff's law, the rate at which rock minerals are decomposed by hydrolysis increases exponentially with temperature; for example clay forms four to five times faster at 30 °C than at 10 °C. Thus soil formation is much more rapid in the tropics than in temperate regions and is not interrupted by a cold winter season.

The material formed by the breakdown of the rocks may remain to form sedentary soils, or it may be blown elsewhere by wind, or transported by water to form alluvial soils. Though alluvial soils occupy only 2 or 3 per cent of Africa and South America, they form the major soils of Asia, occurring on the flood plains, river terraces, coastal plains, deltas and old dry lakes. As these alluvial deposits occur on flat land they are ideal for rice cultivation and irrigation and support a high population density. Frequently the water table is high and the organic matter content of the soil higher than non-alluvial soils of the same physical type found nearby.

Where waterlogging occurs, and anaerobic conditions develop, peat accumulates. In the reducing conditions resulting from the absence of oxygen, and in the presence of organic matter, the ferric iron compounds are chemically converted to more soluble ferrous compounds which leach out leaving colourless minerals behind, giving gley soils their characteristic grey colour (see p. 17).

Tropical soils

With the high temperatures and the heavy rains of the tropics soil formation is a quicker process than in temperate regions. The resultant soils are frequently deep but they are generally poorer than in temperate zones, since the nitrates and bases have often been leached out. The appearance of tropical vegetation is often deceptive, and the new arrival in the humid tropics, like Charles Kingsley, seeing the luxuriant forest growth covering the hills often fails to appreciate that it is maintained on a rapid circulation of nutrients between the leaves and the topsoil. The nutrient status of the different soil types, for example, has not exercised much effect on the distribution of forest vegetation in Trinidad in the way it has influenced the ground flora of cocoa plantations and the weed flora of the cane fields. Once this cycle is broken by clearing for cropping and the nutrients harvested or leached out, deterioration to a low state of fertility follows rapidly, and a long period of fallow is required for recovery. Maintenance of fertility in the large areas of Brazil currently being cleared from forest will be a serious problem where the land is used for annual cultivation.

The tropical farmer, ignorant of soil science, knows the potential of the local soils by their vegetation. A very varied flora is a good indication, a restriction of species such as a grass cover may result from a lack of soil depth, sometimes caused by a pan, or a limitation set by pH or nutrients. Certain grass species are indicative of a poor or a fertile soil. In southern Uganda a good stand of elephant grass (*Pennisetum purpureum*) is only found on a deep fertile soil suitable for coffee and bananas. When the soil fertility has been reduced by overcropping, neither elephant grass nor bananas will grow satisfactorily and the land is colonized by *Cymbopogon afronardus* and other grasses indicative of low fertility. If the land is allowed to rest, elephant grass will re-colonize the area when the fertility level has risen sufficiently.

When studying an area of land it is essential to examine the soil under the surface. Ideally a series of deep soil pits should be dug, but this is a major task, and where the soil type changes over short distances, soil pits will not give an accurate picture. An alternative is the Oakfield soil sampler, an excellent tool for taking slim soil cores quickly to a depth of a metre. If the soil is damp the feel of the soil will indicate whether it is sandy or a clay. A damp clay soil can be polished with the finger nail. A sandy subsoil will not hold sufficient moisture for a tree crop in a dry season, and a heavy clay soil will probably be waterlogged in the wet season. The presence of a red-brown colour in the core shows that the soil is free draining, and a bluish grey mottled or streaked colour shows the drainage is impeded. These are important indicators when choosing a site for tree crops. Cases occur where the soil appears to be good but will not support much vegetation or grow crops due to a high acidity or gross deficiency of nutrients.

Soil profile. During the formation process, material is leached from the surface layers or 'horizons', and either carried away in the drainage water or deposited in the lower horizons. These horizons can be studied by digging a soil pit to expose a vertical section, or profile, down to the parent rock. From the surface downwards the horizons are as follows:

A horizons - The topsoil from which materials are leached - eluvial horizons

B horizons - The subsoil where materials accumulate - illuvial horizons

C horizons - The primary products formed from the parent rock by weathering

There may be one or more A and B horizons but only a single C horizon. Some descriptive systems use suffixes, such as Bg where g denotes a gleyed or mottled B horizon.

Erosion may have removed the A and sometimes the B horizons and in some soils a B horizon never forms. Soil profiles can develop to great depths, 20 or 30 m or more, in the tropics.

A typical soil profile with cocoa roots is shown in Fig. 1.1.

Soil texture. The inorganic or mineral portion of the soil, which results from the weathering of the parent rock, or alluvium deposit, has particles varying in size from inert stones and rocks to colloidal clays. For the purposes of mechanical analysis, the 1927 International Agreement classified the particle sizes as follows:

Gravel	over 2 mm diameter
Coarse sand	2–0.2 mm diameter
Fine sand	0.2–0.02 mm diameter
Silt	0.02–0.002 mm diameter
Clay	less than 0.002 mm diameter

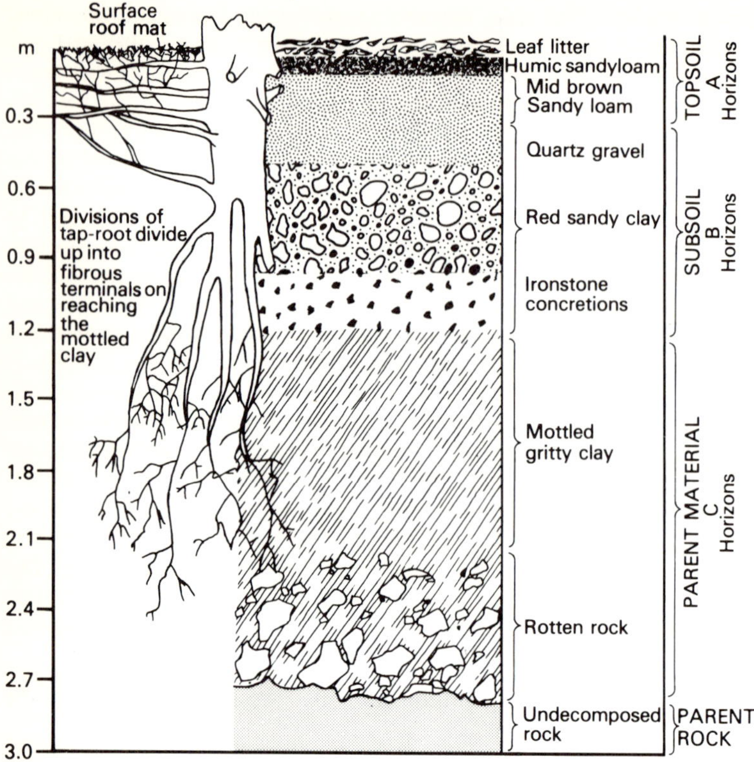

Fig. 1.1 Diagrammatic profile of the red well-drained soil of the uplands showing how the typical cocoa root system exploits the various layers. Ghana. (Charter 1949)

More detailed classification systems have been devised, but this simple system suffices in most situations.

Soils are normally a mixture of sand, silt and clay in varying proportions which determines their characteristics, feel and response to cultivation, rainfall and drought. Soils described as sandy, have a mixture of coarse and fine sand with so little clay that it is loose when dry and not sticky when wet. When rubbed between the fingers it leaves no film on the fingers. A moistened sample of sieved sandy soil cannot be rolled into a ball. Coarse sandy soils are easy to cultivate but drain freely and suffer from drought. They are usually acid. A silt soil feels soapy to the touch and a clay soil is sticky and plastic when wet and gives a shiny surface when polished with the nail. A moistened clay sample can be rolled into a cylinder which can be bent into a ring or can be moulded into any shape. Clays and silts are difficult to cultivate, being sticky when wet and clays bake hard in the dry season. Soils described as loams have a blend of clay, silt and sand; 'sandy loams', 'clay loams' and 'silty loams' being used when one of the three constituents is dominant. A sample of a sieved loam soil can be rolled into a cylinder but if the cylinder is bent it cracks and breaks up.

Clay minerals

The particles less than 0.002 mm in diameter form the very important colloidal fraction of the soil. The smaller the particle size the more active they are both chemically and physically. One reason for the greater physical and chemical importance of the clay fraction is that the total surface area per unit mass or volume of material increases with fineness of division; for example, 0.1 μm clay particles have 200 times the surface area of silt particles, 0.02 mm in diameter (Rose 1966). As the clay mineral particles are so small, the electrical forces of the surface molecules are sufficiently strong to maintain the clay particles in a colloidal state.

In most soils developed in the humid tropics, feldspars and mica have been completely weathered, and the clay fraction is made up of clay minerals with iron oxides and sometimes aluminium hydroxide. The clay minerals are the most important mineral constituents of the soil and are identified by using an electron microscope. They have a layered flat crystalline structure which is revealed by X-ray analysis.

There are three main clay minerals, kaolinite, montmorillonite and the hydrous micas such as illite with transitional forms between each. All the clay minerals are built up from layers of aluminium atoms and silica. The simplest type of clay mineral is kaolinite (kaolin or china clay) which has flat crystal flakes roughly hexagonal in shape. These flakes are built up of double layers, one of a silica lattice and the other an aluminium lattice, bonded together by shared oxygen atoms - a 1 : 1 lattice. Each of these double layers is held to the next by OH ions at small and fairly constant distances in a non-expanding lattice. Kaolinite, which is formed from the weathering of feldspars, is the dominant clay mineral in laterite soils and throughout West Africa. The non-expanding or contracting characteristics of kaolin make it ideal for pottery as the moulded clay retains its shape and does not crack on firing.

Montmorillonite has a 2 : 1 layer lattice structure with two layers of silica on the outside of an alumina layer. The bonds between the three layer units are relatively weak and the lattice can expand and contract on wetting and drying more than the kaolinite lattice.

Illite is a similar 2 :1 lattice clay but with a fixed distance between the units as kaolinite.

The characteristics of these three clay minerals are summarized in Table 1.1.

Table 1.1 Clay Minerals

Characteristic	Kaolinite	Montmorillonite	Illite
Stability	High	Low	High
Swelling and shrinking capacity	Low	High	Low
Cohesiveness	Low	High	Low
Water-holding capacity	Low	High	Medium
Cation exchange capacity	Low (5–15 me/100 g)	High (15–150 me/100 g)	Moderately High (about 30 me/100 g)
K fixing capacity	Low	Appreciable	High
Ammonium fixing capacity	Low	Appreciable	High
P_2O_5 fixing capacity	High	Low	Low
Ca + Mg availability	High	Low	Low
K availability	Low	High	High
P_2O_5 availability	Low	High	Medium
Humus combining	Low	High	Low
Smell	Earthy	None	None

After Hardy (1970a).

Classification of tropical soils

Soil surveys are the essential basis of agricultural development and land use classification. Based on soil-forming factors, Sibirtzev at the end of the nineteenth century suggested the classification of soils into three major groups with a broad application geographically.

Zonal soils. These have well developed characteristics resulting from the influence of the soil forming factors: climate, topography, vegetation and the age of the parent material. Hardy (1970a) divided zonal soils into four groups.

1. Desert soils. These are formed in areas of high temperature and low rainfall. Little organic matter accumulates and little leaching occurs.

2. Chernozems. Chernozems are formed in areas where the temperature is lower than in the deserts and rainfall higher though generally less than the potential evapotranspiration most of the year. The natural vegetation is grass. Organic matter accumulates in the upper layer and as there is no leaching, chernozems are alkaline and black, as for example the black soils of north-west and south-east

Ghana and north-west Nigeria around Sokoto and Lake Chad, the Regurs of the Deccan plateau, and the self-mulching soils of the Darling Downs in Queensland. These soils are generally heavy as montmorillonite is the predominant clay mineral; they are difficult to work with hand tools but are often well suited for irrigated rice and sugar cane (Dudal 1965). They have a high cation exchange capacity.

3. Podzols. Podzols are formed in areas where the temperature is lower and the rainfall higher than where chernozems develop. Such soils are only found at high altitudes in the tropics, as in the Andes and south Mexico. The natural vegetation is forest. As the rainfall generally exceeds the potential evapotranspiration leaching of bases occurs and the upper soil layers are acid and humic. The A2 layer below this is like a layer of ashes owing to the iron being leached out and deposited in the B horizon, often as reddish brown streaks in old root traces.

4. Latosols. These soils are also formed where the rainfall exceeds the potential evapotranspiration but where temperatures are higher than where podzols form. Such conditions are the most common throughout the tropics. Organic matter breaks down as fast as it accumulates, and mineralization is very rapid. Bases and silica are leached out. The colour of the soil profile ranges from yellow, common in Malaysia, to brown or red, so common in East Africa, according to the degree of hydration of the ferric oxide. These soils are free draining. Latosols are sometimes divided into the heavily leached acid oxisols which appear to be developed where the rainfall exceeds 2,000 mm/year; and ochrosols which have a neutral or slightly acid topsoil where the rainfall is below 1,500 mm/year (Williams and Joseph 1970). Latosols vary from coarse sands to fine clays and have a low cation exchange capacity, their fertility depending upon their organic content. As the soils are deep with a good structure and not easily eroded, they are popular for tree crops and general cropping.

Intrazonal soils. These are soils whose characteristics have developed under the dominating influence of some local factor of relief or parent rock, over climate and vegetation. This group includes soils with excessive amounts of soluble salts (halomorphic) as found in the desert regions of East and West Africa; soils with a high calcium content (calcimorphic) as occur on the Deccan plateau; on the East African coast and in Zanzibar, but not in West Africa; and soils associated with wet conditions (hydromorphic).

Hydromorphic soils are widespread in the tropics and include peats and humic gleys which form under conditions of poor drainage, sometimes in low rainfall areas. Hydromorphic soils are common in Africa, including the swamps of Gambia and Sierra Leone and the papyrus-clogged lakes of East Africa.

Poor drainage conditions are frequently caused by a hardpan of an impervious clay layer or a humus/iron deposit. When the soil is waterlogged for long periods, anaerobic conditions are set up, and the effect on the soil profile is called gleying. Under these anaerobic or reducing conditions ferrous compounds are produced which are bluish-grey or greenish grey. Being more soluble than the ferric salts these ferrous salts are often leached out, leaving grey minerals behind. Where the soil dries out for part of the year the grey soil is mottled with rust-coloured ferric oxide.

Coulter (1950) estimated that there are nearly a million hectares of deep peat soils in Malaya.

If drained slowly peat soils can be used to grow sugar cane, pineapples, cassava, vegetables and other crops which can tolerate a high water table.

Hydromorphic soils which have developed in a region with marked wet and dry seasons are called ground water laterites which are the soils of the West African savanna areas and also common in India.

Azonal soils. These soils have no developed profile and are usually of recent origin; they include:

1. Alluvial soils. These have already been mentioned. Generally their fertility is constantly being replenished by fresh water-borne deposits. Their characteristics and fertility are very variable and depend largely upon the soils and parent rock in the catchment area.

Alluvial soils support dense populations of rice growers in South-East Asia as in the Red River, Mekong and Irrawaddy deltas, the central plain of Thailand where the floating rice is grown, and in Indonesia and Burma. Tropical Africa has no delta plains comparable in fertility to those of Asia. The Niger, before reaching the delta, carries very little alluvium. The levees alongside rivers which periodically flood are formed from the heavier particles carried by the flood water and in the uncultivated state are gallery forest. As they are easily accessible by boat, fertile, well drained and moist, they were frequently the first areas settled. They are ideal soils for growing valuable tree crops. The saline alluvial soils along coasts are predominately mangrove swamp. Certain alluvial soils, rich in sulphur, can become very acid after drainage.

2. Regosols. These are shallow soils over unconsolidated parent material deposited by means other than water; for example, fresh volcanic ash and wind-borne loess.

3. Lithosols. These are shallow stony soils formed from freshly and incompletely weathered rock mass. Occurring mainly on slopes, they have low water acceptancy and low organic matter, they are therefore dry and of low agricultural value. They commonly occur where the dense vegetation on sloping land has been cleared for cultivation. The topsoil is washed away and the new soil begins to develop on the exposed parent rock. In the drier tropics they are covered by undulating grassland.

There are at least five other major systems of soil classification widely used in the tropics, but no universally accepted system of nomenclature. These are reviewed by Sanchez (1976) and Young (1976).

In these soil classifications and research reports a number of terms are frequently used, and the more common ones are summarized below. The names in brackets are the approximations in the US system.

Description of soil types in the tropics

Acrisols (Ultisols - US). Soils with clay particles deposited in the B horizon (argillic horizon). Less than 20 per cent base saturation. These soils are midway in weathering between luvisols and ferralsols. Originally classified as latosols. Form the upland soils of South-East Asia.

Andosols (Andepts - US). Soils developed from recent volcanic material.

Arenosols Very sandy soils with a distinct B horizon. Less than 15 per cent clay.

Cambisols (Inceptisols - US). Brown earths without a clay particle deposit horizon (argillic) but with a B horizon that has weathered *in situ*.

Ferralsols (Oxisols - US). Highly weathered soils of the humid tropics. Clay minerals dominantly kaolinite and free sesquioxides. Cation exchange capacity (CEC) of clay fraction less than 16 milliequivalents (me)/100 g which means the virtual absence of 2:1 clay minerals and few minerals for weathering. Can be red (rhodic), yellow (xanthic), normal (orthic), plinthic, humic, have a very low (1.5 me/100 g clay) CEC - acric, may be slowly forming ironstone. Common in the Cerrado and Amazon of Brazil and Central America.

Fluvisols (Fluvents - US). Alluvial soils developed from recent alluvial deposits, the horizons being formed by the layers of deposit. As they are largely on flat land they are very suitable for rice and support high population densities.

Gleysols. Soils where the hydromorphic (poor draining, mottled, dark grey or peaty) properties are dominant. In the savanna zones of Africa mainly permanent grazing. Can be used for rice. Gleyed profiles on recent alluvium are fluvisols.

Histosols. Peat soils which are predominantly organic matter, and muck soils where the organic matter is mixed with mineral matter. Where permanently flooded form the papyrus swamps of Uganda and the Sudd in Sudan, and mangrove swamps of the coast.

Lithosols. Shallow soils with continuous hard rock less than 10 cm below the surface.

Luvisols (Alfisols - US). Soils with clay particles deposited in B horizon, but with base saturation over 50 per cent. Originally classed as latosols, in West Africa, India and Sri Lanka.

Mollisols (US). Soils high in organic matter, red-brown colour, soft when dry and over 50 per cent base saturation.

Nitrosols. Soils of the humid tropics derived from basic rocks. Soil profile is shiny due to clay skins.

Podzols (Spodosols - US). Soils usually sandy with a B horizon where iron sesquioxides and organic matter accumulate.

Rankers. Weakly developed shallow soils on consolidated rock.

Regosols (Entisols or Psamments - US). Weakly developed soils on unconsolidated materials, usually sandy. Horizons usually absent, though A horizon may be pale.

Rendzinas. Shallow calcareous soils on limestones, usually high in humus.

Solonchaks (Aridisols - US). Saline soils, the saturation extract has an electrical conductivity greater than 15 mmhos/cm. Form the deserts of Africa.

Solonetz. Alkaline soils, the B horizon has more than 15 per cent exchangeable sodium. Profile highly saline.

Vertisols (Grumusols - Brazilian system). Dark-coloured deep cracking clays with high CEC. Black cotton soils of Indian Deccan, Sudan Gezira, Ethiopia and Chad. 'Gilgai' in east central Australia. Difficult to cultivate with hand tools and need drainage. Easily eroded, usually ideal for irrigation.

Xerosols. Semi-desert soils. Pale coloured A horizon. Clear humic A horizon, organic matter 0.5 - 1.0 per cent.

Yermosols. Desert soils. Pale coloured A horizon. Weakly developed humic A horizon, organic matter less than 0.5 per cent.

The approximate extent of the major soil types in the tropics is given in Table 1.2.

Laterites (Latin *Later* - a brick)

For many years tropical soils were regarded as red soils and frequently called laterites. Today this term is much less used except as the correct term for 'a massive or concretionary ironstone formation nearly always associated with uplifted peneplains originally associated with areas of low relief and high ground water' (Bridges 1970). The 'original' laterites of South India are now classified as Alfisols, Inceptisols, or Ultisols in the US system.

Laterite is composed of kaolin and the oxides of iron and aluminium together with unweathered minerals particularly quartz. Laterite remains soft while moist and when fresh can be cut into blocks and dried in the sun to form a hard material suitable for house building as the

Table 1.2 Areas of major soil types of the tropics (million hectares)

	Africa	America	Asia	Total area	%
Oxisols (Ferralsols)	550	550	0	1,100	22.4
Aridisols (Solonchaks)	840	50	10	900	18.4
Alfisols (Luvisols)	550	150	100	800	16.3
Ultisols (Acrisols)	100	200	250	550	11.2
Inceptisols (Cambisols)	70	200	110	400	8.2
Entisols (Regosols)	300	100	0	400	8.2
Vertisols	40	0	60	100	2.0
Mollisols	0	50	0	50	1.0
'Mountain areas'	0	350	250	600	12.3
Total	2,450	1,670	780	4,900	100

Source: Adapted from Sanchez (1976), derived from Aubert and Tavernier (1972).

Note: Spodosols (Podzols) and Histosols though important in localized areas do not occupy sufficient areas in total to be mapped.

hardening process is irreversible. Laterite was used to build the temples at Angkor Wat in Cambodia. Hardy (1933) pointed out that laterites represented only a limited part of the soils of the tropics. Although they cover only 5 to 15 per cent of different tropical areas they have attracted much attention, largely due to the difficulty of farming these lands unless the laterite is covered by about half a metre of soil. With less top-soil it is difficult to produce worthwhile crops. Laterite soils are very common in Uganda and adjacent Tanzania. Massive laterite up to 10 m deep, hard and resistant to weathering forms the characteristic flat-topped hills covered with poor grass in the lake shore areas of Uganda (Fig. 1.2). Laterite is used in this area to make murram blocks for building and the murram gravel is used for road making. Where forest has been replaced by grassland the murram in the subsoil builds up more quickly as in the dry season the soil is hotter and drier than under forest. A lateritic horizon is formed near lakes and swamps where the ground water movements within the soil concentrate the iron and aluminium sesquioxides in a layer.

Catena(Latin – a chain)

Milne (1935) defined catena as 'a regular repetition of a certain sequence of soil profiles in association with a certain topography.' Milne used this

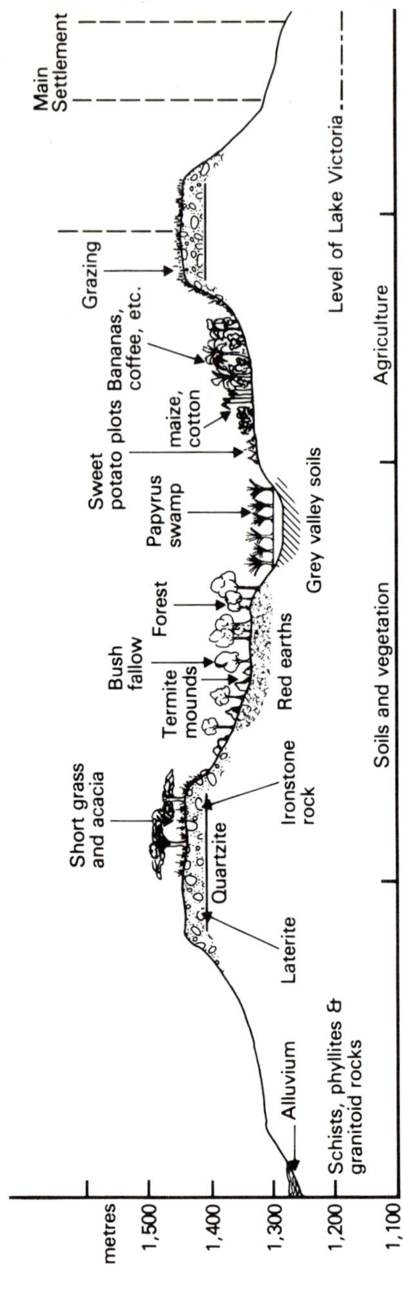

Fig. 1.2 Example of a catena in Buganda. (After Ker *et al* 1978)

concept of topographically controlled variations in soils to describe East African soils but it is applicable for any climate and is widely used in soil surveys. Ahn (1970) illustrates a number of examples of West African catenas. There are different catenas in the same climate varying with the original parent rock, and different catenas develop from the same parent material in different climates.

The lower slopes of a catena may be covered by colluvium, material which has moved down the slope by gravity, and the valley bottoms may be covered by alluvium, material which has been moved by rivers and streams and deposited in the valley. Alluvium may have originated at a distance from the site and is often a source of fertility. These deposits weather and form the local soils. This succession of soils down the slope, from the hillcrest to the valley, is used by farmers in locating their different crops and grazing.

In planning smallholder settlements where the derived soil type is a catena, though the fields must run along the contour to control erosion, each separate holding should run down the slope to include a part of each soil type in the catena.

Soil surveys

The soil surveys of the West Indies are good examples of the many expensive surveys that have been laboriously prepared in the past from a large series of soil samples often collected with great difficulty, mapped in many colours and left unused even though the information is very useful down to farm level. The Barbados soil survey is one exception (Hudson 1965). Besides providing a rational basis for the location of experimental sites, and advice on cane growing and crop diversification, the map was introduced and interpreted to the sugar cane plantation managers, most of whom requested a copy of their own land, and used it as a basis for choosing their varieties, cultivation techniques and reorganization of field boundaries.

Land capability classification

A practical system of classification is one based on its potential uses and limitations. The basis of many classifications is the one outlined by Klingebiel and Montgomery (1961) for the USA, adapted for local use in the tropics. Three broad groups are generally defined:

1. Land suitable for cultivation.
2. Land unsuited to cultivation, useful as forest or pasture.
3. Land unsuitable for productive use.

Each one may be subdivided according to suitability for mechanical cultivation, intensive production, restricted production, etc. The French have used 11 classes for African soils (Aubert and Fournier 1957).

The ultimate unit of each class suggests the suitable crops and management methods, or the system of conservation required. An example of this is Obeng and Smith's (1963) land capability classification for Ghana using seven capability classes. Such a classification is an essential basis for planned agricultural development.

On a world-wide basis Dudal (1966) outlined the classification of land suitable for mechanized rice cultivation.

Aerial surveys

Large areas of the tropics are now photographed from the air and the interpretation of such photographs with limited ground surveys, provides a quick method of mapping and assessing land resources and land use by skilled personnel working mainly in laboratory as opposed to difficult field conditions. From the photographs, which might cover 25,000 square miles (65,000 km^2), can be prepared maps on a 1 : 50,000 or smaller scale, showing the relief for soil and water conservation, geology, soils, vegetation, present land use, population density and distribution and cattle density, all to a greater accuracy than was previously possible and in considerably less time.

Photographs from satellites using special techniques such as infra-red photography can provide such information in greater detail and monitor rapid changes such as crop disease outbreaks.

On a smaller scale, however, say 10,000 ha, where the land is covered by forest such mapping is not easy. Nevertheless such aerial surveys should be used as the basis of all agricultural development schemes, and many expensive failures could have been prevented or planned more successfully with the use of aerial survey data.

Different types of film are used according to requirements.

1. *Panchromatic* - black and white is the standard film used for aerial surveys as it is cheap, easily processed and familiar. Changes in vegetation and soil moisture are not always clearly shown.
2. *Infra-red* - black and white is better for outlining differences in the vegetation and soil moisture, showing up wet soils well. Conifer and broad-leaved trees are distinguished on this film.
3. *Colour film* - distinguishes many more tones than black and white and gives a much clearer picture of the ground cover.
4. *False colour infra-red* - combines the advantages of black and white infra-red with those of colour film, but the colours are unfamiliar and make interpretation difficult. For example, dense

vegetation is bright red, and bare brown soil appears green. It is more expensive than the other films. It is useful for showing up diseased and damaged crops.

Wet paddy soils

There are around 100 million ha in the world devoted to the cultivation of wet rice or paddy of which some 85 million ha are in Asia, predominantly on alluvial soils. The major rice soils of South-East Asia were classified by Dudal and Moorman (1964).

These soils are flooded for a large part of each year and in consequence develop their own characteristic profile as described by Koenigs (1950) (Fig. 1.3). Without irrigation the soil from the same parent material, andesitic tuff, gave an almost homogeneous profile.

Under the water the surface of the soil is reddish brown. This oxidized layer is only a few centimetres or millimetres thick and is covered by a layer of blue-green or green algae which not only maintain the oxidizing conditions but also fix atmospheric nitrogen which becomes available to the rice crop. Below this is the characteristic gley layer.

As described by Koenigs the gley layer was about 20 cm deep, but it may be as deep as 75 cm depending upon the soil type, drainage and cultivation (Dudal 1958). Below this bluish grey layer are oxidizing conditions as occur in a non-irrigated soil.

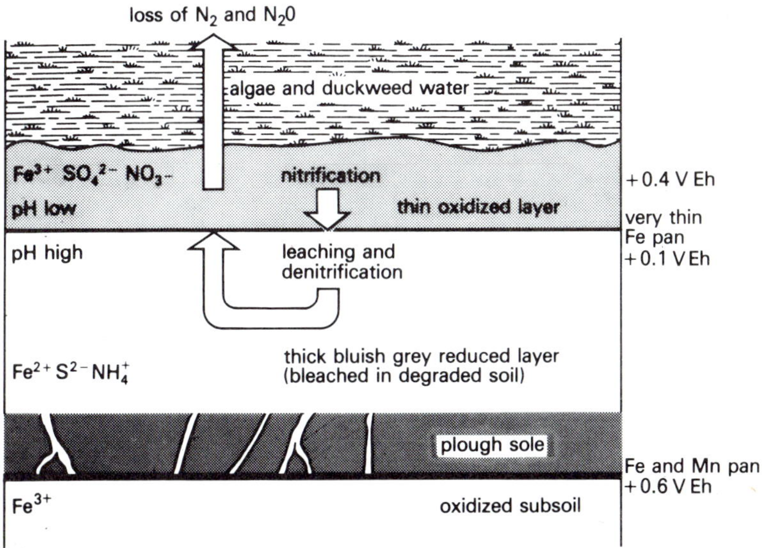

Fig. 1.3 Paddy soil profile (After Greene 1960).

The reduced iron and particularly the manganese compounds in the gley layer are more soluble and more easily leached down to where they are oxidized to insoluble compounds, particularly MnO_2 and Mn_3O_4, forming a cemented concretionary layer of iron and manganese compounds which produce streaks, mottles or a hardpan. When a rice paddy dries out in the dry season or after the irrigation water has been drained off, the differentiation between the three contrasted zones gradually disappears because of increased oxidation as the air permeates the paddy. Under favourable conditions a paddy profile may take 50 to 100 years to develop (Kawaguchi & Matsuo 1956). Such profiles do not occur on alkaline soils or where the parent material is poor in iron and manganese. Thick deposits of silt from flooding may lead to a new profile developing over the original one.

De Gee (1950) gave the oxidation - reduction (redox) potentials through such a profile (Table 1.3). He found that immediately below the brown surface soil of a paddy profile in the blue-grey mud the redox potential fell suddenly over a very small increase in depth from 600 mV to below 100 mV, reaching a minimum at 10 cm depth. Below 17 cm there was a rapid increase within 3 cm, and below the mud in the drier subsoil the redox potential rose again to 600 mV.

Table 1.3 Redox potentials at different depths in a rice profile (Bogor, Java)

Depths (cm)	Millivolts measured	
At the surface	400-600	
1-3	50-100	Reducing conditions
10	0-50	Reducing conditions
17	0-200	Reducing conditions
20	400-500	
30	About 600	

Reducing conditions are present below a potential of 300 mV at pH 5 and oxidizing conditions at 340 mV and above (Pearsall 1950). During the periods the paddy soil is submerged or waterlogged it develops reducing conditions, and the anaerobic bacteria convert ferric compounds to the ferrous state and sulphates to sulphides. The ferrous iron compounds cause changes in the colloidal structure which reduce the permeability of the soil. As the soil surface remains in contact with oxygen carrying water, or the air, or oxygen produced by algae, it has a rusty colour from the ferric iron. The reducing conditions deeper in the soil have a bluish-grey (gley) colour from the iron and manganese compounds in the reduced state. Between these oxidized and reduced layers may be a thin layer of hydrated ferric oxide. This thin bleached layer is formed by nitric acid percolating from the oxidized layer. There appears to be a definite relationship between the thickness of this layer and yield possibly due to this zone forming a barrier to the loss of gaseous nitrogen from the reducing zone below (Dudal 1966).

Fertilizers in wet paddy soils

Pearsall (1950), whose pioneering work in the field of waterlogged soils has led to a greater understanding of their nitrogen cycle and management, explained how these conditions of oxidation and reduction influence the effect of nitrogen fertilizers on paddy yields. Sulphate of ammonia applied to the surface of a paddy field will be converted in the oxidation layer to nitrate, liberating a sulphate radical. The nitrate will be rapidly leached down into the reducing zone where it is converted to nitrite and then gaseous nitrogen which is lost. As much as 80 per cent of the nitrogen applied to the surface of paddy soil may be lost in this way. Nitrogen fixed by the blue-green algae may be lost in the same way. The thicker the oxidized layer, and the more rapidly the nitrate can be taken up by the crop, the less is the nitrogen loss.

If the ammonium sulphate is placed, trodden or worked into the lower reducing zone, the ammonium ions remain unchanged and can be taken up by the crop. One method is to spread the ammonium sulphate on the surface, plough it under about 5.0 to 7.5 cm and then flood the land. This will also kill many of the established weeds. Because of the internal air spaces in the rice plant, typical of a marsh plant, which supply oxygen for cell respiration, the roots of rice can grow and function in the reducing zone. A red-brown oxidizing zone is often found coating the roots of paddy plants which protects them against moderate amounts of hydrogen sulphide. Organic matter which is puddled into the reducing layers is broken down to ammonia by the anaerobic organisms and becomes available to the crop. The sulphate ion left by the breakdown of sulphate of ammonia increases the acidity of the soil and accelerates leaching. In the reducing zone this sulphate is converted to sulphide which combines with the iron. In the absence of iron, free hydrogen sulphide or methyl sulphide is liberated, both of which are toxic. The presence of iron or calcium is essential to prevent this. The ammonium ions will free metallic ions by base exchange which leach away, assisted by the increased acidity from the free sulphate ions. Pearsall considered calcium cyanamide a better fertilizer than sulphate of ammonia. Calcium ammonium sulphate should be preferable to sulphate of ammonia, but should be placed in the reducing zone or applied in split applications in amounts which the crop can rapidly take up.

Work at the West African Rice Research Station in Sierra Leone (Annual Report 1961, 1962), suggests that, assuming nitrogen is the limiting factor, soils of different iron and organic contents require different water regimes for growth: low iron paddy soils should yield best under conditions of less than 100 per cent waterlogging, and high iron soils under standing water. Results suggest that the nitrogen response of paddy grown on a given soil is a function of the state of reduction of the soil (water regime) and not of its level of nitrogen. Sulphate of ammonia dug in, as recommended by Pearsall, gave higher yields than surface applications and there was a marked interaction of phosphate and nitrogen, due to a more rapid uptake of nitrogen in the presence of phosphate.

Pearsall (1950) also pointed out that aeration of a waterlogged soil may have a marked effect on its physical structure. Sub-aqueous muds exposed to air soon lose their semi-liquid nature, due to the precipitation of iron - humus compounds. Greene (1960) states that yields of paddy are increased 10 per cent or more if paddy fields can be dried off between successive crops, due to the greater production of ammonia than in continuously flooded soil. The increase of nitrogen may be as much as 22 kg/ha. Air drying may also increase the availability of phosphate.

Mid-season drainage of rice fields, a common practice in Japan, may not be desirable under most tropical conditions due to the substantial loss of nitrogen, without compensating advantages (Athwal 1972). Keeping the soil submerged for a few weeks before planting gradually improves the soil for growing rice. Flooding increases the availability of nitrogen, phosphorus, iron and silica.

Essential elements

The predominant element in the mineral fraction of the soil is silicon, which occurs either as the oxide, silica, or compounded with other elements such as aluminium and oxygen to form silicates. Iron and the basic elements are also common, occurring either as the hydroxide, carbonate or in combination with other elements, particularly silicon and aluminium.

As shown in Table 1.4 16 elements are considered essential to plants, some in greater amounts than others and a further 3 are essential to animals.

Silicon and aluminium are present in all plants. Silicon is probably not essential for the growth of most plants, though the general vigour of many grasses including cereals and particularly rice is decreased if grown in soils low in available silicon. Silicon appears to increase resistance to paddy blast and borers. Disease resistance of sorghum and the resistance of sugar cane to borers is reported to be improved with silicon applications. Clements (1965) suggested silicon was essential to sugar cane and in the high rainfall areas of Mauritius 7.5 t of calcium

Table 1.4 Essential elements

Essential elements obtained from air and water	Macronutrients	Micronutrients	Elements essential to animals	Toxic elements
Hydrogen	Nitrogen	Chlorine	Sodium	Selenium
Oxygen	Phosphorus	Manganese	Iodine	Arsenic
Carbon	Potassium	Copper	Cobalt	Molybdenum
	Sulphur	Iron		Fluorine
	Calcium	Zinc		Aluminium
	Magnesium	Boron		Nickel
		Molybdenum		etc.

silicate per hectare has increased the cane yield by over 25 t. If present in quantity, certain of the trace elements or micronutrients can be toxic. Molybdenum, manganese and boron have caused toxicity in a number of areas yet give deficiency problems in others. On the lower slopes near certain swamps in Buganda (Uganda), the pH is so low that manganese salts come into solution at a level high enough to be toxic, and none of the standard crops including cassava will grow. It is quite probable that in many other tropical areas, particularly on acid sands, the same conditions occur. The simplest way of correcting this is to lime the soil to raise the pH. The Buganda farmers used to mulch the soil heavily with organic matter. The addition of too much lime and organic matter can make manganese unavailable to crops and cause manganese deficiency symptoms.

In North Borneo small areas have an excessive chromium content in the soil, making it infertile.

Toxicity from an excess of certain minerals can often occur near mining operations.

The physiological functions in crops of the elements in Table 1.4 are dealt with by Russell (1973), and in stock by Underwood (1966).

Tropical soils are generally poor in nutrients. With continued cropping and increased use of NPK fertilizers, other elements such as magnesium or sulphur tend to become limiting at these higher levels of production (see also Ch. 2).

Soil micro-organisms

The cycle of nutrients in the soil is dependent on the micro-organism population, which includes bacteria, fungi, actinomycetes, mycorrhiza, myxomycetes, nematodes and protozoa. The contribution of each of these to the crop growth is still largely undetermined, but between them they are capable of breaking down practically all organic material to simpler compounds which are utilized by plants. It is thought that the more resistant lignified plant residues are attacked by the fungi. Either the organisms must have a wide range of activity or many must remain dormant and increase rapidly with the supply of a suitable source of energy. The same micro-organisms occur in soils of temperate and tropical regions, the soil conditions being less variable than the air conditions. Within the soil they occur in greatest numbers near the organic matter, that is the top few centimetres of soil, and in the root region. It should be remembered that a root system is not static but is continuously growing into regions where the air and water conditions are satisfactory and old roots are dying leaving strands of organic matter. A sugar cane crop produces about 5 t of roots per hectare which maintains the organic content and structure of the soil.

Bacteria are important in the nitrogen cycle and in bringing the C/N ratio to approximately 10. The addition of material to the soil with a C/N ratio over 10 causes bacteria to utilize nitrogen from the soil to the detriment of any crops. This nitrogen is eventually made available again as the C/N ratio narrows. Some bacteria associated with the roots of sugar cane and cereals are now thought to fix nitrogen and benefit the crop.

In rice fields certain of the blue-green algae fix atmospheric nitrogen (see p. 24–25). Algae may also help in maintaining the oxygen content of rice fields.

Many tree crops, particularly conifers but also cocoa, have mycorrhiza associated with their roots but their significance has not yet been explained. They are more prevalent when the crops are growing on poor soil.

Nematodes, which can seriously reduce yields in bananas, tobacco and many other tropical crops, are being increasingly studied, and a number of disease conditions such as 'slow decline' in citrus and 'red ring' of coconuts have been attributed to them. (See Ch. 4).

The benefit of partial sterilization is still inadequately explained, but it is associated with a destruction of the micro-organisms followed by a more rapid recovery of some of the more desirable species with an accompanying rise in soil nitrate. This may help to explain the rapid nitrate rise in tropical soils at the end of the dry season and part of the benefit of exposing soil to the direct heat of the sun as practised in India (Howard & Howard 1910). This process has an effect on soil structure and the availability of iron.

Soil nitrogen

Nitrogen, one of the elements of major importance in agriculture, is considerably influenced by climatic conditions both as to the form in which it occurs and the level at which it is present, and is the element that most frequently limits yield.

Though nitrogen occurs in the soil solution mainly as nitrate, and it is mainly in this form when it is taken into the plant, ammonium ions can be absorbed with equal ease. The nitrogen is then combined rapidly with carbohydrates to form amino acids.

The nitrogen removed from the soil by cropping, leaching or burning is replaced in a variety of ways, apart from that which is left behind in plant and animal remains, or added directly as fertilizer. Rainfall brings a small quantity, an average of 25 kg/ha of N has been estimated in Trinidad, 32 kg in Guyana and 44 kg in India. Roelofsen (1941) recorded a range of 17 to 47 kg over 15 years, with an average of 30 kg/ha N per year in Sumatra. Some of this nitrogen is in the dust or carried by the storms from the sea, some from denitrification in the soil. The formation of nitrate during thunderstorms is estimated to account for only 10 - 20 per cent of the nitrate in rainfall (Hutchinson 1944). The quantity of nitrogen supplied by rainfall is of little agricultural significance.

The most important source of nitrogen in most tropical soils is probably *Rhizobium* bacteria growing symbiotically in the root nodules of leguminous plants. These bacteria can fix atmospheric nitrogen into a compound, which the legume can use. Tropical legumes under favourable conditions may be equal to or superior to temperate legumes for nitrogen fixation (Norris & Henzell 1960). Sen (1958) estimated that root nodule fixation in the pigeon pea added 97 kg/ha of N in the control plots and 149 kg/ha when phosphate was applied. The legume cover crops (*Calopogonium, Centrosema, Indigofera* and *Phaseolus*) used in young rubber were shown in Indonesia to add 64 to 250 kg/ha of nitrogen in six months' growth (Anon 1936). *Glycine wightii* (syn. *G. javanica*) grown as a cover crop in Kenya adds the equivalent to 940 kg/ha of sulphate of ammonia (194 kg N) per year (Jones 1942) Teoh *et al* (1979) suggest this is a low figure, and cover crops contribute about 65,000 t of sulphate of ammonia in Malaysia. When a legume is planted as a cover crop or in a pasture, or introduced into a new area, it should be inoculated with the correct strain of *Rhizobium* before planting. In Malaysia cultures of *Rhizobium* bacteria are available in a compost of coir dust, peat and calcium carbonate, packed and sealed in polythene bags. The correct type is available for each cover crop. The compost is made into a slurry, spread over the seed, dried out of the sun, and then sown with some rock phosphate. Meyer and Anderson (1959) showed that nodules of subterranean clover fixed nitrogen at a maximum rate at 25 °C but hardly at all at 30°C. Mulching a soil planted with legumes aids nitrogen fixation, both by reducing the temperature and increasing the soil moisture. About 11 per cent of all legumes have been shown by Allen and Baldwin (1954) never to bear nodules, and hence never fix nitrogen. Conversely plants from another nine families of which four, Zygophyllaceae, Podocarpaceae, Cycadaceae and Casuarinaceae, are of potential importance in the tropics, have been shown to support root nodule nitrogen-fixing bacteria (Henzell & Norris 1962).

A *Rhizobium* strain extracted from the roots of the non-legume *Trema aspera* found growing as a weed in tea in New Guinea not only forms nitrogen-fixing nodules on *Trema* seedlings, but also on the roots of some tropical legumes including *Macroptilium atropurpureum* (*Siratro*)(Trinick 1973).

In a similar way certain bacteria can fix atmospheric nitrogen utilizing energy sources available in the soil rather than from associated plants. *Azotobacter* has long been thought to be the most important organism in converting nitrogen from the air to a form available to the plant, but it will preferentially use nitrate or ammonia as a nitrogen source rather than fix nitrogen from the air, and the nitrogen contribution from *Azotobacter* in tropical soils is now considered to be small. Certain tropical grasses, including maize, appear to support a high population *Azotobacter* in their root zone. *Beijerinckia*, which can similarly fix atmospheric nitrogen, is more common on the acid tropical soils of low fertility, but there are no claims that these bacteria make a significant contribution to soil nitrogen. *Clostridium*, which can fix nitrogen under low oxygen conditions, makes a contribution similar to *Azotobacter* and probably less than 11 kg/ha per annum.

Observations at ICRISAT suggest that many grasses and particularly the important crops of the semi-arid tropics, sorghum and pearl millet, benefit from nitrogen fixation in the soil. This nitrogen is not from blue-green algae or *Rhizobium* bacteria but from free-living bacteria that fix atmospheric nitrogen independently of a plant. The plants appear to stimulate these bacteria in their root zone.

While no green algae have been shown to fix nitrogen, the blue-green algae (Cyanophyaceae) can (De 1939). As the blue-green algae are widely distributed in the tropics and can tolerate high temperatures and drought they are thought to be the main natural source of soil nitrogen after legumes. Work in India (Allison & Morris 1930) has suggested that these blue-green algae which grow on the soil surface, using sunlight as an energy source, are an important factor in maintaining the yield of irrigated rice paddies which have been cropped continuously for generations without the yield declining to an uneconomic level. Hernandez (1956) gave a figure of 16 to 70 kg of nitrogen per hectare fixed by blue-green algae in rich fields in Hyderabad. The higher figure was in soil where phosphate and lime had been applied and a crop was growing. On a succession of four high-yielding rice crops in Kerala the presence of algae gave a 30 per cent increase in the number of tillers (Aiyer *et al.* 1972). Crop growth and yield were also stimulated even where 120 kg/ha of nitrogen had been applied. There was no increase in soil organic matter or nitrogen but there was a significant reduction in oxidizable matter, total sulphides and ferrous iron in the soil. The last two can be toxic to rice. The aquatic fern *Azolla pinnatu* which grows symbiotically with a blue-green alga *Anabaena azollae* which can fix atmospheric nitrogen is now considered an important source of nitrogen in rice paddies. It is possible that many of the traditional cultivations practised in rice paddies encourage the activity of these blue-green algae to the ultimate benefit of the crop, a fact that must be considered before these systems of land preparation are changed, and particularly before harmful herbicides are used extensively.

At the International Rice Research Institute (IRRI) in the Philippines, the nitrogen-fixing activity in paddy fields is much greater in planted soils (52 kg/ha) than in unplanted soils (28 kg/ha) and it increases as the crop grows (Athwal 1972).

Nitrogen cycle

The organic matter of the soil acts as a nutrient store, particularly for nitrogen, sulphur and phosphate, from which nutrients are slowly released by mineralization. Mineralization is the production of inorganic ions by the oxidation of organic compounds, mainly by bacteria. In this process the carbon compounds of the organic matter are converted to carbon dioxide and lost. The inorganic ions are available to the plants or may be leached out of the topsoil. This process of mineralization takes place in the following stages:

Mineralization →

Organic nitrogen → Ammonia → Nitrite → Nitrate

Ammonification → Nitrification →

← Demineralization

Nitrogen and Oxides of Nitrogen

Under normal soil conditions the rate of each of these stages in mineralization is greater than its predecessor, hence nitrite and ammonia rarely accumulate. Under certain conditions the reverse process can occur, and nitrate is reduced to nitrogen or oxides of nitrogen. This occurs when the oxygen supply is limited, and increases rapidly with temperature, but is slower under acid conditions.

Griffith (1951) suggested that a soil under given conditions of moisture, temperature and possibly insolation has a characteristic nitrogen content specific to that soil, but if any factor, such as moisture

Table 1.5 Effect of fertilizers and mulch on crop yields (Uganda)

Crop	**Sulphate of ammonia** 493 kg NO_3 per hectare	**Nitrate of soda** 493 kg NO_3 per hectare	**Kraal manure** 25 ha	**Control**	**Mulch**
Eleusine, *April–July, total yield (grain, roots and straw), t/ha*	23.75	22.5	51.25	22.5	30
Cotton, *Aug–Feb., seed-cotton, kg/ha*	370	346	448	258	1,053

content, alters, so does the nitrogen change. Conversely the increase of nitrate by a fertilizer application only raises the nitrate level temporarily if other factors remain constant. This replacement of soil nitrogen, as it is removed by the crop, was suggested after studying the effects of mulching where the level of soil nitrogen remained at a low level (about 10 ppm) yet considerable crop increases were obtained, greater than the increases from nitrogenous fertilizer dressings as shown in Table 1.5.

Fluctuation of soil nitrate level

In bare tropical soils the nitrate level rises rapidly at the onset of the rains, and very high high levels have been reported particularly in Uganda (150 ppm; Griffith 1951) (see Fig. 1.4); (200 ppm; Mills 1953) (Birch 1958). The high temperature and low moisture content of tropical soils in the dry season, in addition to making nutrients more available, have a partial sterilization effect which is followed by a rapid upsurge of bacterial activity when the moisture content of the soil becomes more favourable.

As the rains get under way the nitrogen level falls rapidly, partly due to leaching with a corresponding increase in subsoil nitrate. To take full advantage of this nitrogen flush, crops should be planted quickly once the rains start. A similar effect will occur at the start of irrigation. Loss of nitrate from topsoil is minimized if the water storage capacity of the soil is increased which will retain more nutrients in the rooting zone.

Rapid increases of soil nitrate are generally followed by rapid decreases. Microbial nitrate absorption and denitrification account for some of this loss. Near to the tropics there is a single rainy season and a single nitrate cycle with one peak period; at the equator, where there are two dry and two rainy seasons in the year, two soil nitrate peaks occur. The crop has an important effect on the soil nitrogen level. At certain stages in the crop growth nitrogen is being taken up by the plant and at a later date soil nitrogen is being tied up, particularly in the root region, by the organic residues of the crop being returned to the soil.

Nye and Greenland (1960) estimated an annual increase per hectare

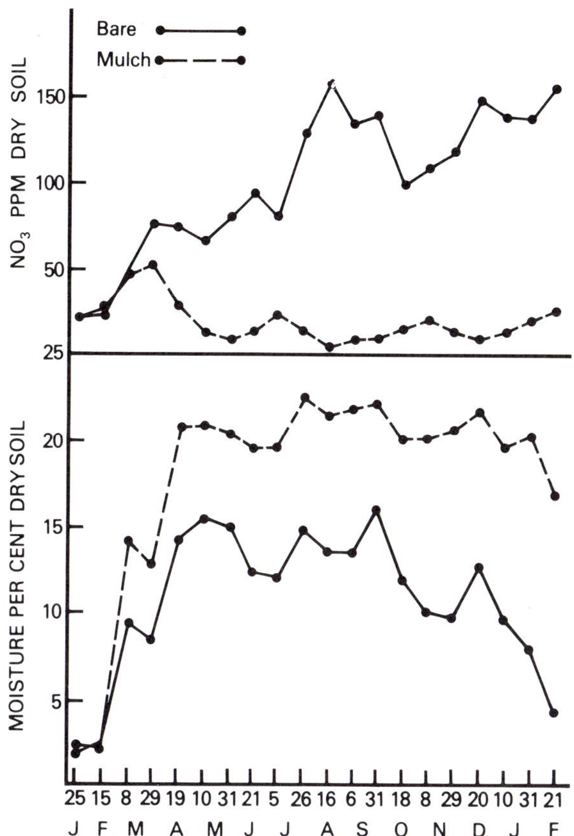

Fig. 1.4 Nitrate in bare and in mulched soil at Serere, Uganda (3-weekly means). (Griffith 1951)

under fallow forest of 39 kg nitrogen in the soil and 67 kg in the vegetation; and under grassland 28 kg in the vegetation which is lost in the annual burn and 11 kg in the soil. On clearing the forest fallow a liberal supply of nitrate from the mineralization of the humus is rapidly available, but the soil may be short of nitrogen the first year after clearing old grassland. This will vary with the species present in the grassland. Mills (1953) found mineralization was rapid after a 3 year resting period under elephant grass (*Pennisetum purpureum*), but slow after *Paspalum*.

Nitrate formation (mineralization) occurs in the soil when the moisture content is between 10 and 25 per cent; 14–18 per cent being the optimum, the 14 per cent level applying to sandy soils. The optimum temperature is 35 °C but nitrate formation can occur between 4 and 54 °C. Maximum nitrate formation occurs in the top few centimetres of soil, which when exposed to tropical sunshine can rise above 54 °C. The greatest nitrogen accumulation occurs under low moisture conditions and high temperature, and the least when the reverse applies. Mulching can keep the soil near the optimum temperature and moisture content for mineralization. As the mulch breaks down to humus the water storage capacity of the soil is increased and leaching reduced.

Cropping, particularly a grass cover, depresses mineralization. The nitrate level of grassland and under a grass fallow is low, all the nitrate present and any nitrate added as a fertilizer being rapidly bound up in an organic form.

Theron (1951) suggested that as the living grass roots of a ley develop they gradually inhibit nitrification, and ammonia appears. This occurs in the second season in South Africa. If legumes are present in the ley or nitrogen fertilizer is applied, humus accumulates. If, however, the ley is grazed from the outset, insufficient carbohydrate is transferred to the soil to combine with the nitrogen to form humus, and there is a drain on the nitrogen originally present in the soil. In Uganda many farmers believe that to graze land resting for a short period impoverishes the soil and subsequent crops are poor. It would appear from South African work (Theron & Haylett 1953) that grazing before the third year would have this effect, particularly if no legumes were present, or fertilizer applied.

Many tropical crops including coffee, tea and sugar cane develop pale green leaf symptoms, similar to nitrogen deficiency when the competition from grass weeds is severe.

Cropping the land immobilizes part of the soil nitrogen, temporarily in the case of annual crops but permanently in the case of perennial crops. If nitrogen fixation exceeds mineralization, there is an increase in total nitrogen and organic matter.

On a continuously cropped area at Samaru in the savanna area in the north of Nigeria, Wild (1972) found the levels of NH_4 and NO_3 at planting time were low unless dung had been applied (Table 1.6), the land bare fallowed, or groundnuts grown. The groundnuts were harvested before the cotton and sorghum, while mineralization was still going on.

The effect on the mineral nitrogen level (ppm) in the soil, of different crops and annual dung applications after 20 years' cropping, are shown in Table 1.7.

Nitrate appears to accumulate under certain tropical forest conditions (Jenny *et al*. 1949). The high proportion of leguminous species may be responsible (Jenny 1950), and the nitrogen is probably circulated between the soil and trees very rapidly (Hardy 1936).

Nye and Greenland (1960) quote nitrogen contents of surface soil samples for different ecological zones (see Table 1.8).

Table 1.6 Nitrogen levels in savanna soils (Nigeria)

	Before start of rains		**After start of rains before leaching**	
Previous crop	**NH_4-N** (ppm)	**NO_3-N** (ppm)	**NH_4-N** (ppm)	**NO_3-N** (ppm)
Cotton	4.5	2.6	0.8	6.3
Sorghum	5.6	1.5	1.3	7.6
Groundnuts	5.0	14.2	1.0	16.0
Cotton after grass *(Andropogon gayanus)*	4.1	3.0	1.3	4.4
Bare fallow	4.0	21.0	2.3	25.8

Note: The total soil nitrogen 100 - 900 ppm (mean 371 ppm).

Table 1.7 Effect of dung and cropping on soil nitrogen

Previous crop			
Dung (t/ha)	Cotton	Sorghum	Groundnuts
	ppm mineral nitrogen		
Nil	2.8	2.5	4.1
2.5	6.0	4.3	9.2
7.5	7.1	8.9	17.0
12.5	13.9	14.9	35.4

Note: The samples were taken about the time of crop planting, 31 May 1970.

Table 1.8 Nitrogen content of surface soil

Area	**Rainfall (mm)**	**Depth (cm)**	**N%**	**Recorder**	**Vegetation**
Southern Sumatra	230	0 - 10 cm	0.29	Hardon	Virgin and old secondary forest
Malaysia	254	0 - 7.5	0.25	Coulter	Virgin jungle
Malagasy Rep.	203	0 - 15	0.12	Pernet	Dense rain forest
Liberia	200 - 460	0 - 15	0.14	Reed	Virgin forest
Trinidad	180	0 - 15	0.13	Hardy and Evans	Young secondary forest
Ghana	125 - 150	0 - 15	0.041	Nye	Long rest grassland (Andropogoneae)
Malagasy Rep.	110	0 - 10	0.083	Pernet	Grassland (Andropogoneae)
Upper Niger	40	0 - 15	0.023	Dabin	'Virgin' grassland

The total soil nitrogen of Queensland is more closely correlated with total phosphate than with annual rainfall (Hubble and Martin 1960). They also consider that among the ways in which nitrogen is lost, by crops, erosion, fire, leaching and volatilization, the last two have been exaggerated.

Soil air

Between the solid particles of soil are spaces or pores, which are filled with soil air and water. As the solid particles have a constant volume, the combined volume of water and air in the soil, the pore space, is also constant in an undisturbed soil. As a soil dries out, the volume of air increases. Conversely a waterlogged soil has insufficient air for root growth. The ratio of soil air : soil water is important for plant growth as shown in Figs. 1.5 and 1.6.

The pore space of a soil varies from about 30 per cent by volume for sandy soils to nearly 60 per cent for clays, but there is no simple correlation between mechanical analysis and pore space. Cultural practices such as ploughing increase the pore space, and rolling or puddling, as carried out in rice paddies, reduces pore space. Though certain plants such as rice can grow in waterlogged conditions where soil air is nearly absent, this is exceptional and the majority of crops need a balance between soil water and air to survive (Fig. 1.6.)

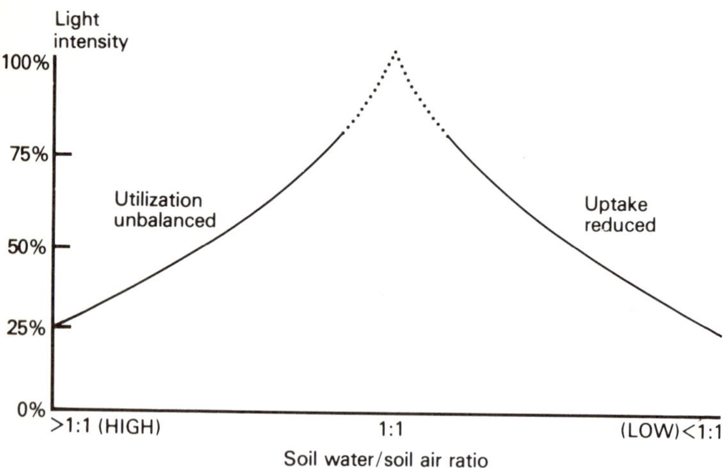

Fig. 1.5 Relationship between optimal growth and light at different proportions of soil water to soil air. (Fennah and Murray 1957)

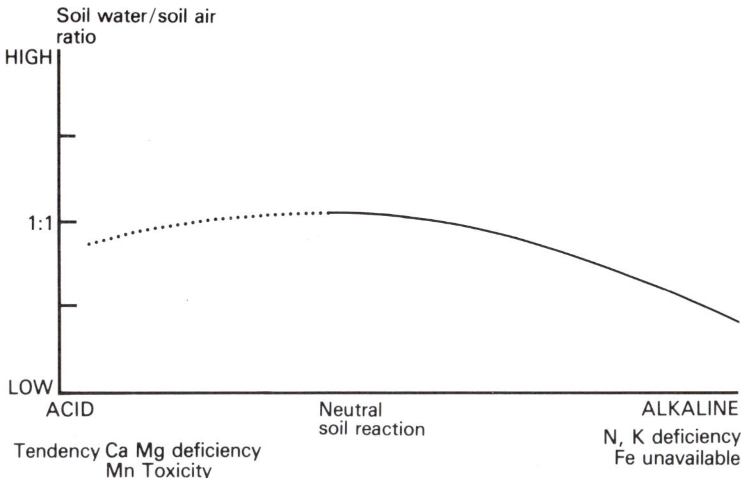

Fig. 1.6 Relationship between optimal growth and soil water/soil air ratio at different levels of soil acidity. (Fennah and Murray 1957)

The composition of soil air is different from the atmosphere, being higher in carbon dioxide and lower in oxygen, as the oxygen has been utilized by the roots for respiration, soil micro-organisms, the decay of organic matter, and the change in the composition of soil minerals. Consequently the carbon dioxide content increases with increasing depth down the profile, and is higher in the wet season than in the dry season as diffusion of the atmosphere into the soil is hindered by the higher water content of the soil, particularly at the surface. If the carbon dioxide exceeds 5 per cent by volume, it is toxic to plant roots, but this figure is normally only reached when water seals the soils surface.

After heavy rainfall or irrigation, oxygen in the soil air may become a limiting factor. Under such conditions the competition by weeds for this oxygen may be more serious than their competition for nutrients, soil water being adequate for both crop and weeds.

In clay soils, the deep cracks which develop in the dry season (Plate 1.1) are important in maintaining satisfactory levels of oxygen and carbon dioxide in addition to assisting in drainage (Hardy & Derraugh 1947). As decaying organic matter and the activities of the micro-organisms are largely confined to the top few centimetres of soil, the carbon dioxide produced by these processes is able to diffuse away rapidly.

The soil air of paddy fields is very different in composition to the at-

Plate 1.1 The first leaves appear on the sugar cane. Note the deep cracks in the soil. Jamaica, West Indies. (Courtesy West Indies Sugar Co.)

mosphere. Harrison and Aiyer (1916) gave the following figures by volume for water-free soil air for paddy fields in India.

Nitrogen	75 to 1%
Oxygen	2.8 to 0%
Carbon dioxide	2 to 20%
Methane	17 to 73%
Hydrogen	0 to 2.2%

Oxygen enters the plant roots in the soil solution, but rice and swamp plants have aerenchymatous tissue in the cortex of the stem or roots to conduct air from the atmosphere to the roots, as the root zone of a flooded or saturated soil is devoid of oxygen.

Soil water

Plants which are transpiring continuously, more during the hot day than the night, transpire 250 to 1,000 kg of water for every kilogram of dry matter produced; 1 ha of tea uses about 25 t of water a day, which is equivalent to 927 mm of rain per annum. The soil acts as a reservoir for the supply of this large volume of water. The roots of most crops cannot grow into the permanent water table, thus, apart from the small amount which may rise, by capillary action, into the metre over the water table, all this water must be obtained from rainfall or irrigation. Balls (1953) suggests that the rising water table in the irrigated Nile delta has restricted root penetration of cotton and has been responsible for a steady decline in yield since 1900. This decline has been masked by improved cultural measures and higher yielding varieties.

If drainage is prevented, rainfall or irrigation water will fill all the pore space to give a waterlogged soil (maximum water capacity) where anaerobic reactions start.

Anaerobic reactions are utilized on the heavy frontal clays in Guyana where at the end of the cropping cycle the structure of the soil is poor and rather difficult to irrigate efficiently. The land is taken out of cropping and flooded to a depth of about 30 cm for 6 months. During this time all the trash from the last ratoon rots down and anaerobic reactions associated with the iron oxides occur. A heavy growth of water weeds and algae builds up on the surface, but the normal canefield weeds are 'drowned'. When the field is drained any excess salts are removed and as the air penetrates, oxidation of iron and manganese compounds proceeds. The final result is a marked improvement in tilth and an increase in fertility. This practice is confined to Guyana, which has its own characteristic system of irrigation and drainage canals (Plate 1.2); these waterways are also used for transporting the cane to the factory. Similar processes operate on irrigated soils, particularly rice paddies as already discussed.

A waterlogged soil has all its pore spaces filled with water to the exclusion of air, and is described as being at saturation capacity. If this soil is allowed to drain freely the non-capillary pores lose their water over a period of 2 or 3 days and it is replaced by air. When no more water will drain away the soil is at field capacity. A minimum of 10 per cent of the soil volume as free capillary pore space is necessary for crops to grow otherwise there is insufficient air for the roots. A soil at field capacity not receiving rain or irrigation water will continue to dry out, losing moisture by evaporation and transpiration. As this water loss continues a moisture level is reached where the plants can no longer draw moisture from the soil and they wilt beyond recovery. This condition occurs regularly in tropical areas with a marked dry season.

Plate 1.2 Aerial photograph of the East coast of Demerara. The first two sections in the foreground show village land, the third section with the buildings is the Government Agricultural Research Station at Mon Repos, and the areas beyond are La Bonne Intention Sugar Estate. (Courtesy H. Evans)

The force with which moisture is held in a soil is referred to as the p*F* (Schofield 1935) which is the logarithm of the length of the water column in centimetres which would exert the same force (e.g. p*F* 4.2 is log 15,000 cm equivalent to 15 atm pressure). The important values are:

Saturation capacity	p*F* 0
Field capacity	p*F* 3.0
Wilting point	p*F* 4.2
Air dry condition (RH 50%)	p*F* 6.0

As the majority of plants can exert about the same force (p*F*) to take water from the soil, for any one soil this wilting point occurs about the same soil moisture content for all crops, but varies between soils according to their ability to hold moisture (p*F*). The ability of a soil to hold water against the extraction pressure of the crop increases with the clay content, thus the permanent wilting may occur on a clay soil with 15 per cent moisture but on a sandy soil at 5 per cent moisture (Fig. 1.7).

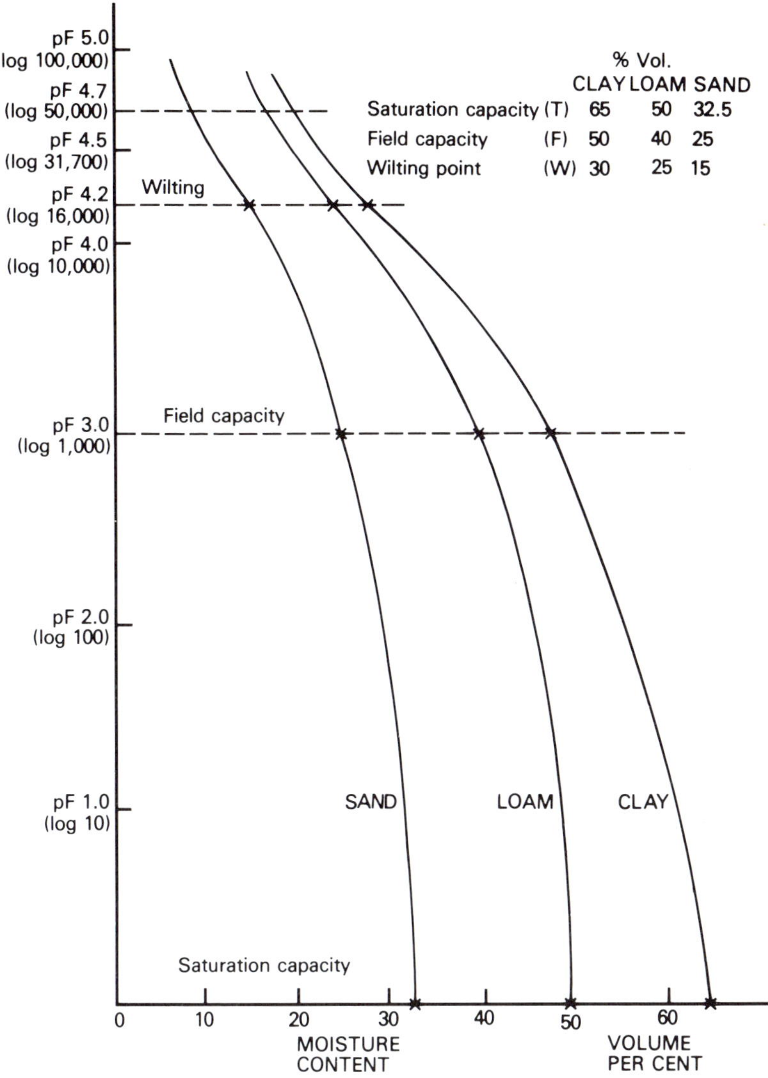

Fig. 1.7 Suction force and soil moisture content. (Hardy 1970)

The presence of dissolved salts in the soil water increases the force needed by the plants to take up water. The importance of soil colloids in relation to the availability of soil water has led to the use of two arbitrary constants, the moisture equivalent which is a measure of the water absorbed by the soil colloids on swelling, and sticky point where the colloidal component of the soil is fully saturated and binds the soil together and with external objects. This latter constant is easily determined with a minimum of equipment and has been used by Hardy as a comparative figure for tropical soils, particularly useful where laboratory facilities are poor.

Non-permanent wilting occurs when the rate of transpiration, as under hot dry windy conditions, is greater than the rate at which the roots can take up water. As a plant wilts, the stomata close and transpiration is thus reduced, but photosynthesis also ceases. Leaf shedding is a normal response of tropical crops to the dry season.

Drought is serious to young newly established crops, hence shade is vital to new plantings of many tree crops. Young cocoa plants have poorly developed leaf cuticles and are particularly sensitive to conditions favouring rapid transpiration. Deep rooted crops have more water to draw upon in deep soils than shallow rooted crops, or crops in shallow soils. Forest will continue transpiring long after grassland has ceased. Sorghum is an important cereal of the dry tropics due to its ability to resist drought conditions which would kill maize. This drought resistance is associated with the difference in stomata structure; other factors are also involved. Tobacco is another crop which can survive periods of drought. When a crop such as lucerne is cut transpiration is reduced as the area transpiring is reduced. There are differences between different cultivars of some crops; for example the roots of the drought-resistant tea clones in East Africa go down to 1.5 m, while those of clones susceptible to drought only go down to 0.8 m.

Penman (1948) has correlated transpiration losses and evaporation losses from open water or wet bare soil, with meteorological observations, though the real and calculated transpiration losses from a crop differ due to the reduction of evaporation when the stomata are closed. Open stomata, although occupying only a small percentage of the leaf surface, behave as an open water surface.

If the surface of a freely drained soil at field capacity dries slowly, the rate of water loss is initially the same as from an open water surface in the same environment. This rate of loss continues until about 25 mm of water have been evaporated, then the evaporation rate decreases rapidly, hence a period of rapid water loss is followed by a period where the loss is slower than under milder surface conditions. In strong sunshine a bare soil is self-mulching. Thus the loss of water by evaporation will depend upon the frequency of re-wetting, hence it is closely linked with rainfall or irrigation (Penman 1963).

Determining soil moisture

Sampling by auger and oven drying is the cheapest method of determining soil moisture but is slow and disturbs the soil.

Electrical resistance methods have been developed for measuring soil moisture under field conditions using either plaster of Paris or nylon resistance units (Bouyoucos & Mick 1940; Bouyoucos 1949). These cannot be used on very wet or saline soils and are not very accurate over a wide range. Where repeated observations in the same location, particularly at depth, are required in experiments, the neutron probe method, which measures hydrogen ions, is the most satisfactory except on organic soils.

Tensiometers are a practical tool, easy to use if not particularly accurate. For a literature review of methods determining soil moisture see Cope and Trickett (1965).

Soil organic matter

In tropical soils there is no direct relationship between colour and organic matter; the soil colour is of no agricultural significance. The organic matter content of tropical soils is not markedly less than similar soil types in temperate regions (Sanchez 1976, p. 162). The percentage organic matter in a soil is normally taken as 1.33 times the per cent by weight of organic carbon determined by the Walkley-Black method.

Typical organic matter contents of topsoils under natural vegetation are:

Lowland rain forest	3-6%
Wet savanna	2-3%
Dry savanna	1-2%
Semi-arid regions	½-1%
Desert areas	down to zero

Under cultivation these may fall to 30-70 per cent of this level.

The nitrogen cycle discussed does not operate in isolation but is inseparable from the carbon cycle and functions mainly on the energy derived from the carbon cycle. The organic matter of the soil, which is formed by the breakdown of plant and animal residues, is an important source and reservoir of nitrogen, phosphorus and sulphur to crops. In many highly weathered soils in the tropics, minerals with a low CEC dominate the clay fractions hence the maintenance of soil organic matter means the maintenance of cation exchange capacity.

The breakdown of leaf litter and other plant debris by insects, fungi, bacteria and other organisms is so rapid, that one rarely finds undecomposed plant matter on the soil surface. The carbon not lost as carbon dioxide forms organic compounds (humus) which are mainly colloidal. This humus is largely retained in the top 5 cm of soil which contains 2 to 5 per cent of organic matter. The 15 cm layer below this contains 0.5 to 2 per cent of organic matter. The soil under permanent grassland burnt each year has a very low humus content.

This colloidal organic matter, in common with the clay colloids, has important effects on soil structure, water relationships and the buffering of the soil pH. Soil organic matter has a high cation or base exchange capacity, 100–500 milliequivalents (me or meq)/100 g (Greene 1961), which is two or three times that of clay, and is thus able to retain calcium, magnesium and potassium cations which would otherwise be leached out (see p. 44). Most tropical soils are very poor in these bases.

Though there are large areas of peat soils in the tropics, as in Malaya and Indonesia, most tropical soils would benefit from an increased humus content, which would improve the acceptance and penetration of rainfall, increase their water-holding capacity, reduce erosion, delay the leaching of nutrients, and improve their structure and ease of cultivation. Jenny (1930) has suggested that the organic matter content of a soil is determined by rainfall and temperature. Rainfall increases the plant growth and hence the volume of residues to be broken down. An increase in temperature stimulates bacterial activity and breakdown. It is therefore impracticable to attempt to build up a permanent increase in organic matter without changing the farming system.

Heathcote (1970) at Samaru found the yields of cotton, sorghum and maize were increased more when the organic matter was burned compared with incorporating it in the soil unburned. This he considered justified the prevailing local practice of burning crop residues, but emphasized that this conclusion may not be valid long term. All the nitrogen and sulphur in the organic matter is lost in burning and these could become deficient.

The build up of soil organic matter during the resting period

Nye and Greenland (1960) estimated that in Ghana about 12.5 t of litter per hectare on an oven dry basis, is produced in an established forest fallow each year. Webb (1956) gives 7.5 t for Queensland, and Jenny (1950) 8.75 to 10 t for Colombia. Dead roots and exudates produce about half as much. A hectare of savanna fallow produces about 10 t of dry leaves and just over 2.5 t from the roots. Only 10–20 per cent of the fresh aerial parts and 20–50 per cent of the roots are converted to humus. Thus 1.25 to 2.5 t of carbon per hectare per annum are added to the humus under a forest fallow and a quarter of this amount under a grass fallow. Where the grass is dug in when opening up the land for cultivation this could add a further 1.25 t of carbon per hectare, but it will be necessary to allow time for the grass to break down before planting. At the same time the humus in the top 30 cm of soil is decomposing at an annual rate of 2 to 5 per cent in forest and 0.5 to 1.2 per cent in the savanna (Greenland & Nye 1959).

The rate of decomposition is greater at higher temperatures, and slower under very acid conditions. Below pH 4.5 most of the bacterial activity ceases. If the mineral plant nutrients such as potassium, calcium and magnesium are at a low level, and the soil acid, peat will accumulate. As the top 30 cm of soil under forest contains around 75 tons of humus carbon per hectare, a balance is reached where the accumulation of humus is equalled by its decomposition.

There are a number of important points about the resting period.

1. The effect of the resting period on the build up of humus will depend upon the original level of humus when cropping stopped. The lower the humus content the faster it will accumulate.
2. The C/N ratio remains fairly constant during the fallow, so the increase of nitrogen will be parallel to the build up of humus in the resting period. Legumes in the fallow help this build up.
3. The humus level does not vary considerably with successions of resting and cropping periods.
4. When the land is cultivated the annual rate of loss of humus may be over 2½ per cent but this rate gets less until equilibrium is reached. The humus content is not all used up even during continuous cropping.
5. Where shade trees are left when the forest is cleared for cropping, or when bananas or a legume cover crop are interplanted the soil organic matter may rise or the rate of loss is very low (du Bois 1957).

The mineralization of the organic matter following a forest fallow provides an adequate or liberal supply of nitrate, but after a resting period under grass the organic matter may be too low to provide adequate supplies for the first year or two of cropping.

Green manuring

The remarkable effects on yield observed in the 1920s in Nigeria when maize followed a green manure crop, caused considerable interest throughout the tropics. Later it was shown that the residues after burning the dried green manure caused a similar yield response, and it became obvious that the addition of organic matter was not the cause of the yield increase. Vine (1953) explains the effect of a green manure crop as a response to nitrate and that short-term green manures have no lasting effect on the soil organic matter or nitrogen content. The effect obtained by burning the green manure he attributes to a phosphate response. Legumes are good extractors of potassium and phosphate from the soil.

Green manuring has given disappointing results in many tropical areas. The loss of a season's cropping, combined with the difficulty of digging the green crop into the soil has resulted in few farmers growing green manure crops. On the other hand it is very common for plant material to be carried on to the plot from nearby unproductive land as in the 'rab' system for rice seed beds, and also in the Chitemene system of Zambia. The full use of crop wastes in the densely populated areas of the Far East, is probably a major cause of the rotation surviving without a long period of rest, as is required throughout Africa. Zimbabwe maize growers are advised to introduce a green manure crop into the rotation every 5 or 6 years and plough it in before it starts to mature.

In marginal rainfall areas the depletion of water reserves in growing the green manure would probably offset the benefits bestowed, though in wetter regions it will reduce the loss of nutrient by leaching. The effects of cover crops and mulching on organic matter are dealt with later.

Animal manure

There is no doubt that many crops including pastures in the tropics benefit from dressings of organic matter such as cattle manure, coffee hulls, grass mulch and compost. Such applications have given very striking results on coffee plantations in East Africa, where a large part of the benefit has been attributed to increasing the rain-absorbing properties of the soil and enabling the crops to survive the dry season better. The effects of such dressings last a number of years. On the poor soil near Piarco, Trinidad, a light dressing of poultry manure has produced a considerable improvement in the Pangola pasture. This Piarco series soil is a sand or sandy loam over a silty clay, waterlogged in the wet season and desiccated in the dry season; acid and poor in nutrients.

Continuous cropping

At Coimbatore, in South India, a series of permanent manurial plots were laid down in 1909 and after 81 crops had been harvested by 1950 the results were summarized (Sanyasi Raju 1952). The major nutrients were added at the rates shown in Table 1.9. The plots were irrigated up to September 1937, fallowed until November 1939, and subsequently rain-fed. The 46 cereal crops grown in the period (1910–51) gave yields in kg/ha as shown in Table 1.10.

The microbial population increased most rapidly with the cattle manure, but the complete NPK fertilizer mixture was second in encouraging microbial activity.

The pH of the soil was most affected by cattle manure; this treatment in 1950 caused a pH of 7.6 compared with pH 7.9 in the control plots. A parallel experiment started in 1925 confirmed these results. The fertilizer treatment does not appear to have adversely affected the soil though the manured plots were more friable.

This strongly suggested that it is possible to farm on a continuous cropping system provided that sufficient nutrients are returned to the soil. The manner in which these are added is a matter for the individual

Table 1.9 Nutrients applied in Coimbatore permanent cropping trial

Nutrient	Dressing (per hectare)	Nutrients (kg/ha)[a]		
		N	P_2O_5	K_2O
Ammonium sulphate	125 kg	25.0	—	—
Superphosphate	375 kg	—	72.6	—
Potassium sulphate	125 kg	—	—	60.5
Cattle manure	12.5 t	67.2	42.6	100.8

[a] Two sets of figures are given in this reference.

Table 1.10 Crop yields in Coimbatore permanent cropping trial (kg/ha)

	Eleusine coracana 15 crops	**Sorghum** 15 crops	**Wheat** 6 crops	**Panicum miliaceum** 7 crops	**Pennisetum americanum** 5 crops	**All 46 crops as % of control**
N + K + P	1,732	2,102	1,144	1,146	348	225
N + P	1,680	2,112	1,017	1,253	357	223
Cattle manure	1,491	2,218	903	1,198	304	213
K + P	1,472	1,811	861	1,098	236	191
P	1,016	1,197	632	1,038	164	141
K	821	958	576	764	188	115
N + K	673	918	600	668	248	108
N	618	885	516	752	207	106
No manure	531	802	417	964	157	100

farmer and is largely dictated by costs. Where the population density is low, as in parts of Africa and upland areas of the Far East, after a short period of cropping the land is abandoned to a bush fallow. During this resting period the deep rooted trees and shrubs transfer plant nutrients from the deep subsoil to the topsoil. Nye and Greenland (1960) estimate that in a mature high forest 40 kg N, 2.8 kg P, 58 kg K, 62 kg Ca and 12 kg Mg per hectare are 'pumped up' from below the first 30 cm each year, and that the figure for younger forest fallows should be similar. Where the population is more dense and land cannot be spared for resting, the fertility has been maintained by carrying on organic waste from other places. There seems to be no reason why this cannot be replaced by mixed fertilizer dressings, which raises the question of replenishing the soil organic matter. It should not be forgotten that a good crop has a good root system and leaves a lot of residues both from the roots and aerial parts. A fertilizer application not only increases the crop but also the residues, and, as the Coimbatore experiment showed, maintains a high degree of bacterial activity. It therefore appears that the soil organic matter which remains constant for a given temperature and rainfall condition is maintained by the decomposition of plant residues. If the nutrients removed by the crops are not replaced either by natural decomposition or by fertilizer application, the crops get poorer, and the soil cover decreases. This causes a rise in soil temperature and more compaction by raindrops decreasing the receptivity of the soil to rainfall. The process of decline goes on and the soil produces a poorer crop each year until a balance is reached. The reverse process occurs if the nutrient status is built up.

The possibility of continuous cropping under African conditions is supported by the results of the fertility experiment at Serere in Uganda (Annual Report 1959). This experiment was started in 1936 to compare different rotations and variable resting periods, together with a dosage of 0, 6.25 and 12.5 t/ha of farmyard manure (FYM) every 5 years. Unfortunately no fertilizer treatments were included. The decline in production of the heavily cropped plots was surprisingly small. This was attributed to strict soil conservation measures. Another surprising feature is the cumulative effect of the FYM. Each 12.5 t application yielded an increase of 78 kg of cotton per hectare, thus the crop immediately after the fourth application yielded 312 kg more than the unmanured plot. The cumulative increase 2 years after a single application was 62 kg, and 37 kg 4 years after application. Groundnuts, finger millet and sorghum respond similarly. Hartley and Greenwood (1933) in Northern Nigeria obtained a striking response to a small dressing of farmyard manure. Even 2.5 t FYM per hectare gave an increase of over 50 per cent on a poor crop of Guinea corn (237 kg grain per hectare). Further work (Hartley 1937) demonstrated that this was a phosphate response which could be reproduced by an application of superphosphate equal to that in the FYM, but rock phosphate did not give the same response (see Table 1.11).

Table 1.11 Response of Guinea corn to fertilizer (Northern Nigeria)

Treatments per hectare	**Yields Guinea Corn –1936** (kg/ha)	
	Grain	**Leaves and stems**
No manure	558	3,774
168 kg Nitrate of soda	718	7,358
58 kg Superphosphate	1,051	8,770
168 kg Nitrate of soda, 58 kg Superphosphate	1,103	12,320
116 kg Superphosphate	1,167	11,514
168 kg Nitrate of soda, 116 kg Superphosphate	1,254	12,992
2 tons FYM (1.01 N, 0.30 P_2O_5 (Equivalent to 168 kg Nitrate of soda plus 58 kg superphosphate.)	1,095	10,080
Significant difference	157	1,926

Source: (Hartley, 1937).

In previous experiments potash had given some response alone but depressed yields when combined with other nutrients. The response was not confined to Guinea corn. It is interesting to see that doubling the grain yield trebles the yield of leaves and stems, and a similar relationship must hold for the roots. This may explain why sugar cane soils have been cropped annually in the West Indies since the eighteenth century and are today yielding higher than ever. These soils are well fertilized, and the roots and trash return a large quantity of organic matter and nutrients. About half of the root residues are converted to soil organic matter.

It is frequent practice in Malaya when replanting rubber, as a precaution against root diseases, to cultivate the soil thoroughly to remove all the old roots. This cultivation destroys much of the soil structure and organic matter, but this is corrected over the next 2 or 3 years by the cover crops which are well fertilized with phosphate.

While the use of fertilizers on many peasant-grown crops, particularly food crops, remains uneconomic the best method of increasing the soil organic matter is by maintaining a good soil cover and returning to the soil the maximum amount of crop and household residues. The soil cover is maintained by mixed cropping or intercropping, planting shade trees for perennial tree crops, close planting or using crop varieties of a spreading habit, mulching, the use of cover crops, and avoiding bare fallows. Many of these practices also break the impact of raindrops and facilitate rainfall acceptance by the soil. These practices are part of the traditional farming systems of the peasants in the tropics.

Cattle manure, where it is available, and can be transported to the cultivated land, is of great benefit in maintaining soil fertility, though most of the benefit is from nutrients added in the manure as is shown in the trials at Matuga in Kenya (see Table 1.12).

Over a much longer period of cropping other differences may show up. Others, including Dennison (1961) in Northern Nigeria, Doughty (1953) in East Africa, and Djokoto and Stephens (1961) in Ghana obtained better yields with cattle manure than with fertilizers providing similar amounts of NPK. Traces of other elements as well as the organic

Table 1.12 Response of grain and root crops to fertilizer and manure, Kenya (Yields in kg/ha)

Crop	No. of crops	No fertilizer or manure	Fertilizer each year	FYM 7.5 t/yr.	FYM 22.5 t/3 yrs.
Maize	6	583	970	874	943
Sorghum	6	1,057	1,928	1,648	1,586
Cassava	6	6,303	10,223	9,596	9,266
Sweet potato	4	2,584	5,155	5,429	4,685

Source: Grimes and Clarke (1962).

matter may have been beneficial. Heathcote (1970) showed molybdenum was important at Samaru.

The organic matter in the cattle manure benefits markedly certain crops such as bananas and coffee by improving the rainfall acceptance and retention. Cereals and root crops, particularly sweet potatoes and yams, respond much better to dressings of cattle manure than do cotton, sim-sim, and groundnuts. Cattle manure should not be used on tobacco as it produces an uneven crop and may spoil the leaf quality.

On soils not liable to compact where cattle are grazed under coconuts, as in Tobago, the southern part of Trinidad and the South Pacific provided soil moisture is adequate the coconuts yield more, partly due to the benefit of the manure and partly due to the grazing keeping down grasses and weeds.

A heavy dressing of organic matter corrects the infertility condition 'lunyu' (manganese toxicity) which occurs near valley bottoms in Buganda, by binding the manganese and making in unavailable to the crop. Soil moisture is also important in immobilizing manganese (Chenery 1954).

Retention of cations by the soil

The colloidal particles of humus and clay in the soil have negative charges which attract and hold the positively charged cations such as calcium, potassium, hydrogen and in acid soils, aluminium.

This taking up and releasing of cations is referred to as cation exchange or base exchange, and the cation exchange capacity (CEC) of a soil is a measure of the negatively charged sites on the surfaces of the clay and humus molecules expressed by the milliequivalents of ions 100 g of soil will absorb. Total exchangeable bases (TEB) are the total exchangeable Ca^{2+}, Mg^{2+}, K^+ and Na^+ ions adsorbed on these surfaces. In some very acid soils, some of these exchange sites may be occupied by Fe^{3+} ions but these are not included in the TEB. Base saturation is a simple percentage of TEB/CEC. Cation exchange capacity is determined in the laboratory of pH 7.0, and this figure is usually higher than the true CEC in the field, as most tropical soils are acid and the CEC falls with increasing acidity.

A high CEC is a desirable feature of a soil as this protects a higher proportion of nutrient cations from being leached out. They are removed from the colloidal particles by replacement with other ions. Roots appear to do this by exchanging with hydrogen ions. Each nutrient ion leached away takes with it an anion which is often NO_3. Some typical CECs are:

Vermiculite	120 me/100 g
Montmorillonite	100 me/100 g
Illite	30 me/100 g

Kaolinite 8 me/100 g
Organic matter 200 me/100 g

Sand, understandably, has a very low CEC.

There is a close correlation between the CEC of a soil and its productivity. A CEC of 8–10 me/100 g in the top 30 cm of soil is regarded as minimum for irrigated soils, although rice may tolerate lower values. A satisfactory soil at pH 5 may have only 20, 5 and 2 of its CEC saturated with Ca, Mg and K respectively and these are mainly in the top 20 cm of a tropical soil. Ca^{2+} is not the dominant cation adsorbed on the colloidal particles of tropical soils as it is in temperate soils.

Organic matter can also adsorb anions, but not chloride, sulphate and nitrate. The power of cations to displace other cations is in the following descending order:

H^+ Ca^{2+} Mg^{2+} K^+ NH^{4+} Na^+

According to Marshall (1949) hydrogen does not remain fixed but moves in and out and does not neutralize the negative charge on the soil particles like the other cations. Cations with the lowest displacing power tend to give sticky or soapy dispersed soils, as produced from sea flooding which occurs in tidal estuaries such as the Mekong delta in Vietnam, or by irrigating with saline water. Sticky sodium soils can be corrected by heavy dressings of calcium sulphate or lime.

For a permanent irrigation system the total salt content of the soil, and particularly the proportions of exchangeable sodium and possibly magnesium must be kept below a certain level. More than 12 to 15 per cent of exchangeable sodium ions reduces the permeability of the soil, and with more than 40 to 50 per cent sodium the plant may not be able to take up sufficient calcium for its needs. Magnesium has a similar but lesser effect. During irrigation, sodium ions in the downward percolating soil water, replace the calcium and magnesium ions by base exchange. As the calcium ions are held more tightly than the sodium ions, this exchange is not serious until the concentration of sodium ions is two or three times that of the calcium. With a higher total ion concentration and the same ratio of sodium to calcium ions, the greater the entry of the sodium ions, hence with more saline irrigation water a lower sodium to calcium ratio is necessary. It is the concentration of ions in the soil solution leaching down rather than in the irrigation water which is important. On very permeable soils the two concentrations will be very similar but with reduced permeability the concentration in the soil water may rise to as much as 10 times that of the irrigation water. Thus on impermeable soils the sodium ion concentration must be kept low otherwise the permeability will be reduced further (Russell 1973).

Sodium soils occur occasionally under arid conditions. Anions such as NO^{-} and SO^{2-} tend to stay in solution, but phosphorus is held very strongly by calcium, iron and aluminium cations in the soil, often referred to as phosphate fixation. This leads to difficulties in supplying crops with adequate phosphate on certain soils. Conversely nitrate and sulphate are easily leached out of the soil. Root hairs appear to be able to take up nutrients direct from the soil colloidal complex.

Soil pH

The relative proportions of hydrogen and basic cations on the soil colloids largely determines the pH of the soil. The acid condition of waterlogged soils is due to the production of organic acids including carbonic and sulphuric acids.

The acid nature of many tropical soils results from the hydrolysis of aluminium clays and the subsequent ionization of exchangeable aluminium, releasing hydrogen ions.

Acidity in tropical soils is not a symptom of low fertility as in temperate regions. Many tropical crops are adapted to acid conditions.

The influence of soil pH on crop growth is indirect, and results from the imbalance of nutritional elements under acid or alkaline conditions as shown in Table 1.13. The wide pH range that certain crops can tolerate is illustrated by the oil palm which in Indonesia will grow well between pH 4.0 and 8.5, provided that other conditions are satisfactory. In general the oil palm prefers a pH below 7.0.

Table 1.13 The effects of excess soil acidity or alkalinity

Effect on	**Low pH below 5.5**	**High pH Above 7.5**
Uptake of nutrients	Retarded	Retarded
Aluminium toxicity	Develops due to free Al ion	Develops due to aluminate anion (AlO_2)
Iron		Lime-induced chlorosis (Fe deficiency)
Phosphate	Fixed	Fixed
Calcium	Deficient	Often in excess
Availability of anions Mo, Bo	Reduced	Increased even to toxic levels
Availability of cations Cu, Mn, Zn	Increased even to toxic level	Reduced
Certain plant pathogens	Increased	Increased

Source: Hardy (1970b).

Note: Molybdenum (Mo) is important for the *Rhizobia* of leguminous plants.

Usable soils lie between pH 3.5 and 9.5. A pH less than 3.5 only occurs on acid sulphate soils (see p. 000) and above 9.0 only in solonetz (see p. 000). Below pH 5.5 nutrients, particularly phosphate, are less available. The desirable micro-organisms thrive best between pH 5.5 and 7.8. The root nodule bacteria normally prefer a neutral soil but *Rhizobia* of tropical legumes can fix nitrogen in an acid soil, as in the case of soya beans which can fix nitrogen at pH 4.0 (see p. 000).

If all other factors are favourable, crops tolerate a greater degree of acidity under more fertile conditions. Tropical legumes grow on soils more acid than temperate legumes but they may suffer from iron deficiency. Tolerance to more acid conditions can be increased in certain crops by breeding.

A very acid soil is generally considered essential for tea, but in East Africa tea is growing well in soil with a pH of 6.0 or more, considered too alkaline in Sri Lanka or India. On hut sites in East Africa the pH may be as high as 7.5 or more. Sulphur is applied to such areas to reduce the pH below 6, otherwise the tea dies and is replaced by invasive weeds. At the other extreme tea fails in East Africa with a pH below 3.3. Smith (1963) showed that the Comber test, which is the production of a red colour when an acid soil is shaken with a saturated alcoholic solution of potassium thiocyanate, was the most useful, and a simple test for choosing soils suitable for tea. Chenery (1966) considers the low pH requirement for tea is related to iron availability. At a pH over 6.0 the available iron in the soil solution is only 1 per cent or less of that at pH 4.0. In Sri Lanka where very heavy rates of sulphate of ammonia have been used regularly to produce very high yields of tea, planters are concerned about the very acid conditions, around pH 3.8, being developed and the lack of calcium in the soil. Calcium ammonium nitrate is to be preferred on such soils, but this should not be used on tea once the pH has risen to 5.0. In Sri Lanka ammonium sulphate nitrate is recommended for soils with a pH between 5.0 and 5.5. Urea should also have a place with its neutral reaction.

Nematodes appear to be more serious on acid soils, but this may be a result of crop resistance being lowered.

Soil temperature

Soil temperature, with its effect on soil formation, solubility of nutrients, the activity of soil micro-organisms and the accumulation and breakdown of organic matter, is an important ecological factor in the tropics. Except at high altitudes the air temperature of the agricultural regions of the tropics is normally between 16 and 37 °C in the shade but there are wide variations. The mean soil temperature is higher than the surrounding atmosphere (Mohr *et al.* 1972). In Zaire it has been estimated that a rise in temperature of 1 °C over 36 °C causes a loss of 28 kg of nitrogen per hectare and unprotected soils lose 1,120 kg per hectare of nitrogen annually. Hardy (1970a) has suggested that the optimum soil temperature for absorbing water and nutrients is about 5 °C higher for tropical crops than temperate crops.

Soil temperature is dependent on the sun and is influenced by factors influencing air temperature (see later), particularly latitude and altitude and also:

1. Aspect and slope which influence the amount of the sun's energy reaching the soil. This is less important at the equator than at the tropics.
2. The shade effect of vegetation and mulching. This also reduces the daily and seasonal variation.
3. Soil colour - black soil may be 6 °C hotter than an adjacent white soil. At Poona (India) where the surface temperature of the soil often rises to 75 °C, black cotton soil absorbed 86 per cent of the total heat radiation, grey alluvial soil 40 per cent and grass-covered soil 60 per cent (Ramdas & Dravid 1936).
4. Soil moisture. Evaporation from a moist soil reduces the soil temperature. Also a wet soil reflects more of the sun's heat. Drainage raises the temperature of a wet soil.
5. As the specific heat of most soil constituents is about 0.2, it takes about five times as much heat to raise the temperature of water each degree, as is required for the dry matter of the soil.

The variation in soil temperature is mainly confined to the top 30 cm of soil and depends upon the thermal conductivity of the soil.

At a depth of 1.5 m the mean annual temperature of the soil remains constant (Vageler 1933).

Soil temperature at 1.5 m

Equator	26.2 °C
10° latitude	26.5 °C
20° latitude	25.0 °C
30° latitude	19.4 °C

In the sun, soil surface temperature can reach 85 °C, but if water is available for evaporation it will not rise above 50 °C. In tall grass the shade restricts the temperature to 40 °C at the maximum. Under forest or dense tree crops the soil receives very little direct radiation and the soil surface is about the same temperature as the air.

The temperature of topsoil if unshaded will rise above the temperature tolerated by many soil bacteria and microbial activity ceases. Activity at lower levels where the temperature is lower will depend upon the supply of oxygen and nutrients. Thus deep cultivation aerates the deeper soil layers, raises the temperature, and accelerates the breakdown of organic matter and mineralization of nitrogen, giving increased yields which the soil reserves may not be able to maintain. Deep cultivation is seldom practised by peasant farmers but has been used successfully on estates growing perennial crops such as sugar cane to increase the root room through which the roots can forage for water and nutrients.

Soil structure

Soil structure is the mode of arrangement of the units of which a soil is built up. It is important to good crop growth and the prevention of soil erosion. Ideally the small particles should be bound together into water-stable aggregates or crumbs, a process which takes place when land is rested under a grass or bush fallow. On cropping the reverse occurs and the structure quickly deteriorates. The process of aggregation has not been worked out, though it is certain there is no simple explanation. Organic matter binds the particles together, and the aggregation bears a direct relationship to the numbers and activities of micro-organisms. It is known that the greater and more readily available is the organic matter in the soil, the better is the aggregation. The clay content of the soil, and the iron and aluminium oxides may also have an important part in the aggregation process. Soils with a good crumb structure show less tendency to cap, have a greater receptivity to rainfall, drain better, are better aerated, cultivate more easily, have less physical resistance to the growing crop (hence plant roots penetrate deeper and thus have a larger reservoir of water and nutrients) and soil microbes are more active - a mark of a more fertile soil.

In East Africa a minimum of 3 years' resting period under grass has long been recommended in the cropping cycle. Since Martin (1944) showed that such a resting period had a beneficial effect on crumb structure, this has been considered to be the main gain from the resting fallow. It is now clear that this improvement of soil structure largely disappears in the first year of cultivation, and there is no clear evidence that the declining yields under continuous cropping are due to the loss of soil structure.

Mulches have been shown to increase aggregation (Alderfer & Merkle 1943). In a Kenya coffee plantation a better structure was produced by mulching 2 m of the 3 m rows than by resting the land under grass (Pereira *et al.* 1954a). Mulching supplies both organic matter and nutrients, it also reduces the soil temperature and pH and improves the soil moisture conditions, including rainfall acceptance, pore space and the amount of water retained. Soil conditions under a mulch, provided that rainfall is not excessive, are conducive to microbial activity and soil aggregation. The mulch also protects the soil from the pounding by the rain which would otherwise break down the crumb structure (see also Fig. 1.4). The results with mulching have shown that there is nothing unique about the roots of the bush fallow plants in relation to restoring soil structure which cannot be achieved by dead organic matter.

In the Coimbatore continuous cropping experiment (see p. 40) there was no evidence of loss of soil structure when production was maintained by artificial fertilizers rather than organic materials. Over much of tropical Africa with present peasant farming conditions, farmers growing annual crops use no fertilizer and must rest their soil regularly, mainly to restore the available phosphate and other nutrients, and increase the humus. Mulching their crops would delay the time when their land must be rested (see Plate 2.13 - Mulched banana garden).

TROPICAL RAINFALL

Within the tropics lies approximately 40 per cent of the earth's surface, three-quarters of which is covered with water and the tropics receive over half of the world's total rainfall. As much as two-thirds of the total precipitation has been estimated to fall between 30 °N and 30 °S. Annual rainfall varies from 10,000 mm to zero in the tropics. As the temperature in the tropics is relatively uniform, except where irrigation is practised, rainfall, particularly its monthly distribution, determines the vegetation and agricultural system of each tropical region (see Figs. 1.14 to 1.18). The farmer's year divides into 'dry' and 'rainy' seasons or 'monsoon seasons' rather than winters and summers.

The main characteristic of tropical rainfall is the intensity of the storms, much of the rain falling in a few heavy downpours usually accompanied by strong winds. (A fall of 25 mm (1 inch) of rainfall represents 250 t of water per hectare pounding on the soil or ground cover.) As the intensity of these storms is often in excess of the infiltration capacity of the soil, much of this rain is lost by run-off, rather than building up the soil moisture reserves. The run-off takes with it fertile topsoil and causes flash flooding of the streams and rivers. These effects are most marked where the ground cover is scanty rather than dense as with forest cover. The existing vegetation is maintained by the proportion of rainfall that becomes available to the roots.

The source of rain is a large mass of moist air which has travelled from the sea or a large lake or a large forest. This is forced upwards by a mountain range, or by meeting a cooler, drier and more dense air mass. As it is forced upwards the air cools and the moisture in the warm air mass condenses to form rain. Dew is often heavy in the tropics and may contribute up to 250 mm of water per year.

The axis of rotation of the earth is not stationary but moves rhythmically through an angle of 23½° in each direction during each year. As a result it appears from the earth as if the overhead position of the midday sun is migrating from tropic to tropic. The sun is overhead at midday at the tropics on 22, June and 22 December (solstices) and at the equator on 21 March and 23 September (equinoxes). Following this movement of the sun is a belt of low pressure, the Inter-Tropical Convergence Zone (ITCZ) which reaches the Tropic of Cancer in the northern hemisphere in July - August, the Tropic of Capricorn in December - January and the equator about October - November and also April - May. The rainfall in the tropics is associated with this passage of the sun overhead, which gives rise to a double rainy season at the equator and a single one at the tropics (see Fig. 1.8).

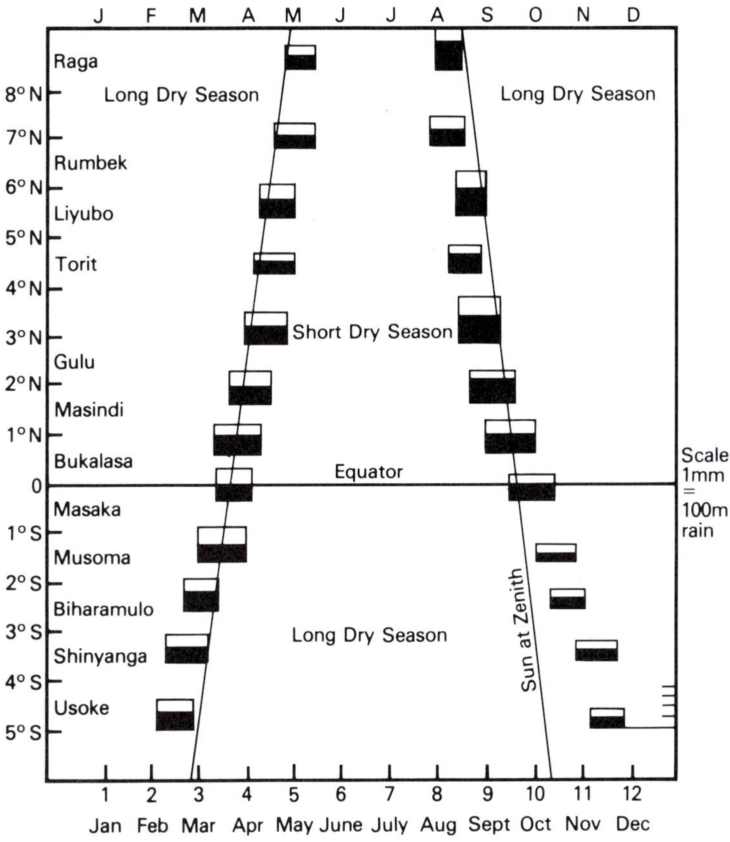

Fig. 1.8 Association of rainfall with passage of sun overhead. East Africa. (Manning 1956)

This simple explanation is complicated by other factors such as the earth's rotation which creates an east wind, and the influence of high and low pressure areas associated with land and water masses.

At the equator the period of the year when the sun is overhead at midday is divided, and the days during which it is overhead are shorter than at the tropics. Thus the regions at the tropics where the sun turns are hotter in summer than at the equator where there is more cloud cover and more rainy days. Thus the northern regions of both East and West Africa are hotter and drier (Figs 1.14–1.16; Table 1.14) than the southern regions, as in the north there is a single rainy season and a long dry summer.

The length and severity of the period between the rainy seasons determines the boundary between the rain forest and deciduous forest, and the suitability for certain tree crops such as coffee and plantains. Figure 1.25 shows the relationship of the cocoa area of Ghana to the rainfall received between November and March.

Rainfall patterns depend upon the circulation and precipitation of water-laden air. The maximum evaporation and transpiration on the earth's surface occurs near the equator, where the convectional air current rises to a high altitude and flows out towards the poles in a direction influenced by the earth's rotation (Fig. 1.9). This is north-east in the northern hemisphere. Just outside the tropics it has cooled sufficiently, meets high pressure cells, and falls to the earth's surface where it divides, part continuing in the same direction and the rest returning to the equator to replace the air that is rising (Fig. 1.9). Between 2 ° and 4 ° of the equator where the air is rising, are the doldrums, usually north of the equator due to the greater land surface of the northern hemisphere (Fig. 1.10). This generalization on air circulation is greatly influenced by land masses which have a wider range of temperature change. The air over a land mass, particularly if arid, gets very hot,

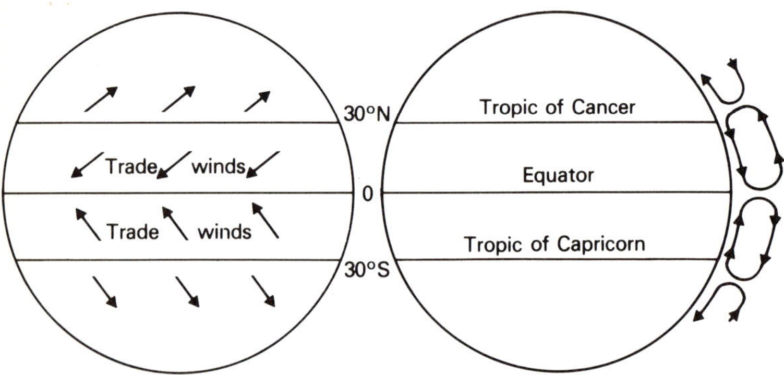

Fig. 1.9 Air circulation in the tropics.

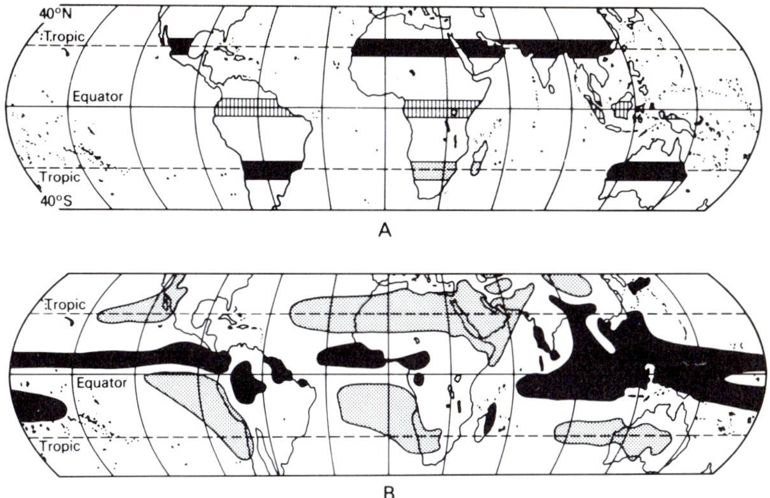

Fig. 1.10 (a) The relative position of land masses athwart the tropics and the equator. Black denotes no meridional overlap of land at the opposing tropic. (Beckinsale 1957)
(b) General distribution of wet areas and dry areas in the tropics. Black - over about 2,000 mm in precipitation annually; shaded - under about 250 mm in precipitation annually. (Beckinsale 1957)

rises, and draws in large quantities of cooler air from the sea. Before midday, if cloudless, an island is warm enough to draw in moisture-laden sea breeze. Over flat land this breeze may travel 130 km (80 miles).

Monsoons

The creation of high pressure areas over land masses in winter and low pressure areas in the heat of summer is of vital importance to the rainfall of India and South-East Asia. The average wind currents in January and July over India and the surrounding countries are shown in Figs. 1.11 and 1.12. In January the mass of Asia is cold and the air flows out, deflected by the Himalayas it modifies the trade wind system and produces the cool north-east monsoon. Rainfall occurs on the north-east side of high land, particularly where the air currents have passed over the sea, as in Madras, north-east Sri Lanka and Malaysia. Towards the end of May the land mass of Asia is warming up and creating a low pressure system which draws air from the oceans in the south. Early in June the south-west monsoon breaks on the Malabar coast and Sri

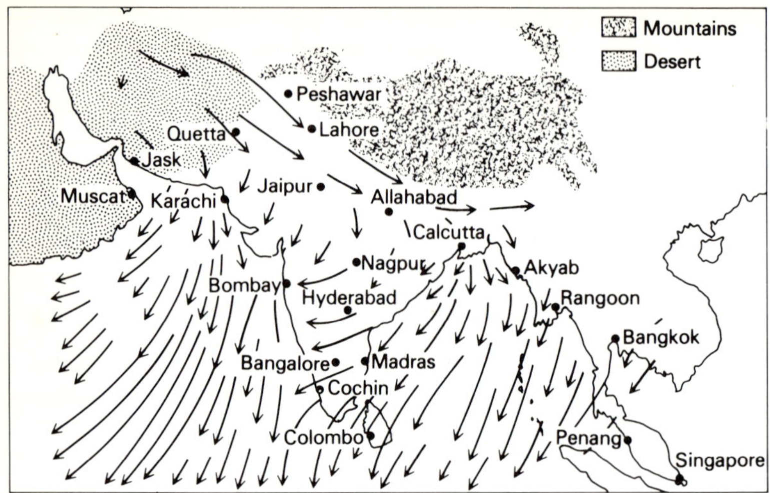

Fig. 1.11 Average wind currents, January. (Normand 1938)

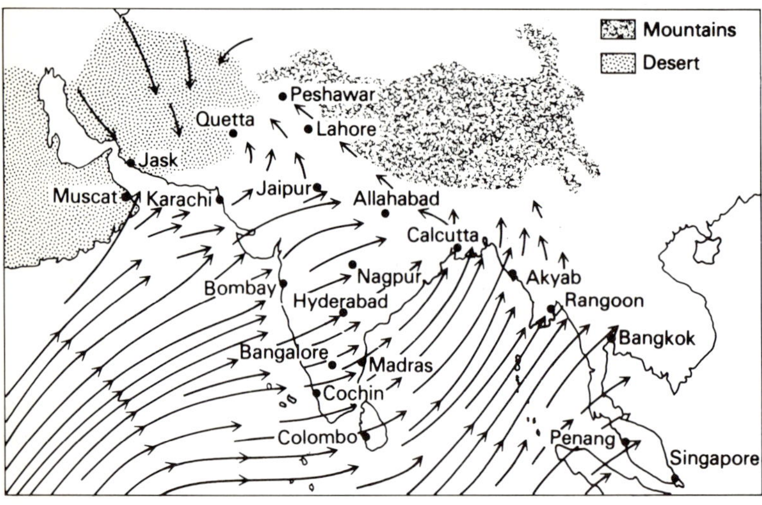

Fig. 1.12 Average wind currents, July. (Normand 1938)

Lanka. The wind currents (Fig. 1.12) and rainfall are influenced by the Ghats on the east coast of India and the Himalayas to the north, bringing heavy rain to the south-east coastal area, and along the Ganges – conditions inducing the monsoon to die down at the end of August. Thus most of India has its major rainy season between June and September, Malaysia and territories in South-East Asia, Indonesia, the

Philippines, Borneo, etc., benefit from both monsoons, and in Sri Lanka due to the high central region the effects of each monsoon are most pronounced on the windward side of the island.

The rainfall of South-East Asia is also influenced by the land mass of Australia. Situated on the opposite side of the equator to Asia, the high pressure in the cool winter of Asia corresponds with the opposite conditions in Australia and the induced wind currents over the intervening seas bring rain. Consequently, no other region of the earth gets such a high rainfall over such a large area. All Malaysia expects more than 1,500 mm of rain a year and generally receives over 2,000 mm well distributed through the year, though in the Cameron Highlands, April and October are the wettest months, following the passage of the sun at its zenith.

Unfortunately the rainfall of India is not so reliable. The monsoon may be late and bring no rain in the July-August growing period; it may finish early before the crops are fully developed; or it may rain excessively to cause floods in one region and be insufficient in another. These are the causes of the desperate famines of India. The rain which falls in the north-east monsoon in north-east India is essential for the winter wheat crop.

The cooling down of the Sahara desert in the same manner gives rise to the harmattan of West Africa, an unpleasant north-east wind, hot, dry, dusty and desiccating. Transpiration from the tropical forest of West Africa ameliorates the effect of the harmattan. As the population increases and reduces this forest cover it is likely to mean a more severe harmattan, even less rain in savanna country, and more frequent and more widespread famines like the Sahel disaster of 1972/74.

Altitude

With a rise in altitude there is generally a rise in rainfall of approximately 333 mm for each 100 m. Certain coastal belts may be wetter than the high hinterland, and in other parts the rise in rainfall may be greater, as in Assam where as much as 833 mm increase per 100 m is recorded. On high peaks the rainfall rises to a maximum at a certain altitude and falls above this height. Mountains, by intercepting rainfall, cast large rain shadows and tremendous rainfall differences occur between the windward and leeward sides of mountains as in Mauritius and the Windward Islands.

Geographical distribution of rainfall

Figure 1.13 indicates the distribution of the mean annual rainfall by

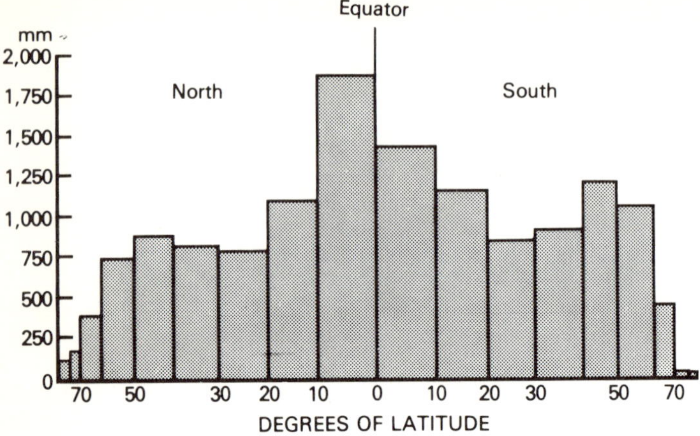

Fig. 1.13 Distribution of mean annual rainfull by latitude. (Beckinsale 1957)

latitude. Maximum rainfall occurs about 5 °N. Figure 1.13 shows the distribution of the rainfall in the tropics. Large wet areas and 'deserts' occur over the ocean. Certain deserts such as the Kalahari are landward extensions of large oceanic regions having a low rainfall. Total rainfall and its distribution in West Africa varies as one moves northwards from the Guinea coast, and the zones become increasingly dry (see Table 1.14). Exceptions to this include the very dry coastal strip centred round Accra (Kendrew 1953).

Table 1.14 Rainfall distribution in West Africa associated with latitude

Zone	**Number of rainy months** (50 mm or more)	**Number of rain-days in a rainy month**	**Mean annual rainfall** (mm)
Coast to 9°N	8–7	20–15	1,270
9–15°N	7–4	15–10	1,270–254
North of 15°N	3	10	254

The change in anticipated rainfall as one travels north in West Africa and its effect on vegetation and agriculture are well illustrated by Harrison Church (1974). Figure 1.14 shows the number of months with less than 25 mm of rainfall which is a measure of the length of the dry season.

Figure 1.15 shows the number of months the areas of West Africa can expect at least 100 mm of rainfall, the minimum requirement for agricultural needs, and is a measure of the length of the rainy season. These two parameters determine the climatic (Fig. 1.16) and vegetative zones (Fig. 1.17) of the region.

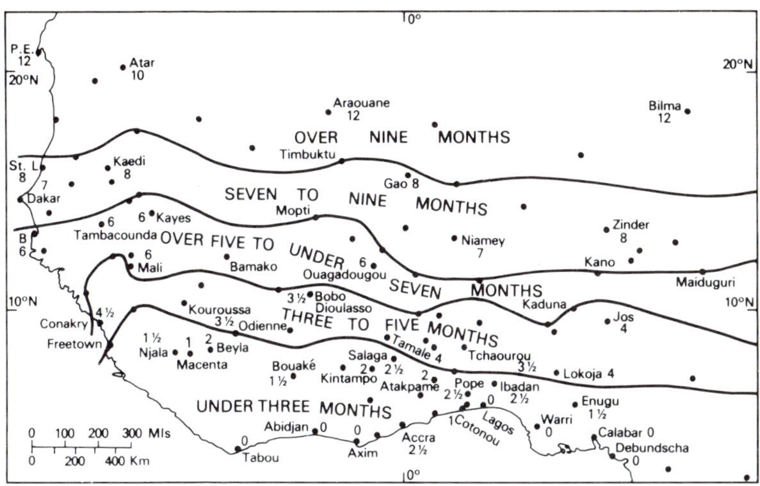

Fig. 1.14 Number of months with less than 25 mm of rain. This figure is critical for tree growth. Based on readings from 103 stations. West Africa. (Harrison Church, 1974)

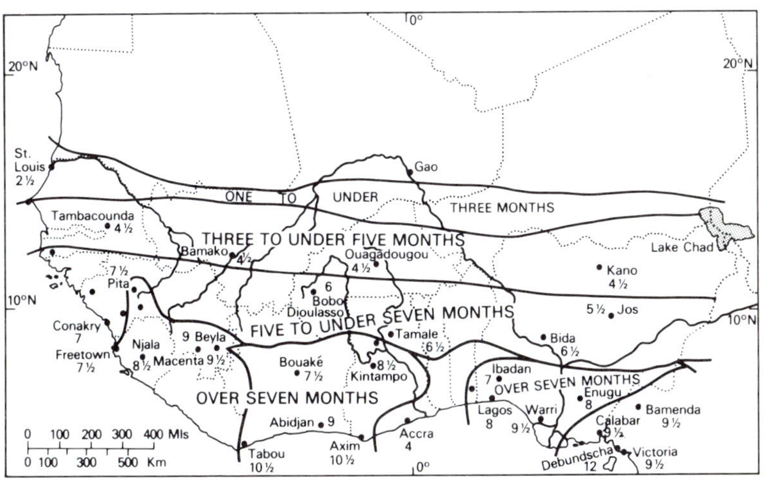

Fig. 1.15 West Africa. Number of months with at least 100 mm of rain. (Harrison Church, 1974)

As tree crops require a rainy season of at least 7 months, and the most drought-tolerant short-term millets at least 3 to 4 months with 100 mm rain, the generalized crop belts shown in Fig. 1.18 have a close relationship to the length of the dry and rainy seasons shown in Figs. 1.14 and 1.15.

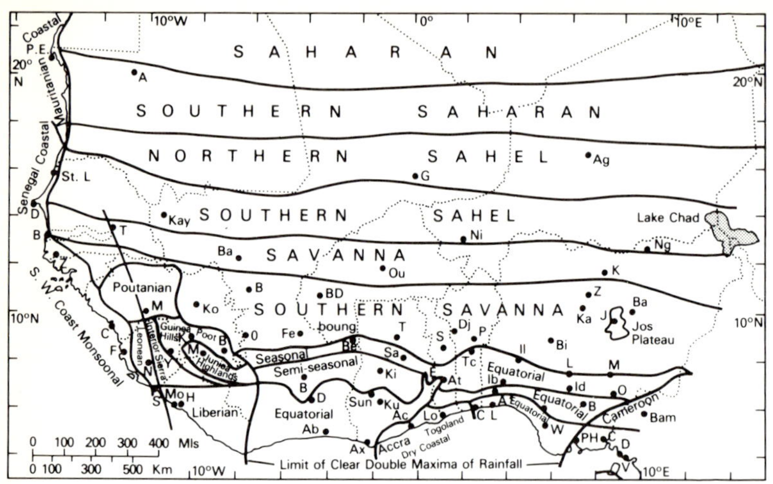

Fig. 1.16 Climates of West Africa. (Harrison Church 1974).

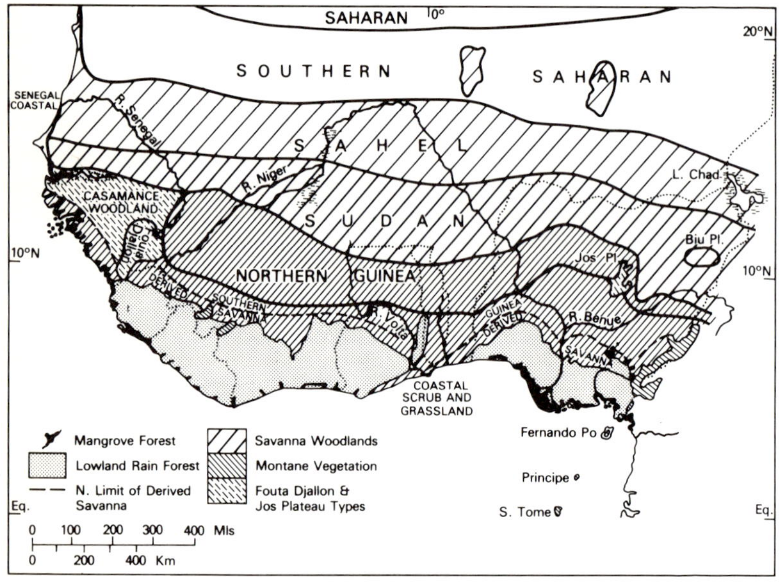

Fig. 1.17 Vegetative zones of West Africa. (Harrison Church 1974, after Keay)

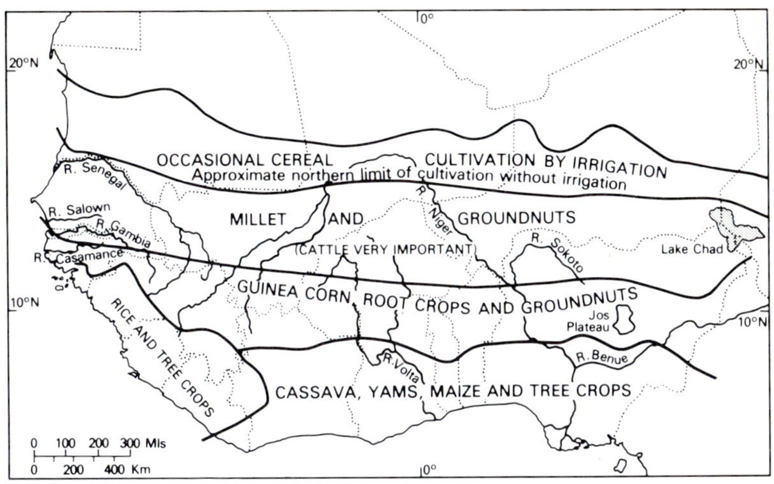

Fig. 1.18 Approximate crop belts of West Africa. (Harrison Church 1974)

Rainfall for rice regions

The water requirement for about 80 per cent of the world's rice comes directly from rainfall, and the irrigation water is the monsoon rainfall which has been stored. Rice appears to use a minimum of 1 m of water for each crop. Most of South-East Asia, including the important rice areas of Burma, Thailand, Indonesia, Cambodia, South Vietnam and the Philippines has over 2 m of rainfall and water supply is not limiting. Where the average rainfall is 1,200 - 1,500 mm a year, provided this falls mainly in the monsoon season, a single rain-fed crop can be grown. However, large areas of China, India and Pakistan have less than the required rainfall.

As explained later the variability between years in total rainfall is a characteristic of the tropics, which limits rice production in certain areas to the deltas and basins of the great rivers including the Mekong, Brahmaputra, Ganges, Indus, Irrawaddy and the Chao Phraya of Thailand, where drainage is a major problem.

The start of the monsoon determines the planting time, and during the monsoon, typhoons often occur when one-third to a half of the average rainfall may fall in 24 hours. Once the rice farmer has water in his paddies, he is reluctant to drain even when drainage is possible. This introduces a major practical problem into the crop management (Moomaw & Vergara 1964).

Forest effect on rainfall

Forests affect the local rainfall pattern. The air-stream reaching East Africa from the Congo is laden with moisture, maintained by the transpiration of the forest. Forest belts around a river delta such as the Amazon may cause localized heavy rainfall.

The destruction of forests is widely believed to reduce the local rainfall, but as far as can be shown the influence is more on distribution than total precipitation. In Malaysia when large blocks of old rubber were being felled and replanted, many planters believed that the newly replanted areas with only small seedling trees and germinating cover crops did not receive as many showers as the surrounding mature rubber. Regions near to forest appear to have more short showers which have little effect on soil moisture, though together with the cooling effect of the forest they may reduce transpiration and evaporation losses. Such showers result from the influence of the forest cover on the air above not from the moisture transpired by the forest.

Intensity of rainfall

The high annual rainfall common in tropical areas results not from an increase in the number of days or hours when rain falls, but from a considerable increase in rainfall intensity in comparison with temperate regions. Within a tropical region, differences in rainfall result from more frequent showers of longer duration rather than higher intensities, though rainfall intensities do vary considerably when compared over short periods. Few tropical areas receive rain on more than 200 days of the year, but an average of 12 - 20 mm falls on these wet days. As much as 50 - 75 mm of rain frequently falls in 1 hour, and even 125 mm in 1 hour has been recorded in exceptional cases. In 1 day 250 - 500 mm of rain may fall and twice this amount has been recorded in certain hill stations. As the storm conditions persist over a number of days more rain may fall in 1 week than Manchester receives in a year. During August 1841 Cherrapunji received 6,120 mm of rain, 3,810 mm of which fell in 5 days. This station has recorded over 1,016 mm during a single day (Beckinsale 1957). The wettest place in Africa is Debundscha at the foot of Mt. Cameroon (4,070 m) where (60") 1,500 mm often fall in August alone. In Indonesia the mean number of rain hours per rainy day is fairly constant throughout the archipelago at about 3½ hours. The mean intensity in Djakarta is fairly constant between 4.57 mm and 6.85 mm per rain-hour throughout the year. The highest intensity observed did not exceed 4 mm minute. About 22 per cent of the rainfall in Indonesia falls as cloudbursts, that is in showers with an intensity of at least 1 mm/minute for 5 minutes (Mohr *et al*. 1972). The distribution of storms at Namulonge in Uganda, a station receiving 1,000 - 1,500 mm of rain a year, is shown in Fig. 1.19. At Shinyanga in Tanzania where the average annual rainfall is 635 mm, a quarter of this has been recorded in two storms.

The heavy intensity of the tropical storms is frequently in excess of the receptive capacity of the soil and the beating of the soil by the rain-

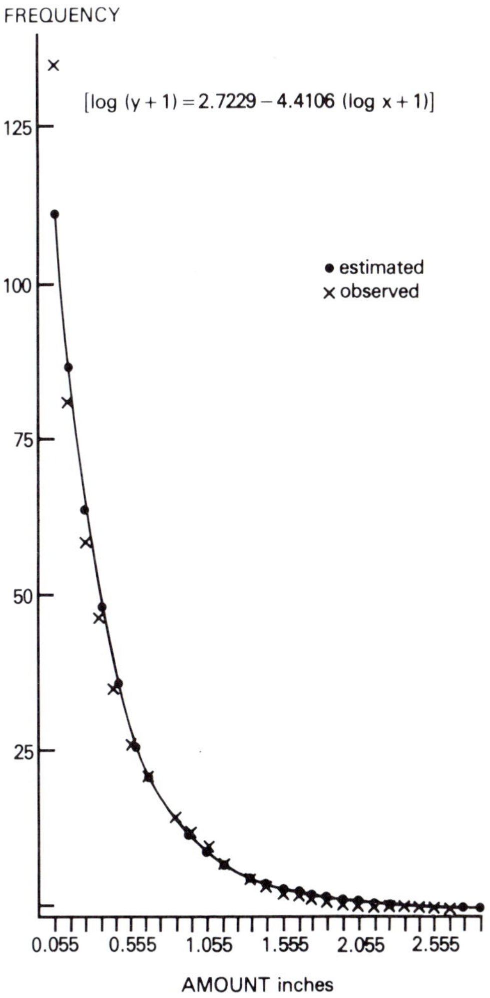

Fig. 1.19 Distribution of storms at Namulonge, Uganda. (The Cotton Research Association)

drops seals the soil surface, hindering penetration. Run-off is thus increased (see Table 1.16) causing erosion, flooding, the rapid rise of rivers, and landslides. The concentration of the rainfall, however, leaves many hours of sunshine, particularly outside the humid tropics. Medium showers, 6 mm in 30 minutes, are absorbed best by most soils.

Variation of rainfall from the mean

The total annual rainfall shows not only a variation from place to place but also a wide difference from year to year for the same rain gauge. The rapidity with which the average total rainfall can change over a short distance is well seen in Mauritius, where a single sugar estate may over a short distance cover the three climatic zones with an average rainfall ranging from 900 mm on the coast to 3,200 mm or more at about 360 m above sea level. This is due to the mountain ridges causing a sudden elevation of the south-east trade winds. Variations in rainfall from year to year of 50 per cent from the mean are not uncommon and usually a result of cyclones in the South Indian ocean.

Average rainfall is a commonly used expression of little practical

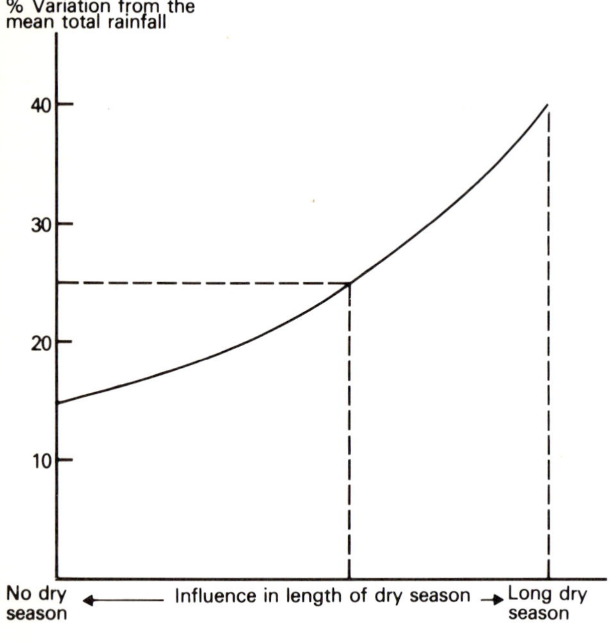

Fig. 1.20 Variation of rainfall from the mean.

value, except as a broad generalization, indicating the type of vegetation and cultivation likely to be found. Distribution and reliability are more crucial factors. Manning (1956) has explained the importance of fiducial limits and the association of two rainy seasons with the sun's movement between the tropics. In East Africa, distant from coastal influence, the two rainy seasons may become continuous at the equator (Fig. 1.8). The variation of rainfall from the mean is greater where the dry season is longer (Fig. 1.20).

Manning's analysis of the rainfall data for Uganda shows the regions where double cropping is feasible. It also analyses the chances of establishing a crop like maize at an early planting date when there is no traditional knowledge available for guidance. The probability map (Fig. 1.21) shows an association of low population (Fig. 1.22) and low rainfall expectancy, a minimum expectancy of 840 mm appears necessary for a high population density. No irrigation is practised in this area.

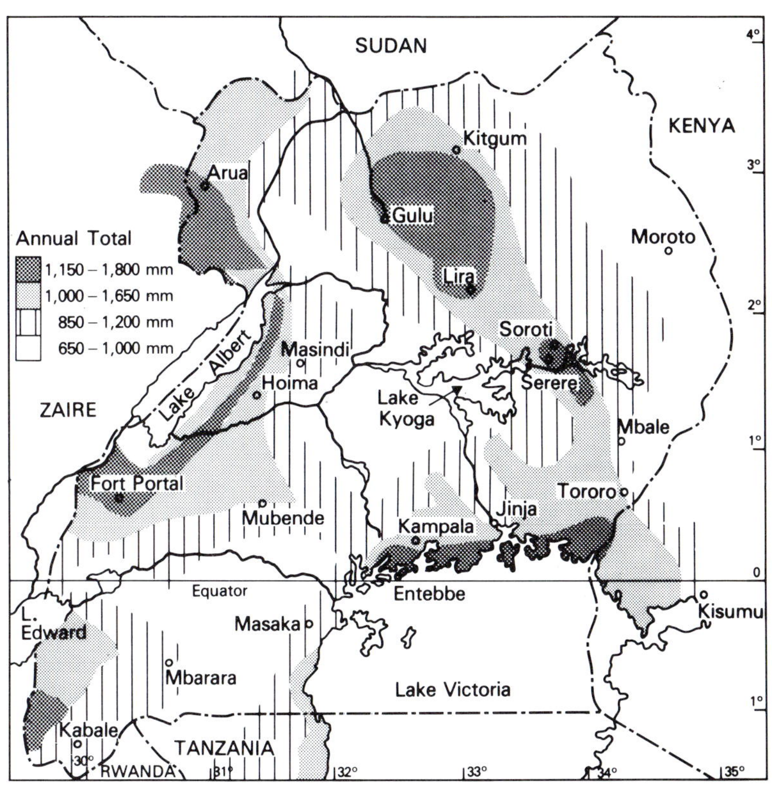

Fig. 1.21 Rainfall probability map, Uganda. (Manning 1956)

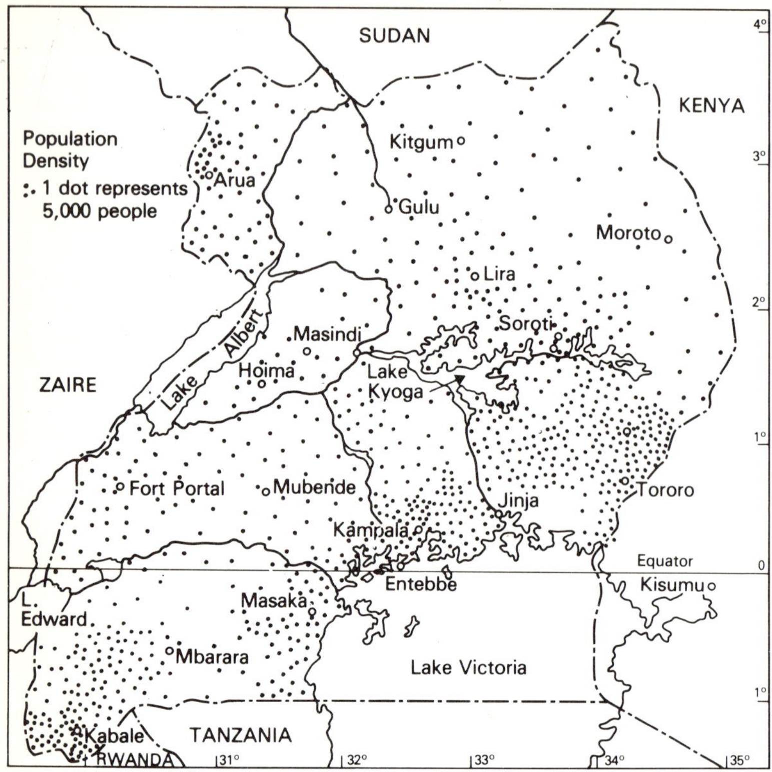

Fig. 1.22 Population distribution in Uganda. (East African Royal Commission 1951)

What happens to rainfall*

While the previous section explains the cause, distribution and need of rainfall, it is important to see what happens to the rain, particularly as the amount of available rainfall may be very critical for the density of population supported.

Some of the rainfall never reaches the ground, being intercepted by vegetation and evaporating before it can reach the soil. Mohr *et al.* (1972) estimate that 4 mm of each rain shower is intercepted by the tropical jungle cover, and that the amount of rainfall reaching the soil surface varies with the cover as follows:

* For a full discussion see Penman (1963).

Dense forest or jungle	70-80%
Dense high grass	80%
Cereals and other crops	80-85%
Bare soil	100%

They also quote data from Surinam showing that the amount of rainfall retained by the forest canopy rises with the amount received, though the percentage retained falls (Table 1.15). On average 75 per cent of the rainfall reaches the soil in a tropical forest. As this vegetation cover reduces the pounding effect of the raindrops on the soil, and influences the soil moisture and temperature, this is a loss which must be accepted.

Table 1.15 Retention of rainfall by forest canopy

	During a rainfall per day (mm)								
	1	2.5	5	7.5	10	15	20	30	40
Retained in the canopy (mm)	0.8	1.2	1.6	2.0	2.5	3.2	4.2	6.0	7.8
% of the rainfall retained in the canopy	80	48	32	26	25	21	21	20	19.5

Penman (1963) has pointed out that the evaporation of rainfall or irrigation water from the wet leaves will be greater during the day than the transpiration of soil water, and can go on at night. But as the same energy cannot be used twice, while the intercepted water is being evaporated the drain on soil water is checked. Thus the intercepted water is not wasted.

The rain which reaches the soil surface is partly absorbed by the soil, or enters the soil through the cracks in a dry soil (see Plate 1.1); and part runs off over the soil surface. The rate at which water enters the soil determines the amount lost by run-off and affects the amount lost by evaporation. The proportion which enters the soil depends upon the rate at which the soil will accept rainfall in comparison with the rainfall intensity. A soil having an open surface and good structure will take up rainfall more rapidly than overcropped soils whose surface has been compacted and sealed by the pounding of raindrops in heavy storms. Subject to the modifications by aggregation, the finer the soil texture the lower the rate of infiltration and the greater the force with which it can retain the water. Soils at Samaru in Northern Nigeria which are sandy loams in the upper layers, have a very high infiltration rate, up to 130 mm/hour under a grass fallow. This means the soil will take in nearly all the rain in a heavy thunderstorm. This high infiltration rate is associated with large conducting channels produced by the activity of worms and ants.

An infertile fallow has little worm activity and a slower infiltration rate, but as the fertility rises, so does the worm activity and the infiltration rate, and the growth of the fallow become more lush. Cultivation after a fallow reduces and clogs up these channels, and cuts down the infiltration rate. With the hotter, drier, less vegetative conditions under the crop, and the sealing action of the raindrops, the infiltration rate at the end of the first cropping season may be down to 7 or 10 mm/hour which is the level recorded on these soils after continuous cropping. This is insufficient to take up rain in heavy storms and soil erosion results.

Worm activity and rainfall infiltration can be improved by applying dung (Wilkinson 1970). Organic matter in the surface soil increases the rate of rainfall acceptance. A cover crop or surface mulch intercepts the raindrops, disperses their kinetic energy, and reduces compaction. As mulching also improves the soil structure, it increases the amount of rain absorbed by the soil. This absorbed water, if sufficient in amount, penetrates deeper and deeper, bringing successive layers to field capacity in excess of which the water drains away.

An indication of the loss of rainfall by run-off in a wet period is given in Table 1.16 (see also p. 91, Table 1.24).

Table 1.16 Amounts of rainfall, run-off and soil loss on mission slopes (Namulonge, Uganda, April 1957)

Type of storm	**Rainfall** (mm)	**Run-off** (mm)	**Soil loss** (t/ha)
Heaviest storm (1)	82.6	34.8	17.6
2nd heaviest storm (1)	56.6	23.6	13.8
All other storms (7) with measured losses	93.0	16.0	8.8
All other storms (13) with no appreciable loss	74.7	Nil	Nil
Total for month (22)	306.9	74.4	40.2

The data in Table 1.16 refer to a field at Namulonge Cotton Research Station in Uganda which was planted with beans prior to the eight consecutive cotton crops. The slope has a 5 per cent gradient and the crops are sown 'on the flat'.

Evapotranspiration

The rainfall absorbed by the soil is either lost in drainage, if it exceeds the field capacity of the soil, or is transpired by the crops and weeds, or evaporated from the soil. As it is very difficult to estimate the water loss by evaporation from the soil and transpiration through the plant independently, this total loss is usually assessed as evapotranspiration (E_T). Evapotranspiration depends upon humidity, temperature, solar radiation, wind velocity, soil moisture, soil type and structure, crop, mulch and weeds. A better understanding of plant and soil moisture relationship depends upon obtaining a correlation of E_T with meteorological observations under tropical conditions. Penman (1948, 1950) working with Schofield, correlated evaporation from an open water surface (E_O) and E_T for a grass surface, with meteorological observations for conditions at Rothamsted. They found that $E_T:E_O$ was closely related to day length and for a 12 hour tropical day E_T should be about 0.7 E_O. This is largely because the evaporation from a bare surface and from the surface of a growing crop is primarily determined by the radiant solar energy. As leaves reflect more than a free water surface the rate of evaporation from leaves is less, but as the crop develops the leaf area is greater than the soil surface area and $E_T:E_O$ can rise to 1 and over 1 where water is plentiful as in rice paddy. Except where soil moisture or cloud cover is limiting this implies that mean evapotranspiration E_T is nearly constant for all types of vegetation and crop cover and independent of rainfall for one latitude. For maximum plant growth rainfall should at least equal potential evapotranspiration, and irrigation should provide the difference between potential evapotranspiration and rainfall. In East Africa the efficient use of water is critical as annual rainfall is less than half the potential evapotranspiration on more than half the land surface, on only 3 per cent of which does rainfall reliably exceed potential evapotranspiration. With the closure of leaf stomata at night transpiration is much reduced.

Drainage and leaching

The water loss by drainage is not usually so important in itself as for the nitrate, sulphate and cations (bases), which are leached out. Continual leaching over the centuries has produced the acid soils low in bases and general fertility, so common in the very wet tropics. Clearing of the forest for cultivation accelerates the process, for example when the forest at Lomaivuna in Fiji with a rainfall of 3,556 mm a year was cleared for bananas, the pH fell within 3 years from about 4.2 under the forest to 3.8, and all the important bases and nitrogen became deficient, making the soil too poor for bananas without liming and fertilizers. It is probable that after applying lime and fertilizer a dense stand of bananas could be established. Mulched with split stems and leaves, the water-holding capacity of the soil would be increased, leaching losses would be reduced and the productivity restored. This is a problem common to all the South Pacific islands when forest is cleared for cropping under conditions of high rainfall, and it is essential to keep the soil temperature down and the soil organic matter up. In Zimbabwe losses of nitrate from ridges by leaching is less than with crops on the flat, as the ridges are drier, more water being lost by evaporation than drainage.

Inter-relationship of crop and water

Roots absorb water and nutrients from the soil along their whole length provided the outer layers of cells in the root remain alive. However, nutrients are not withdrawn equally from the entire rooting zone. The absorption and transport of phosphate and potassium take place through the entire root system of plants several weeks old. In contrast calcium is translocated to shoots from the relatively young roots, hence a continuing supply of calcium near the growing root apices is essential for growth (Scott Russell 1977). The plant can adapt itself to a locally favourable condition and an increased concentration of nutrients can cause the root form to change. More and longer laterals grow in the zones where nutrients are present in a higher concentration irrespective of their position on the root. Ethylene produced in waterlogged soils inhibits root extension, more seriously in crops less tolerant to waterlogging.

Nutrient uptake capacity is more closely related to root volume than either root length or surface area. Nutrient uptake from the upper soil layers can drop markedly during quite short periods of dry weather.

The root systems of most tropical crops have been studied. Characteristically they have a mass of surface roots which permeate the fertile and microbially active top 15 cm of soil, and a series of deep penetrating roots that can tap deep supplies of water and minerals. Cereals, sugar cane, and other grasses do not have a tap root like the tree crops, but can still penetrate deep into the subsoil. The important feeding roots are in a circle around the main shoot which is important for fertilizer placement. Where a hard pan or undecomposed rock occurs or a high water table, the root depth is reduced (see the diagrams of coffee root systems in East Africa (Nutman 1933)), but the trees remain productive if the topsoil is very fertile and the moisture adequate. In a severe drought the root hairs of the surface roots die, and when the rains break or the land is irrigated their regeneration takes about 3 days and absorption recommences. Nutman considered that the dryness of the hard pan rather than its physical resistance was the cause of lack of root penetration. Neither the roots of cotton (Balls 1951) nor coffee (Nutman 1933) will survive in the water table, and this applies to other crops known to suffer from waterlogging, including the coconut. Roots of some crops can survive in waterlogged soil provided that it is not stagnant and is continually being replaced by water with dissolved oxygen. Die-back of coffee caused by overbearing is also accompanied by a die-back of the roots, associated with a depletion of carbohydrate reserves in the roots. Starch reserves in the roots are important in the recovery of tea from pruning (see p. 211).

The depth to which roots can penetrate under suitable soil conditions is rather surprising. A cotton plant will send its roots down 2 m or more and the tap root of a cocoa tree to 1.5 m. The tap root of coffee tree is short, perhaps 0.3 to 0.5 m, but the finer roots can penetrate 3 m. The 'rope system' of sugar cane roots described by Evans (1936) penetrates 3–3.5 m if given the opportunity (Fig. 1.23) but this 'rope system' is rare the superficial roots absorbing the moisture and nutrients. The roots of elephant grass on a deep fertile soil in Uganda, were found going down to the bottom of a newly dug deep latrine pit. Thus the root system of a crop is often larger in size if not in volume than the aerial parts.

The influence of available water on crops

Knowledge of the water requirements of tropical crops is generally lacking. Water availability at certain periods in the growth of the crop has a very marked influence on the yield though the exact correlations have not yet been worked out for many crops. Cotton is one that has been studied intensively and a number of relationships are known for certain regions.

In Northern Nigeria, King (1957) found a highly significant correlation between rainfall after the end of September and the bales of lint for export produced from every ton of seed issued for planting. While much

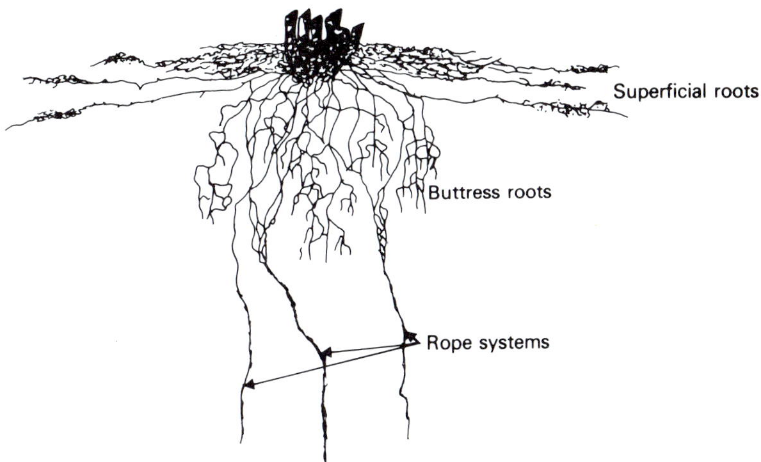

Fig. 1.23 Root system of sugar cane. (After H. Evans)

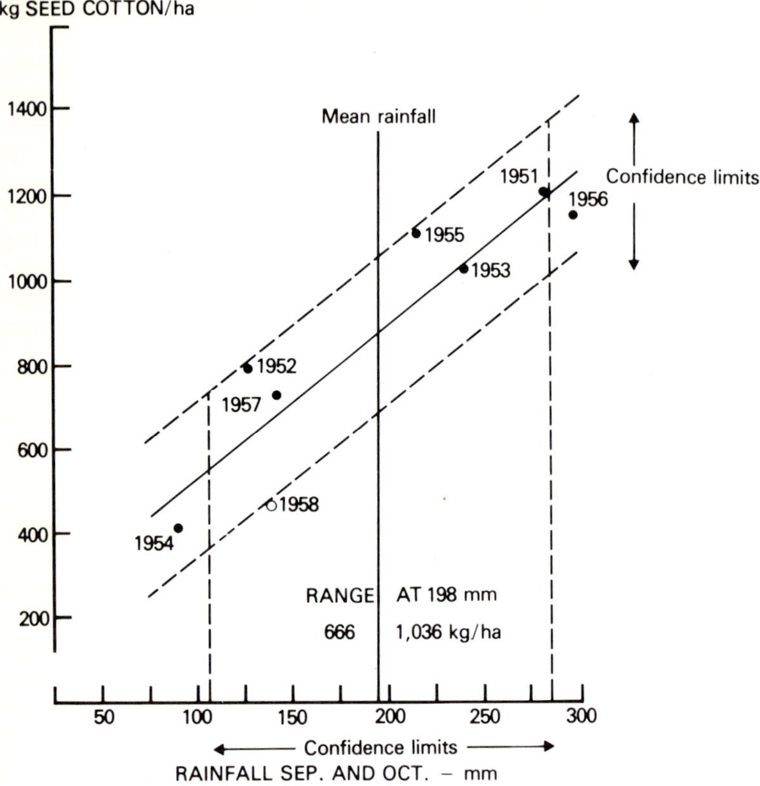

Fig. 1.24 The effect of rainfall during September and October on large-scale Namulonge cotton yields. (Manning and Kibukamusoke 1960)

of the cotton is planted late every 25 mm of rain missing in October was equivalent to a loss of 20,000 bales of lint worth about £1 million at 1957 prices.

A similar association applies to Namulonge (Fig. 1.24), which is probably true for a large part of Buganda. The time of onset of water stress is critical to ultimate yield, which shows an increase of 90 kg/ha of seed cotton for every 25 mm of rainfall received in September and October. The effects of rainfall variability from year to year are best minimized by early planting, or even dry planting, and rainfall conservation.

The yield of cotton in the Sudan Gezira is positively correlated with the rainfall between 1 July and 15 August, which is the period prior to sowing (Crowther 1944). High total rainfall in the year prior to sowing has a detrimental effect on crop yield. This has been attributed to the weed growth on the fallows utilizing the nitrogen and decreasing the soil moisture. Thus the cotton yield is largely determined prior to planting. A formula has been developed for calculating the expected crop. The incidence of severe pest outbreaks can upset all calculations, as in the 1957–58 season, a year of unfavourable rains, when the average yield from the irrigated Sakel cotton in the Sudan fell to 170 kg of lint per hectare, about one-third of the normal yield. Rice must have adequate water from panicle initiation to full flowering.

The influence of rainfall on tropical tree crops is more obvious. Tea dislikes a period without rain and ceases to put out new growth, causing a decline in production. Gadd (1935) estimated that a tea plantation transpires 900 mm of rain a year. Considering that the crop prefers areas with 1,500 mm of rain it would appear that only about 60 per cent of the rainfall is utilized by the crop.

Rainfall is not the sole factor controlling flushing in tea, as was illustrated at an irrigated estate in Zimbabwe where crop production fell when the temperature dropped, and rose at the end of the cool season (Harler 1956, p. 40). A combination of high moisture content and temperature is needed for flushing.

Flowering of coffee trees is induced by the showers at the start of the rainy season (see Ch. 2 p. 199). This is fortunate as the succeeding rainy season is available for the cherry to fill out and the crop is usually harvested as the dry season starts.

The length and severity of the dry season frequently determines the geographical distribution of tropical tree crops including tea, coffee, rubber, cocoa and bananas. Figure 1.25 illustrates this for cocoa in Ghana, where the crop is restricted to areas receiving more than 254 mm of rain between November and March.

SOLAR RADIATION

In the past solar radiation, which provides the energy for both photosynthesis and evapotranspiration, has been largely neglected as a factor in tropical agriculture. Under farmers' cropping conditions water and nutrients are normally more limiting to crop yield than solar radiation.

A large part of the solar radiation reaching the atmosphere is absorbed in the atmosphere by dust, clouds, water vapour and the air, so that the maximum received at the ground is around 200 or 220 kcal/cm^2 per annum in the desert regions. Tropical forest receives about 120 or 160 kcal. On reaching the vegetation or soil part is reflected, vegetation reflects 10, 15 or even 30 per cent. Consequently only about half the solar energy entering the atmosphere is ultimately available at the ground, and only a quarter of this is of the correct wavelength for photosynthesis.

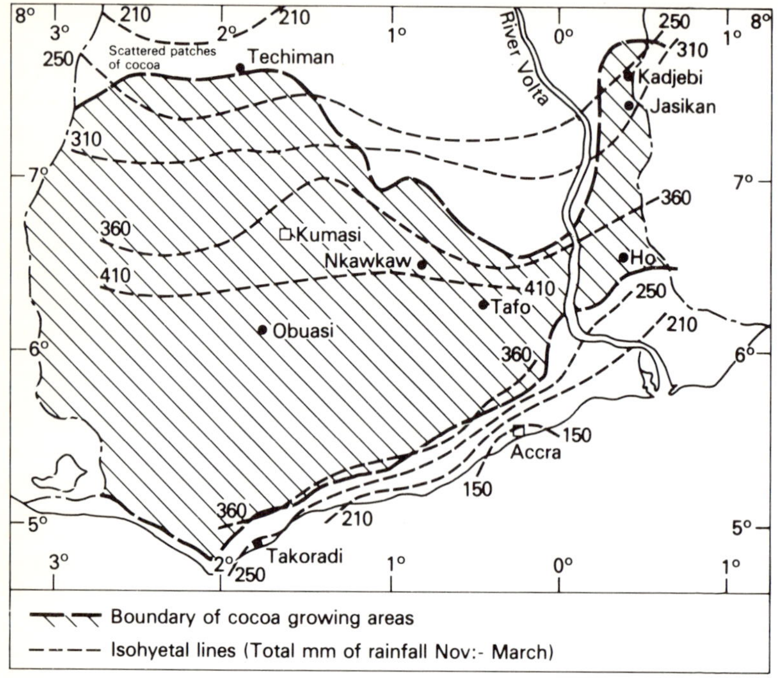

Fig. 1.25 Relationship of cocoa area of Ghana to the rainfall received between November and March. (Adams and McKelvie 1955)

Loomis and Williams (1963) calculated that with a crop receiving 500 cal/cm^2 per day, 40 per cent was used in evapotranspiration and the potential net production of the crop was 250 t of dry matter per hectare per year, which is far far more than with any agricultural system yet devised. At the best crops are only utilizing 1 to 2 per cent of the incident solar energy. With a leaf surface fully exposed to sunlight, the CO_2 concentration in the atmosphere is the limiting factor in photosynthesis. Leaf surfaces within a crop, shaded by higher leaves, receive less radiation than they can utilize and this limits photosynthesis. Breeders are working towards a plant form with a minimum of mutual shading to obtain maximum photosynthetic production. The crop can then be planted at a spacing which will give as complete a canopy as possible at the growth stage when maximum photosynthesis is required so that all the incident light will be intercepted.

Sunshine is more intense in the tropics due to its more vertical and shorter passage to the earth. In the rainy seasons and monsoons the sun is often obscured by the heavy cloud as is shown by the reduction in solar radiation recorded in India during the monsoon (Fig. 1.26).

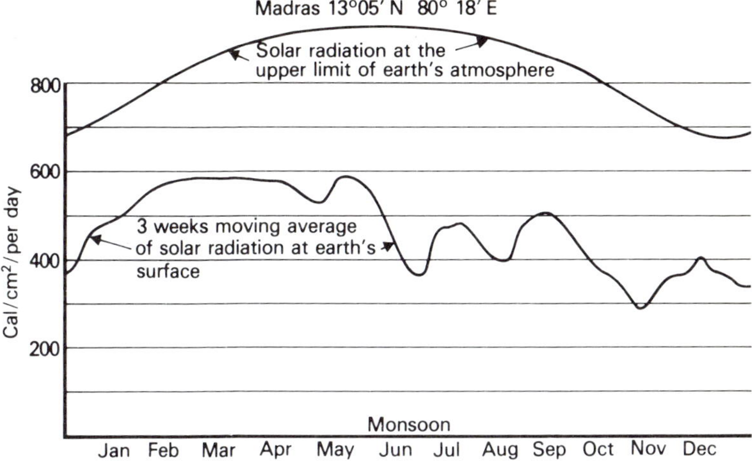

Fig. 1.26 Variation in solar radiation received at earth's surface in Madras, showing effect of monsoon. (Huxley 1963)

Similarly during the harmattan months in West Africa, the appearance of the afternoon sun in the heavily dust-laden atmosphere is reminiscent of a foggy afternoon in Northern Europe, and can be looked at directly without discomfort (Plate 1.3).

The variation in solar radiation throughout the year at Los Baños, Laguna in the Philippines is shown by a comparison of the dry season in April and the end of the rainy season in December (Table 1.17).

Table 1.17 Variation of solar radiation at Los Baños, Philippines

Month	**Rainfall** (mm)	**Number of rainy days**	**Mean daily temperature** (°C)	**Solar radiation** (gcal/cm^2) (1959-65)
April	41	5	28.5	14,850
December	147	17	25.4	8,140

Source: Annual Report IRRI (1965)

The importance of this variation is shown from the calculation that 100 cal/cm^2 per day increase in solar radiation could at a minimum estimate increase dry matter production by 25 kg/ha per day using 1 per cent of the available incoming radiation (Huxley 1965).

Prior to the advent of the relatively cheap and straightforward Gunn-Bellani radiation integrators, sunshine hours were generally recorded. Solar radiation and sunshine hours are not directly related as a lot of the sunshine may be absorbed by dust and humidity. Sunshine hours vary from just over 1,000 hours per year in the coastal areas of

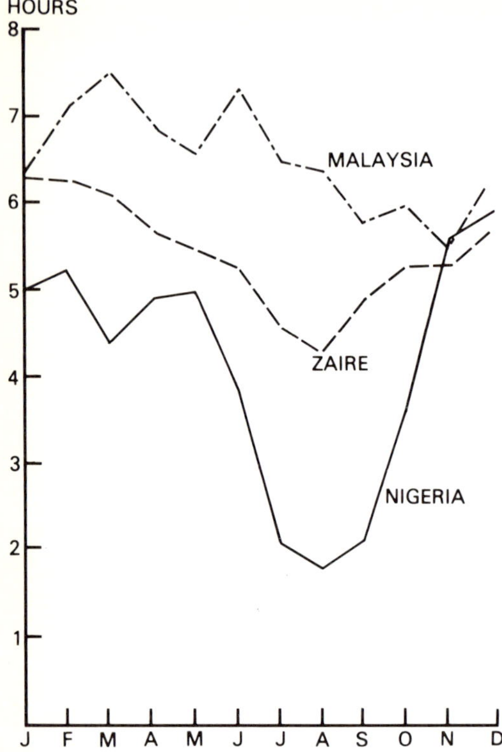

Fig. 1.27 A comparison of the average daily sunshine of three oil palm regions. (Hartley 1958)

equatorial West Africa to around 3,000 hours in Queensland. The latter figure helps explain the high sugar yield of Queensland cane. In Hawaii the difference in light intensity between two otherwise identical environments is responsible for one site producing 140 t of sugar cane per hectare and the other 400 t (Evans 1955a). Figure 1.27 shows the comparison of daily sunshine hours of three oil palm regions. Yields are highest in Peninsular Malaysia which has the highest average number of sunshine hours.

Hardy (1960) quotes results for cacao showing the effects of increased sunlight on the assimilation of dry matter (see Table 1.18) and comments that these figures are only about one-quarter of the value obtained with temperate crops such as potato and apple, but high for tropical crops. Coffee is about two-thirds as efficient, and coconuts which have motor cells along the midrib of each leaflet which keep the edges of the leaf towards the sun, only one-sixth as efficient as cocoa. The growing

Table 1.18 Effect of sunlight on dry matter assimilation - cacao

	% of full sunlight	g.dm²hr*
Trinidad	10	0.042
	20	0.099
	75	0.188
Ghana	20	0.072

* grams of dry matter per square decimeter of leaf surface per hour

Table 1.19 Response of rice varieties to solar radiation

Variety	**Grain produced per unit of sunlight** (g)
Taichung (Native) 1	15.5
Milfor	11.5
Peta	8.4

Source: IRRI Reporter, March 1966.

of a tall and a short crop together, such as coffee in bananas, is a 'two-storey' cropping system which makes good use of the solar energy. In the early stages of growth a ground crop like taro (*Colocasia*) is usually planted to use the solar energy that penetrates the two upper canopies. Brown (1965) suggested that for cotton grown at Samaru in Northern Nigeria there appeared to be a yield barrier at 2,464 kg of seed cotton per hectare, which might be a result of the limitation in the quantity of available radiation as this crop forms a high unbroken canopy.

Studies on three varieties of rice in the Philippines showed that the yield of each variety was correlated with the amount of sunshine measured in g cal/cm^2 per day, received in the 4 to 6 weeks before harvest. The utilization of sunlight varied between varieties (see Table 1.19). Taichung (Native) 1 and to a lesser degree Milfor were more efficient during cloudy weather. Taichung (Native) 1 produced about 6,500 kg/ha in the dry season, and about 5,040 kg/ha in the wet season, but the yield per unit of sunshine was greater in the cloudy wet season. Other studies at the same Institute (IRRI 1966) have shown that light is often limiting for the rice crop. In the monsoon season the heavy clouds may reduce the light intensity below the level needed for the plant to gain weight, losses from respiration being greater than the carbon assimilated (Fig. 1.28). Mutual shading, of one leaf by another, may also mean that on sunny days some leaves receive inadequate light. Mutual shading only becomes important at the stage of maximum tiller number. The addition of 100 kg of nitrogen per hectare by increasing the leaf area so increased the mutual shading effect in the tall indica variety Peta that the yield was almost halved, compared with a 10 per cent yield reduction by mutual shading when no nitrogen was applied.

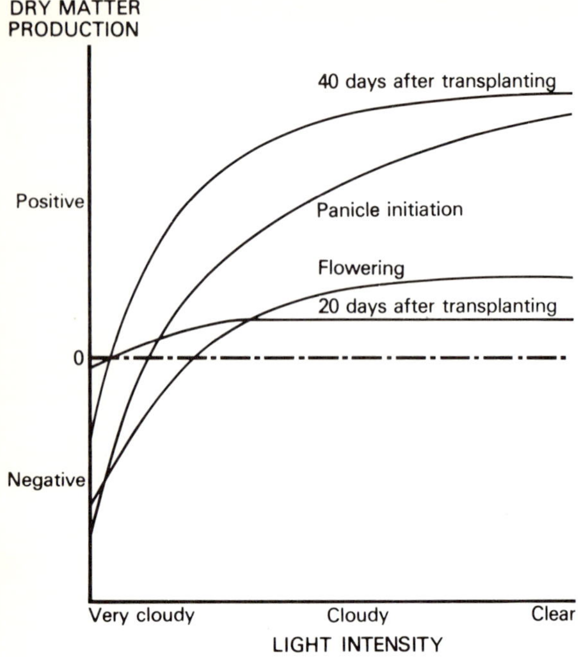

Fig. 1.28 Effect of light intensity on dry matter production at different growth stages. (IRRI Reporter May 1966)

The ability of shaded leaves to photosynthesize is reduced, and they die prematurely. Shaded rice plants grow taller and have a greater tendency to lodge. The reduction of yield with closer spacing, frequently recorded in rice spacing trials, is often a consequence of mutual shading, which is less with varieties having short upright leaves than the tall varieties frequently grown.

Full utilization of solar radiation is important in all crops, not just rice. Where the soil is poor, or rainfall inadequate, the restricted leaf production cannot fully utilize the solar energy. Plant breeders are now consciously designing plants with the right leaf number, size and angle to intercept as much sunlight as possible, provided they are planted at a spacing to give a complete canopy when fully grown. In perennial crops, such as rubber, this may be achieved by top grafting. The traditional practice of intercropping makes the best use of solar radiation throughout the year. Cocoa is often shaded to reduce the amount of solar radiation taken up because the plant cannot obtain adequate water and nutrients from the soil to maintain a higher level of photosynthesis. If shade is removed nutrients must be applied.

Day length

Days are shorter in the tropics than in the summer growing season of the temperate regions. At the equator the sun is above the horizon for 12 hours 4 minutes every day of the year, but at the tropics (23 °N and S) the day length varies from 10½ to 13 hours. Tropical plants are considered 'short day' plants, and some are very sensitive to photoperiod. Day length and solar radiation are not well correlated in the tropics. Certain plants, or more commonly certain cultivars or varieties of certain plants, require a definite length of daylight, either long days or short days to induce flowering or form tubers. Many temperate crop varieties requiring a very long or short day to flower will not flower in the tropics. This could be an asset in crops such as pasture where flowering is undesirable. Rice varieties occur which are short day, others long day, and others day-length insensitive.

It is important when introducing a new variety of a crop such as maize or sorghum from outside the tropics, to make certain that the variety will flower in the day length of the new environment. Some crop varieties such as soya beans are very sensitive to small changes in day length, others like groundnuts are not. Day length may determine the optimum planting date of a photosensitive crop variety as illustrated in Table 1.20.

Table 1.20 Influence of sowing date on a photosensitive rice variety at Bogor, Java

Sowing date	**Crop duration** (days)	**Yield** (kg/ha)
10 Nov.	147	4,000
24 Nov.	140	4,480
8 Dec.	132	5,040
22 Dec.	125	4,210
5 Jan.	118	3,730

Source: Siregar (1954).

Craufurd (1964) comparing different sowing dates of swamp rice at different latitudes in West Africa, showed the most photosensitive varieties had the longest period between sowing and flowering and were least influenced by climatic conditions, the converse being true for the least sensitive varieties. He concluded that in order to obtain the highest yields, the sowing date should be as near to the longest day as possible; and that the relationship between sowing date, yield and duration was valid wherever the sensitive varieties were grown.

Vegetable seed grown in tropical latitudes is usually more satisfactory than seed introduced from Northern Europe, largely on account of this response to day length.

Air temperature

The following factors influence air temperature:

1. Latitude. The amount of heat received annually is greatest between 15 °N and 15 °S. At higher latitudes the high intensity of solar radiation is combined with a long day and temperatures in summer will exceed those nearer the equator. Winters will be cooler.
2. Altitude. There is a fall of approximately 6 °C for every rise of 1,000 m.
3. Distribution of land and water. Near water, as on islands or lake shore areas, the diurnal range is much reduced and lower maximum temperatures occur, but humidity will be high which reduces solar radiation.
4. Distribution of water and air currents. They may have a cooling effect as the sea breezes in the Caribbean, or the opposite effect as with the later stages of the harmattan in West Africa.
5. The amount of dust and water vapour in the air to absorb the heat of the sun. About 20 per cent of the sun's energy is transformed this way (Plate 1.3).

Within the tropics the temperature at sea level rarely falls below 16 °C. Where humidity is high the maximum shade temperature rarely exceeds 38 °C, but where the humidity is low 55 °C may exceptionally be reached. It is therefore difficult to dissociate excessive temperature from insufficient moisture or humidity. The seasonal fluctuation in ambient temperature is less in the tropics than in temperate regions. Low temperature, which is often a limiting factor in plant growth in temperate agriculture, is only limiting in high altitude areas of the tropics.

Tropical crops have a higher optimum growth temperature than temperate crops but the temperature range over which the crops are successfully grown varies not only between crops but between varieties of a crop. Hardy (1970b) gives the following cardinal temperatures in °C for three tropical crops.

	Minimum	*Optimum*	*Maximum*
Cacao	15	25.2	38
Maize	10	30 - 35	45
Arabica coffee	10	20	30

Bananas, cacao, coconuts, oil palms, rubber, sisal and sugar cane are true tropical perennial crops requiring a high minimum temperature for growth, whereas citrus, arabica coffee, tea, and avocado will grow over a wider range of temperature conditions provided that frost does not occur. Sugar cane makes little growth when the temperature is under

Plate 1.3 Ivory Coast, end of January. Air laden with dust from burning vegetation nearly obscures the sun. Time 3.0 p.m. Palms killed by tapping for wine.

22 °C. The traditional tea growing areas of China and Darjeeling lie outside the tropics and quality tea is produced in the tropics at altitudes of 1,500 m and over, suggesting that quality is associated with lower temperatures.

Varieties of annual tropical crops which are adjusted to the longer hours of daylight can be grown well outside the tropics in the hot summer. Tobacco and maize are grown in the Niagara peninsula of Canada about 45 °N. Rice, which is always considered a typical tropical crop, gives higher yields grown outside the tropics as in Italy, Japan or Australia (Copeland 1924), but this results from a combination of a higher standard of farming, growing varieties responsive to fertilizers in longer days with greater solar radiation during the growing season, rather than a temperature effect. Olives, which thrive in hot summers, are not grown in the tropics as low winter temperatures are necessary for the initiation of flowering. This need for a cold period during the year applies to most of the tree fruits grown in temperate regions. Vines like a cold winter, but certain varieties are grown successfully on the Venezuelan coast around La Guiara, and a sport of a certain apple variety will grow well in sub-tropical South Queensland suggesting that varieties of temperate fruit trees might be developed for the tropics. The maximum temperatures in the tropics on the other hand are too low for flowering and fruit setting of the date palm (*Phoenix dactylifera*), which requires the very high summer temperatures of North Africa and Arabia.

The important influence of temperature on tropical crops is illustrated by the work done on cotton. In Florida, optimum germination occurs at 35 °C. At 16 °C germination is very slow, ceasing altogether at 14 °C (Camp & Walker 1927). Prior to emergence cotton grows between 14 and 35 °C with optimum growth at 24–29 °C. In colder soils the germinating seedlings are more prone to soil-borne diseases.

Balls (1916), in Egypt, obtained the best continuous growth at 32 °C and correlated the number of flowers produced on a given day with the growth of the plant 29 days before when the flower primordia were differentiating. As cotton likes abundant sunshine, the daily temperature was a major factor in this initiation. Balls found growth of the main stem to be correlated with night temperatures.

Cocoa in Ghana grows under conditions where the diurnal range 35–13 °C is large, but when the weekly mean of the maximum temperature falls below 28 °C flushing is suppressed. When the mean daily temperature falls below 29 °C it appears to be followed by a reduced number of flowers 2 months later (Adams & McKelvie 1955).

Arabica coffee grows well between 13 and 27 °C, the optimum being about midway. In many areas the temperature rises above 27 °C but when the temperature reaches 35 °C the trees appear to suffer. They can tolerate a reduction in temperature to freezing point provided that no wind occurs. The Brazilian crop in some years has been severely reduced by frost. At high temperatures the absorption of nutrients is reduced. The starch content of tea roots in Sri Lanka rises by 2 per cent for every 300 m rise in altitude (Gadd 1949). This has had an important influence on the recovery of tea plants from pruning (see Ch. 2, p. 211 2.22). The green colour associated with ripe oranges in the tropics is largely due to the high air temperature as carotin develops best below 21 °C (Hardy 1970b).

Shade

The annual crops of the tropics, cotton, tobacco and the short-term perennial sugar cane, can utilize the maximum amount of sunlight available, so much so that excessive amounts of cloud can cause yield reductions. Certain perennial crops, including coffee, tea and cocoa whose natural environment is the lower storeys of the forest, usually welcome shade at planting out and frequently the mature crop is grown under shade.

Investigations started at the Imperial College in Trinidad in 1950 on the effect of shade and fertilizer, showed that young cocoa grew best

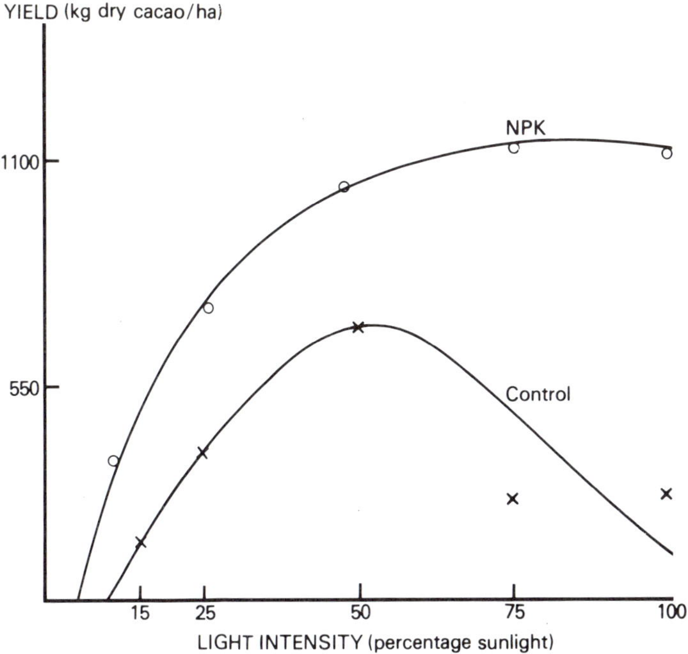

Fig. 1.29 Response of cacao to fertilizer at increasing light intensity. (Murray 1958)

with 50 per cent full sunlight. As the trees came into bearing there was no response to fertilizer in the shaded plots but with increasing light intensity the yield increased logarithmically up to 75 per cent of full sunlight provided no other factors remained limiting (Fig. 1.29). In the first few years there were some spectacular effects. With 15 - 25 per cent of sunlight the trees grew 1.8 to 2.4 metres high irrespective of fertilizer application and retained the fan type of morphology. At 75 - 100 per cent sunlight the trees grew to 1.8 m with fertilizers but only 0.3 to 0.5 m without. In low light intensities the leaves lived for many months, but in full sunlight they remained small, but hard and with brown discolorations and lasted only a few weeks, suggesting that toxic products had accumulated in the leaves at high light intensities. (See reports of Cocoa Research at the Imperial College of Tropical Agriculture, 1945 - 53).

The argument as to whether tree crops should be grown with or without shade, is really on a fundamental principle in tropical agriculture, the balance between carbohydrate nutrition and mineral nutrition. The effect of shade is to reduce the production of carbohydrates by photosynthesis. If one wishes to keep a cocoa tree alive under unfavourable soil conditions, lower the photosynthetic rate and it will remain alive indefinitely. Conversely in areas of high light intensity with no shade the tree will have to take up minerals at the maximum rate to balance the accumulating products of photosynthesis, which are otherwise toxic. This implies both an adequate supply of mineral nutrients and permeable soil which does not become waterlogged in the wet season, and allows the roots to forage. In the volcanic islands of the Pacific and in Grenada, provided adequate minerals are available cocoa produces its maximum yield in full sunlight. In Trinidad the cocoa trees cannot cope with more than 75 per cent sunlight due to waterlogging in the wet season, and in a severe dry season windbreaks are necessary. Both cocoa and coffee in Trinidad are usually grown under very heavy Immortelle shade. The leaves of coffee are said to scorch if the shade is removed, indicating that fertilizer must be applied (Evans 1961).

The same ability to use full sunlight if the nutrients and air/water balance of the soil are not limiting, has been shown to apply to arabica coffee (Kona) grown on the volcanic soils of Hawaii in full sunlight with 10 applications of 250 kg/ha of balanced fertilizer throughout the year. In Central America this is applied in the sun hedge system of coffee growing (Cowgill 1958). Die-back, frequently seen in other tropical tree crops, is often a result of an inadequate supply of minerals to a plant growing in a high light intensity.

The same principle applies to sugar cane, though here the carbohydrate is removed from the leaves and stored as sugar. With high light intensities leaves become pale green and respond to higher rates of fertilizer.

A comparison of economic rates of fertilizer application is given in Table 1.21. Little response is obtained with increased fertilizer rates in

Table 1.21 Relationship between solar energy and economic rates of fertilizer application to sugar cane.

	Solar radiation (gcal/cm^2)	**N**	**P_2O_5**	**K_3O**
Hawaii	600	600	300	400
Guyana	350	110	80	100

Source: Evans (1961).

Guyana. The interaction of shade (*Gliricidia sepium* = *G. maculata*) and fertilizer in the field is shown in a 3 ha experiment at the Cocoa Research Institute at Tafo, Ghana (Table 1.22). Amelonado cocoa was planted in 1947 under shade, and the fertilizer applied in 1957. The trial was sprayed regularly to control capsids. The yields in pounds of dry cacao per acre are as shown in Table 1.22.

Table 1.22 Shade and fertilizer experiment on cacao. (Tafo, Ghana)

Treatments	**Pre-treatment 1956/57**	**1957/58**	**1958/59**	**1959/60**	**1960/61**	**1961/62**	**1962/63**	**1963/64**	**1964/65**
Shade, no fertilizer	205	586	958	756	702	494	657	1,036	1,183
Shade, with fertilizer	211	849	1,211	906	969	679	1,040	1,398	1,640
No shade, no fertilizer	170	1,100	2,348	2,187	2,449	1,516	2,093	2,018	2,259
No shade with fertilizer	173	1,602	3,091	3,088	3,474	2,435	2,945	3,023	3,275

Source: Annual Reports WACRI and CRI CRIG.

The application of the results this experiment and other fertilizer trials in Ghana at 1963 prices for cocoa and fertilizers was outlined by Cunningham (1963). Where the yield is only 280 kg dry cocoa per hectare, shade should not be removed, where the yield is 560 kg shade should be reduced to two-thirds of that considered best for cocoa. Under full or thinned shade only the best farmers will profit from applying 125 kg of triple superphosphate each year. To all high yielding and unshaded cocoa must be applied 250 kg urea, 125 kg triple superphosphate and 500 kg of potassium sulphate at the beginning of the main rains, which should maintain a yield of 2,240 kg dry cocoa per hectare. To justify the extra work with clear felling and growing unshaded cocoa, yields of 3,360 kg/ha are necessary. Vernon (1967) has pointed out that the maintenance of a closed canopy is essential to the success of deshaded cocoa. Suppression of weed competition he regards as a major function of the closed canopy, and herbicides must be used to control weeds as cocoa (or coffee) is deshaded. Foliar urea sprays help to increase the ground shade by increasing the leaf size.

Shade in tea grown at low altitudes may be important in reducing leaf temperature. High leaf temperatures over 35 °C will reduce photosynthesis and increase respiration. This may also help to explain the association of quality tea with altitude where leaf temperatures will be lower. Without shade, leaves of a tea bush may reach 40–50 °C on clear hot still days in Assam (Sharma 1968).

Van Dierendonck (1959) explains that tea has a low nutrient requirement relative to other tree crops and is generally grown under shade on well drained hillsides of low fertility where other crops would fail. If the shade is removed, there is usually a noticeable yield increase for some years, after which, depending on the soil fertility status, a decline may occur. On the rich volcanic ash of certain humid tea areas in West Java, shade may be omitted without much decline in productivity. The yield under such conditions is almost directly proportional to the light intensity (see Table 1.23).

Though the effect of time and the need for fertilizer could not have been recorded when reported, it should be noted that on the fertile well

Table 1.23 Yield increase in West Java after removal of shade from tea

Treatment	Yield pre-trial April–July 1950		Trial yield Oct. 1950–Oct. 1951	
	Fresh leaves (kg)	Index	Fresh leaves (kg)	Index
Without shade	360	100	2,200	100
25% shade	358	99.4	1,591	72.3
50% shade	365	101.4	1,395	61.3
75% shade	324	91.4	1,063	48.3

drained soil the tea bushes would be able to maintain the balance between mineral nutrition and carbohydrate synthesis even at the high light intensity. This might not have been maintained on a poor hillside without fertilizer. At Tocklai in India even high rates of fertilizer could not stop the decline in yield where the shade was removed from the tea. By the fifth year the yield was only that of the shaded tea.

In East Africa unshaded tea has responded to nitrogen, but no response is obtained in shaded tea (Annual Report 1958). This same interaction has been observed at Tocklai, India (Wight 1958). At two Kenya estates a linear increase in yield followed the reduction of the *Grevillea robusta* shade, whose shade patterns reduced incident radiation from 1.4 cal/cm^2 per minute in full sunshine at noon to 0.1–0.2 cal/cm^2 per minute. Contrary to certain opinions, shading reduced the quality of the tea (McCulloch *et al*. 1966).

The Kenya Tea Development Authority does not permit smallholders to grow shade trees except as a windbreak. Windbreaks can be more effective than overhead shade in maintaining humidity.

On good permeable soil with adequate rainfall or irrigation and correct use of fertilizers, the complete or partial reduction of shade will give considerable increases in the yield of tree crops. Where these ideal conditions are not fulfilled as in the present level of peasant farming, the ultimate result of removing shade might be desolation, once the soil can no longer provide the necessary balancing minerals for the high rate of photosynthesis. Where tree crops suffer from die-back and 'stag headedness' the introduction of a limited amount of shade, particularly a leguminous tree, would reduce or eliminate this. Many of the observations of the susceptibility of a crop to certain pests and diseases with or without shade, no doubt follow from the incorrect balance of photosynthesis and mineral supply.

The advantages and disadvantages of high shade are listed below but these are secondary to the function of regulating photosynthesis.

Advantages of high shade

1. The shade trees act as a windbreak. Apart from the physical damage from the wind, hot dry winds (e.g. harmattan in West Africa) can increase transpiration to a serious level.
2. The range of the air temperature within the crop is less, e.g. on Mount Elgon, Uganda, 'hot and cold' disease of *arabica* coffee is attributed to this change and occurs less frequently in shaded crops. (This could be a result of high light intensity.)
3. The range of the soil and leaf temperature under shade is less.
4. The humidity within the crop is higher.
5. The surface soil moisture is higher. Particularly important where a dry season occurs.

6. Shade trees act as a drain in removing excess rainfall.
7. Shade trees add organic matter to the soil.
8. Shade trees bring up nutrients from the deep subsoil and add it to the surface leaf mould. The deep penetrating roots open up the subsoil.
9. Leguminous shade trees add nitrogen to the leaf mould. *Albizia chinensis* shade in North India often increases the tea yield by an amount equivalent to adding 100 kg of nitrogen per hectare (Annual Report Tocklai, 1959).
10. Weed growth, particularly of undesirable grasses, is less under shade. A well grown crop canopy should do this equally well.
11. Shade may reduce the incidence of pests and disease. Cocoa capsid and coffee thrips prefer unshaded crops. In Costa Rica, unshaded cocoa has a more severe *Phytophthora* infection due to the heavier dew.
12. Shade may also be a crop, e.g. *Ficus* (bark-cloth) in coffee, bananas in coffee and cocoa, timber species, rubber in coffee, oil palms in cocoa.
13. Unshaded forest soil rapidly loses its structure. A good crop canopy and mulching will prevent this.
14. Shade tends to reduce overbearing, and 'die-back' of coffee. This can be prevented by the use of manure and fertilizers, provided the soil is deep and permeable.

Disadvantages of high shade

1. Shade reduces the light intensity which the crops can utilize fully on well fertilized, well drained soils adequately watered.
2. Shade competes with the crop for nutrients.
3. Shade competes with the crop for water in the dry season. This can be evaded by using a deciduous shade tree such as *Terminalia superba*, which drops its leaves to form a mulch at the start of the dry season. Such trees will have a reduced effect as a dry season windbreak.
4. Shade competes with the crop roots for oxygen in the wet season.
5. Some trees are incompatible with the crop.
6. Some trees are alternate hosts for pests and diseases. *Albizia falcata (molucanna)* is a host for *Xyleborus* borer.
7. Shade trees or their branches may fall and cause serious damage.

Seedlings should be grown in shaded nurseries, as full sunlight would reduce the leaf size and cause too frequent leaf shedding. If no shade is used in the crop it may often be advisable to protect the crop with windbreaks. Cocoa in Grenada on fertile volcanic soil is often in full sunshine without shade but with a surrounding box of trees to act as a windbreak.

Tree species for shade

Shallow rooted species such as palms and *Musanga*, and trees such as *Ceiba pentandra* which are leafless for long periods, are undesirable.

Erythrina spp. are popular for cocoa in the West Indies and Sri Lanka and are being used in the Pacific. The sight of these Immortelles flowering in Trinidad alone justifies their existence, and when these flowers fall they contribute to the soil about 22 kg/ha of nitrogen which has originated in the root nodules. This is more nitrogen than is removed in 360 kg of dry cocoa beans. Unfortunately one species, *Erythrina poeppigiana*, is very susceptible to pests and diseases and causes serious damage to the cacao crop when they die and fall. Murray (1966) considers that the disease of the swamp Immortelle (*Erythrina poeppigiana*) shade has been beneficial by reducing shade density. In the Congo, *Terminalia superba* is used for cocoa. *Albizia* spp. are popular for tea, and also among certain coffee growers. In Assam *Albizia stipulata* is most common but it has drawbacks. *Albizia procera*, an erect long-lived tree, is less attacked by canker than *A. stipulata*. For tea in Sri Lanka *Grevillea robusta* and *Erythrina indica* (Dadap) are common, but not in Assam where Dadap is considered a source of root diseases. Dadap in Sri Lanka is cut back frequently and the prunings left in the tea. The non-leguminous *Ficus* (bark-cloth) is recommended for coffee in Uganda. The soil under fig trees tends to remain moist. *Erythrina* has been replaced by *Leucaena leucocephala (glauca)* in Indonesia. *Grevillea robusta* has been removed from a number of tea estates in Kenya to increase yields.

Under peasant cultivation, partial clearance of the forest, leaving a number of established forest trees and planting ground shade such as bananas, tannias or cassava, is the common practice. In the Congo (Zaire) this system was followed in a more organized manner, and alternative strips were cleared and planted, the debris of the clearing being heaped in the uncleared strips which are cleared and planted in a subsequent season.

Most tree crops on planting out from the nursery require shade the first year, and this may be provided by a crop such as bananas or tannias in coffee and cacao; a legume such as *Tephrosia* in tea; or a shade of dry grass, palm leaves, or a clay ring. The living shade may be allowed to grow on and form a cover for the mature crop. As *Tephrosia vogelii* is susceptible to *Poria* root disease of tea it has been recommended as an indicator crop for locating the presence of a source of infection as well as a cover crop for tea. Where soil moisture is marginal, a living shade or cover crop might be detrimental and in such circumstances non-living shade is used, followed by mulching of the mature crop.

Plate 1.4 Young oil palms with legume cover crop. Ivory Coast.

COVER CROPS

Cover crops, usually creeping legumes, are frequently interplanted in young perennial crops including rubber, oil palms (Plate 1.4), coconuts, sisal, coffee and cacao, and should be considered for other tree crops, even when there is no previous experience. These cover crops, besides protecting the soil from the pounding effect of the raindrops, soil wash, and the undesirable effects of sunshine, build up the organic matter in the soil and maintain the soil surface in an ideal condition for the spread of the surface feeding roots of the trees. Such cover crops will reduce leaching and the roots of deeper rooting species bring up nutrients from the subsoil which become available for the crop. Leguminous cover crops are generally preferred as they increase the soil nitrogen. Over ten years of tapping, rubber trees in legume covers may produce 20 per cent more yield (Teoh *et al*. 1979).

The ground covers must be easily propagated by seed, grow rapidly without competing with the crop, tolerate some shade and cutting back from around the crop, be resistant to pests and disease, not act as an alternate host to pests or diseases attacking the crop, suppress weed growth particularly grasses, and be free from irritating hairs or spines which would be unpleasant to cultivators without shoes or leg covering. The weeds can be kept out of newly sown cover crops until they are established by spraying oxyfluorfen (Goal^{R}) or metolachlor (Dual^{R}) immediately after planting. *C. pubescens* is sensitive to Goal. For quick germination the seed should be scarified, or treated with hot water and soaked overnight and inoculated with *Rhizobium* before sowing.

A number of legumes have most of these attributes including: *Calopogonium mucunoides, Centrosema pubescens, Dolichos hosei, Glycine wightii = (G javanica), Indigofera spicata, Pueraria phaseoloides* (syn. *P. javanica), Stylosanthes gracilis. Mimosa invisa* is a vigorous legume, and a good soil improver which crowds out weeds, but it has sharp thorns and is a fire hazard in the dry season. *Mucuna cochinchinensis* planted with more persistent legumes on steep land provides a quick cover to stop erosion.

Calopogonium tolerates poor soil and partial shade but not heavy shade. In rubber it is often planted mixed with *Centrosema* which is more competitive to the roots of young rubber but persists longer in the shade of older trees. *Pueraria* sheds its leaves in the dry season and is often planted mixed with *Centrosema. Pueraria* is commonly used in Cameroon where the dry season is not serious. These three legumes are the ones commonly used when replanting rubber in Malaya. The planted strips along the contour rows are kept bare and weed free, and the area between the planting rows sown with these cover crops. As the soil at replanting is usually very poor, liberal amounts of phosphate are applied to both the crop and covers. The phosphate used by the covers later becomes available to the rubber as the leaves fall and rot. Where a creeping leguminous cover was used in trials in Malaya, the trees came into tapping 12 months earlier than where a natural ground cover had been allowed to regenerate (Mainstone 1960). Compared with natural regeneration or grasses, these legumes increased the nitrogen in the soil and in the rubber leaves where the magnesium level was also increased and sometimes the phosphorus (Watson *et al*. 1963).

Leguminous cover crops reduce root diseases in young rubber (Newsam 1963). To maintain some soil cover under the mature rubber canopy, the deeper rooting shrubby legumes may be interplanted with these low growing covers. *Moghania macrophylla* (syn. *Flemingia congesta F. macrophylla*) is often used in Malaya, and *Desmodium ovalifolium* is popular in Sri Lanka for its tolerance to shade and drought. The Rubber Research Institute, Malaya (RRIM) Planters' Conference, Kuala Lumpur, June 1963, reviewed the role of cover plants in rubber in Malaysia (Anon 1963). *Calopogonium caeruleum* a creeping legume indigenous to Central America, grows well in association with rubber in Malaysia. It is more shade tolerant and vigorous than *Calopogonium mucunoides* and appears to return more nitrogen to the soil. It may persist for ten years after planting. In the Philippines *Calopogonium caeruleum* and *Pueraria phaseoloides* have proved to be excellent cover crops in coconut plantations, crowding out the weeds including *Imperata cylindrica* in the first year. *Centrosema pubescens* does not eliminate this weed even after four years. The planting of cover crops on neglected plantations greatly increases the coconut yield after only 1 year (Cabato 1970). *Pueraria* also gives good weed control and increases the yield of coconuts in Fiji (Leather 1972). *Stylosanthes* in Indonesia and Malaya has been found valuable as a 'drain' in the wet season, and is a useful cover crop for coconuts in Malaya. *Centrosema* is a popular cover crop for oil palms in West Africa (Plate 1.4) and is planted to leave a circle of bare soil round the young palm. In Surinam a continuous cover of *Pueraria phaseoloides* is sown prior to planting citrus, coconuts or shade trees. This system is seldom used for coffee or cocoa due to the high cost of keeping the cover away from the young trees.

Certain sisal estates in East Africa plant in double rows and sow cover crops in the wide inter-row. Established covers must improve the fertility of the poor sisal soils. *Desmodium sandwicense* is the most promising cover tried in sisal in Kenya, except for the drier areas where *Macrotyloma uniflorum (Dolichos biflorus)* is better (Richardson E.F. 1965).

Leucaena leucocephala (glauca) has been used as a leguminous cover in some robusta coffee estates in the fertile crescent of Uganda bordering Lake Victoria. It must be cut back regularly otherwise it grows into a shade tree, but if properly managed maintains the soil and coffee in good condition, and the soil stays moist in the dry season. It will not thrive as a cover or shade tree on soils of low base status as occur on many of the rubber estates of Malaysia (Ruskin 1977).

The basic seeding rates (80 per cent germination) are

Pueraria phaseoloides	3.5 kg/ha
Calopogonium caeruleum	0.5–1 kg/ha – Double if planted alone.
Centrosema pubescens	2.0 kg/ha
Mucuna cochinchinensis	1.0 kg/ha at 1.5 m spacing in centre of inter-row.

MULCHING

Areas of marginal rainfall, such as the robusta coffee areas of Uganda, should respond better to mulching with dead organic matter than to cover crops provided that the weeds are controlled. Indeed dry elephant grass (*Pennisetum purpureum*) has proved to be consistently a most useful ground cover. Many coffee farmers, however, do not spread it in their small holdings as, apart from the heavy work of cutting and carting, mulch introduces a serious fire risk in the dry season. The effect of such a mulch on run-off and drainage is shown in Table 1.24. With a total rainfall of 576 mm the figures in Table 1.24 were recorded.

Table 1.24 Effect of mulch on run-off

	Run-off	Drainage	Retained
Bare soil (wetted)	368 mm	10 mm	198 mm
Bare soil (undisturbed)	224 mm	8 mm	345 mm
Mulched	28 mm	117 mm	432 mm

Source: Farbrother and Manning (1952).

The effect of mulching on the soil nitrogen level for the same zone is shown in Fig. 1.4.

The increase by mulching of available moisture in a dry year could prevent crop failure, both for annual crops such as cotton and maize, or perennial crops such as bananas or coffee. While the benefit of mulching is often obvious in the perennial crops, the expected increase is not consistently realized with cotton.

Gilbert (1945), studying the effect of mulching coffee in Tanzania, showed the soil temperature had a diurnal range of 12 °C at a depth of 50 mm, mulching reduced this to 1.5 - 3 °C. Banana leaves increased the reception and retention of rainfall so much that even after a prolonged drought the moisture percentage in the top half metre was 50 per cent higher in the mulched plots than where clean weeding was practised. Surface rooting crops benefit particularly from mulching as the mulch maintains the temperature and moisture constant at a level suitable for root growth, and often the roots are found in the partly decomposed interphase between soil and mulch. Bare soil may become so hot that these surface roots are destroyed.

The effect of organic mulches on the yield and quality of coffee in East Africa has been studied for more than 20 years, and grass mulch is now widely used in Kenya. Mulching reduces soil acidity, and increases carbon, nitrogen, exchangeable potassium and available phosphorus in the soil. Exchangeable calcium and manganese are decreased. These changes are reflected in the leaf analyses (Robinson and Hosegood 1965). The added organic matter improves the rainfall acceptance and penetration. Alternate row mulching with 25 t/ha of elephant grass adds about 1,250 kg of potash to the soil which can induce magnesium and iron deficiency by antagonizing their uptake (Robinson and Chenery 1958). Magnesium deficiency is easily corrected by adding Epsom salts (magnesium sulphate) to the routine fungicide spray. At Ruiru Coffee Research Station, Kenya, between 1950 and 1959 the mulched plots yielded significantly higher than the unmulched plots, but in certain seasons overbearing followed by a low crop occurred, which suggests that a modified system of pruning together with fertilizers are necessary for mulched coffee.

In coffee experiments in Brazil (Medcalf 1956), where heavy mulch applications increased the yield by 72 per cent in the 1956 crops, phosphorus and potassium levels were increased, but nitrogen and manganese availability was reduced. When tea is pruned 30–50 t/ha of vegetative matter are cut from the bushes. This is too valuable both as a source of nutrients and organic matter and as a protection for the soil, to allow it to be removed for domestic fires. At the Tea Research Institute in Kenya, mulch and sulphate of ammonia had opposite effects on the pH and the level of the bases of the soil. Both were increased by mulching and decreased by applying sulphate of ammonia as shown in Table 1.25

Table 1.25 Effect of mulching and sulphate of ammonia on the concentration of bases to a depth of 0.9 m over 5 years

	Sulphate of ammonia 224 kg/ha per annum[(a)]	**Equivalent**	**Mulch** 37.5 t/ha per annum	**Equivalent** to
Ca	−19%	716 kg/ha	+52%	1,960 kg/ha
Mg	−25%	157 kg/ha	+47%	295 kg/ha
K	−17%	249 kg/ha	+30%	440 kg/ha

Note: 100 kg of sulphate of ammonia 110 kg of calcium carbonate (Smith, A. N., 1959).

[(a)] It was estimated that 1,134 kg of sulphate of ammonia were actually applied to the area sampled during the 5 years.

Thus the application of sulphate of ammonia to mulched tea has no effect on the level of the bases or pH. In the experiment, double superphosphate (50 kg/ha per annum) and muriate of potash (37 kg/ha per annum) applied with the sulphate of ammonia maintained the level of the bases. Mulching does not normally increase yields so considerably unless it is preventing run-off or erosion.

The Rubber Research Institute (RRI) Sri Lanka (Advisory Circular No. 66 – 'Manuring of Rubber') states: 'It has been established that mulching is advantageous to the growth of rubber and it is of greatest benefit during the early years of a plantation. Mulching should preferably be done with easily decomposable material of low C/N ratio, but mulching with any material is better than no mulching at all. Loppings of leguminous cover crops and weeded *Mikania* are suitable materials for mulching.'

Mulch is more effective if applied at the start of the rains as it intercepts the rain and increases the take-up, but it is frequently more practical to mulch towards the end of the rains when the grass is available. Alternative row mulching is sometimes preferred to full mulching, this also reduces the risk of fire. Grass for mulching should be allowed to dry before applying as this not only reduces the weight to be carried but also the chance of the grass rooting.

Mulching of sugar cane in Barbados with sour grass (*Andropogon pertussus*) at the end of December as the dry season is starting, increased the yield of cane by 10 t/ha. Combining the mulch with a heavy fertilizer dressing increased the crop by 30 per cent and produced the same effect as 100 t/ha of pen manure (Saint 1930). Mulching has significantly increased the soil organic matter during the 20 years since it was encouraged (Robinson 1951).

Ratooned sugar cane in Mauritius is mulched along alternate inter-rows with the trash from the previous crop. In the wetter areas complete soil cover with mulch would cause excess water to be held in the soil, creating anaerobic conditions and nitrogen deficiency through increased leaching. The trash mulch breaks down during the year, and after the cane is cut the new trash is collected on the inter-row previously bare. Some sugar estates have a large covering of volcanic rocks (basalts) which are piled between alternate rows of sugar cane as a stone mulch. After the last ratoon crop these walls are moved to the previously unmulched area. While these stone walls maintain the soil moisture and collect dew they also harbour perennial creeping weeds including *Paederia foetida*.

Mulching of seed beds for tree crops is quite common and this has been successfully used with tobacco (Garmany 1956). The tobacco seedlings can penetrate a considerable amount of mulch, and the mulch reduces the amount of watering necessary.

In Northern Nigeria increased yields have been obtained from mulching annual crops with heavy applications (62.5 t/ha) of groundnut shells. Sunflowers on the mulched plot were 50 per cent taller and yielded 348 kg/ha of threshed seed compared with 95 kg without mulch. The yield of cotton grown on metre ridges was increased from 543 to 724 kg of seed cotton. Later trials with 10 to 12.5 t of mulch, manure, fertilizer and insect control raised yields to around 2,240 kg/ha of seed cotton in three successive seasons with markedly different rainfall patterns. Such yields had not been produced previously using manure, fertilizer and insecticide as the lack of moisture and bad aeration were limiting (Lawes 1962). Different mulches on sorghum in Northern Nigeria gave the yields shown in Table 1.26.

Mehlich (1966) has suggested the use of maize stover which has a high potash content as a mulch in coffee. Nitrogen fertilizer would also be needed to correct the high C/N ratio.

Table 1.26 Effect of different mulches on sorghum yield. - N. Nigeria

Mulch	**Yield of grain** (kg/ha)	
	1957	1958
Groundnut shell	1,249	1,688
Dead grass	1,170	1,702
Chopped sorghum stalk	857	1,396
None	980	1,289
SE	44	54
Rainfall	1,394 mm	886 mm

In the Ivory Coast two legumes, *Tithonia diversifolia* and *Macrotyloma uniflorum (Flemingia congesta)* cut at regular intervals have been tried as a coffee mulch (Verliere 1966).

Yams planted in the dry season in West Africa are always mulched along the ridge or the mounds 'capped' with leaves, dry grass or sorghum stalks. Without mulching the high soil temperature severely reduces germination. When the rains start, the tops of unmulched ridges are washed away, exposing the developing yam tubers. Exposed tubers develop chlorophyll and are generally unpalatable, besides being exposed to rats and other pests.

Pineapple growers in Hawaii use a black polythene mulch which is laid by machine immediately after the soil fumigant and fertilizer have been incorporated. Py and Tisseau (1965) reported no advantage with polythene in Ivory Coast and Martinique where the rainfall is well distributed, but in Guyana in a very dry year there was a 78 per cent increase. Any form of readily available herbage or crop waste should be suitable as a mulching material provided that it is not a carrier of a pest or disease.

At the International Institute of Tropical Agriculture (IITA) in Nigeria a wide range of plant residues have been compared as mulches for maize, cowpea and soya bean. The yields of all these crops whether planted early or late were increased by all the mulch materials though the increase was not statistically significant with every type of mulch (IITA Annual Report 1976).

Care must be taken when mulching in areas receiving heavy rainfall that anaerobic reactions do not occur. If the mulch is rich in potassium, a foliar spray of magnesium sulphate may be needed, and in the first year of mulching nitrogen deficiency might occur.

Live mulch

The harvesting, transporting and spreading of mulch takes a lot of labour and experiments have been carried out at IITA growing tall crops such as maize in a live legume mulch or cover crop which grows close to the soil and gives a good ground cover. Another approach at the same research centre has been to grow a cover crop such as *Pueraria phaseoloides* for 30 months and then destroy it with paraquat. This leaves a 14.5 t/ha dry weight of mulch into which tomatoes are planted. This dead mulch prevented weed growth, and by protecting the fruit from ground rots produced high quality fruit (IITA Annual Report 1976). *Stylosanthes* chemically killed also produced excellent mulch for sod seeding of maize, cowpea, soya bean, pigeon pea, and cassava.

Dust mulching, the keeping of a loose layer of soil at the surface, has little to commend it other than removing the weeds which are drawing on the soil moisture. The idea was based on a misconception that water rises by capillary action in the soil and breaking up the 'capillary tubes' would reduce this loss. Evaporation losses from a soil, which is in a condition suitable for cultivation to form a dust mulch, are at a minimum. Thus it does not prevent the loss of moisture, but the constant movement may severely damage the surface roots. The finely divided soil surface produced by dust mulching, seals rapidly when rain falls, and there is an increase in run-off and a loss of topsoil by erosion. The beneficial effects attributed to dust mulching were due to weed control.

2 Crop Culture

AGRICULTURAL SYSTEMS*

The object of farming is to provide food, or cash, or both, and in the widely varying ecological and economic situations of the tropics many different farming systems are practised and many farmers operate more than one system. There are still a few food gatherers who collect much of their food in the wild, and the collectors of wild oil palms in West Africa and gum arabic from the wounded bark of *Acacia senegal* in the Sudan who sell their surplus. Hunting, fishing and the collecting of wild honey make important supplements to food supply and income in many regions but this is not strictly agriculture.

Sanchez (1976) lists five main tropical farming systems:

1. Shifting cultivation covering	45%	of the tropical land area
2. Settled subsistence farming	17%	
3. Nomadic herding	14%	
4. Livestock ranching	11%	
5. Plantation systems	4%	

The remaining 9 per cent is non-agricultural. This division is based on an old estimate (Whittlesey 1936). A recent estimate suggests that shifting cultivation, the predominant farming system in the tropics, is practised over some 36 million km^2 (14 million square miles) or 30 per cent of the cultivable soils and is a means of livelihood of 10 per cent (200 millions) of the world's population, particularly in the Amazon basin, West and Central Africa, and in the hilly areas of Vietnam and neighbouring countries. However, much of this cultivation might be more correctly considered as the fallow system described below.

Ruthenberg (1976) has six major cultivation systems and three grazing systems based on the type and intensity of land use as given below.

Shifting cultivation**

Shifting cultivation is where the community crops the land and when yields decline to an unprofitable level, abandon the land and houses and move to a new site. The woody vegetation is felled and burnt in the dry

* For a detailed survey see Ruthenberg (1976).

** For a full account see Nye and Greenland (1960).

season and the ash disturbed as little as possible in planting the crops. The main planting tool is a digging stick, though a hoe may be necessary where savanna land is cleared as the land is harder. Livestock are not usually a part of this system. In the Far East, hill rice is frequently grown. The crops are seldom planted in rows. Usually an interplanted mixture of crops is grown. Yields tend to fall off in the third year, in some cases the weed invasion, particularly 'lalang' (*Imperata cylindrica*) in the Far East, becomes unmanageable, and no fertilizers are used. When cropping ceases, the forest regenerates. It takes 5 to 10 years for a forest fallow to become established. Originally the forest fallow could have been 25 years, but population pressure has in many cases shortened this until the resting period may be as little as 4 years, and this is the minimum needed to restore any worthwhile degree of fertility. Shifting cultivation as practised in South and Central America, Africa and Asia has very many variations, and types of vegetation cleared; from forest, to bush, to savanna. The shifting cultivators may only shift the cultivated land and have a more permanent house with a nearby garden plot kept in permanent cultivation with organic waste from the house, and droppings from the chickens and the other animals kept there. The boundaries of the holdings under shifting cultivation are not clearly defined and the land rights vague. In Uganda where land tenure is to a degree secure, shifting cultivation has practically everywhere been replaced by the fallow system.

The general system followed in West Africa is described by Nye and Greenland (1960). During the dry season the lianes and small trees are cutlassed, and the larger trees axed. Trees with a diameter more than 60 cm are left and provide a light shade. When the mass of vegetation is dry, it is burnt and maize is planted into the forest topsoil with a digging stick, without moving the surface cover of ash. In the wet zones in Liberia and Sierra Leone upland rice replaces maize as the first cereal crop. Weeds are slashed with a cutlass. While the maize is growing or after harvest, cassava, cocoyams, tannia and bananas are planted, and a small area planted with a wide range of vegetables. These crops can be harvested in the next and subsequent years until the plot is abandoned to allow the forest to regrow. Once the first crop has developed the ground is covered and protected until it is abandoned. The regenerating 'bush' grows from stumps and large roots left at clearing; and from the seeds in the soil and from adjacent areas. After 5 years the forest may be 6 m high and 15 m after 10 years and as time allows, slower growing tree species become more dominant. As the boundaries of the new clearing do not necessarily coincide with the old the resting period on a single patch may vary from 5 to 20 years, though 8 years of fallow appear to be needed to restore fertility after 3 years' cropping.

In the savanna country where the grass may be 3 to 4 m high and the fire-resisting trees 15 m tall, to deal with the grass roots after burning the soil needs more cultivation than in the forest. The soil when moist is scraped into mounds which are planted with climbing yams and interplanted with maize, beans and vegetables. The following year the yam mounds are broken down and sorghum and maize planted on narrow ridges. In the final year of the cycle groundnuts are interplanted with millet. The soil is exposed during the dry season and for a month when the rains begin. If *Imperata cylindrica* ('lalang') invades, keeping the plot clean is difficult. Its sharp pointed leaves can penetrate yam tubers which subsequently rot. The land is not cleared again until other grasses have crowded out the 'lalang'. In short grass areas where there is less rainfall sorghum, millets, groundnuts, beans, sweet potatoes and cotton are grown, often on ridges. Much erosion could be stopped if the ridges made for annual crops ran across, rather than down the slope.

Fallow system

In this system the land is cultivated for 3 or 4 years until yields fall. It is then allowed to revert to natural vegetation for a similar number of years to restore the fertility, after which the whole cycle is repeated. The length of the cropping and resting periods depends upon the pressure of population, and is not the same for each plot in the holding, or standard for the area. Where a 4 years' cropping and 4 years' resting rotation is common, certain plots will be cultivated 7 or 10 years before being rested and there will be a fertile 'garden plot' near the house under permanent cultivation, producing vegetables and plantains on the organic refuse of the holding.

The fallow system, which is often confused with shifting cultivation, is much more important in Africa, grass fallows being common in Central and East Africa, and bush fallows in the wetter parts of West Africa; and also in South America. This system has characteristic differences from shifting cultivation. The holdings and frequently the rights to the holding are clearly defined with field boundaries. The house compound has a permanent location. The shifting cultivator burns the bush in the dry season and plants as the rains start, but in the fallow system cultivation cannot start before the rains have softened the soil, and planting cannot take full advantage of the nitrogen flush at the start of the rains (see Ch. 1). Even though cultivation is mainly with the hoe the area cultivated is often larger than under shifting cultivation. In Uganda where the fallow system is the main farming system, the average area cultivated per family varies from 2 ha in the fertile crescent of Buganda, to 4 ha in Karamoja where the rainfall is less reliable and the dry season very long. In other countries up to 8 ha may be cultivated, but averages mean very little as there are inequalities and no set sizes for holdings. In some areas hired labour help at weeding and harvest, the two major peaks of labour demand. Arable farming is the main source of food for the cultivator and his family, animals are supplementary. A lot of labour is required for cotton where this is the main cash crop. Tasks are divided between the sexes, the men clearing the land and doing the heavy work including thatching, and women cultivating the food crops, preparing and cooking them. Work in the field starts at dawn in the busy season, finishing just before midday when it gets too hot. A further 2 hours' work may be done in the evening making a 5 or 7 hour day. At certain times of the year little field work can be done and there are family and tribal occasions which take up much time.

Cattle are money and also have an important social function, providing the bride price and security in old age, as well as milk, meat and skins. Grazing is normally communal, in the swamp, or on hills too rocky or steep for cultivation, and as these areas are unfenced the cattle must be herded, or if few tethered. The young boys that once herded the cattle are now often at school.

As the human population increases or the soil fertility declines, more grazing is cropped so less grazing is available for cattle. As there is no control over the numbers of cattle grazed or the area or location grazed, cattle numbers may be increasing, overgrazing results and unpalatable thorn bushes take over, run-off and consequently soil erosion increase, and the carrying capacity of the grazing declines.

Mixed cropping is standard in the fallow system. Often the swamp bottom is used either to grow rice or cassava as a famine reserve. Away from the swamp there may be a tree crop such as coffee.

The peaks in labour needs have stimulated interest in labour saving devices such as tractor services and ox-drawn implements. Weeding early planted food crops often delays the planting of cash crops such as cotton, with a subsequent yield loss. Failure to weed at the correct time is a major cause of reduced crop yields. Farmers are showing active interest in a simple safe system of chemical weed control, particularly where nutgrass (*Cyperus* spp.) and perennial grass weeds like 'lalang', which are difficult to remove by hand, are serious competitors.

Regeneration of fertility under fallows

In both the fallow system and shifting cultivation there is increased pressure on the fallow period due to increases in population, and there is a need to manage the fallow to improve and speed its effectiveness. Traditionally the land is just abandoned and natural vegetation takes over. Pendleton suggested that the value of the resting period in Thailand would be increased if seeds of nearby leguminous trees and shrubs were scattered on the land when it was abandoned. In Eastern

Nigeria when the cropping cycle is finished in the fallow system, growers may plant *Acioa barteri* a deep rooting rosaceous shrub which thrives under poor conditions, and *Anthonota macrophyllum* (Ruthenberg 1976), and it is very likely that in many areas certain indigenous legumes are valued for the restoration of fertility. When clearing forested land for cropping many trees are left either because they are too difficult to remove or on account of local superstitions. It is more than probable that many of these trees protected by superstition or folklore have a value in the fallow.

Ending the cultivation cycle with a shade producing crop such as cassava or plantains aids the regrowth of tree species. Pigeon pea is a particularly useful crop with which to end a cropping cycle as it will survive in the encroaching bush for 2 or 3 years, and provide some nutritious food. As well as building up the soil nitrogen pigeon pea is a very good accumulator of potassium, magnesium, calcium and phosphorus.

In the more fertile parts of Uganda, the government recommended that elephant grass (*Pennisetum purpureum*) should be planted at the start of the 3 year fallow but this was seldom done by the farmers due to the labour involved both in planting and clearing.

Research has shown the benefit of grazing the grass resting period; and also where plenty of kraal manure is available that crop yields can be maintained without the resting period.

Where it is economic, some areas such as the West Africa savanna zone would benefit from an application of phosphate to encourage legumes in the resting period.

Ley system

In this system several years of cultivation are followed by several years of a grass and legume ley grazed by cattle. The ley if unregulated, in that the grasses are self-sown and the grazing communal, is a short-term grass fallow. The regulated ley is planted with a grass legume mixture, the grazing fenced and regulated. This system is rare in the tropics (e.g. Kenya Highlands) as it is a high cost system dependent on a high price for meat or milk, and conditions good for pasture growth most of the year. This system was widely used on the Rhodesian tobacco farms to allow the nematode population to decline before the next tobacco crop was planted. Equipment for ploughing the leys is necessary. Even in Queensland, where much of the tropical pasture work has been pioneered, less than 2 per cent of the pasture land is planted to improve pastures and much of this is treated as a long ley or permanent grassland.

Permanent upland cultivation

As the pressure on the fallow increases, farmers have no alternative but to crop their land each year without a break, as occurs in most of India. The fields are clearly demarcated and the grassland is permanent. Land is scarce but labour often plentiful. Permanent cultivation may only apply to part of the farm, irrigated rice being grown in the valleys and tree crops further up the slope. This system is mainly practised in semi-arid areas, and to minimize the drought risk, planting is delayed until the rains have definitely started otherwise the seed is lost. A crop may be planted at various times in the season as insurance. The crops, and even varieties, tend to be chosen more for their drought resistance than yield. In such a high risk farming system little use is made of fertilizers or crop protection sprays. Mixed cropping and, where the rainy season allows, relay cropping, that is intercropping in a standing crop just before harvest, is the normal practice. Where there is sufficient land, cattle are kept for cultivation, and land has become the status symbol rather than cattle. With the heavy pressure on the land for cropping, trees for firewood are scarce and in India nearly half the cattle dung is dried to provide fuel for cooking. Whenever possible cultivators like to have an irrigated area as a drought insurance. To maintain fertility in India the rotations include many legumes, groundnuts, grams, peas, lentils (see Table 2.5) or green manure crops such as *Crotalaria juncea* (Sann hemp), *Sesbania aculeata* or *S. speciosa*, which are cultivated into the soil 2 or 3 months after sowing.

In Senegal, groundnuts are rotated with millet, sorghum or maize. The acacia trees (*Acacia albida*) (Plate 2.1) left when the land is cultivated, provide dry season fodder for cattle, and increase the yield of millet sown under the spreading trees – 'five acacias fill a store with millet' is a Sérèr proverb (quoted by Ruthenberg 1976). In the drier parts of Africa and India a sorghum–groundnut rotation is common.

In the humid and semi-humid tropics fertility cannot be maintained for permanent upland cultivation unless fertilizers and cattle manure are generously applied. In addition weeds tend to get out of control without a fallow. The 'no-tillage' system developed at IITA in Nigeria may offer a solution for these difficult areas (see p. 107).

Arable irrigation farming

This is a high-cost, high-management, high-yield system. It is labour intensive but supports the highest population densities in the tropics. As yields vary little from year to year, the farmer can risk expenditure on fertilizers and weedkiller, and spray his crop against pests and diseases.

Plate 2.1 Fertility of arable land maintained by *Acacia albida*. (Rose Innes)

The irrigation water may be brought to the area from the river or dam by a system of canals and ditches; or the rainfall impounded by bunds, sometimes supplemented by irrigation. Where irrigation water is available in the dry season, yields are higher than in the rainy season, owing to the higher solar energy during the cloudless days and the absence of waterlogging of the roots.

More than one crop a year is harvested; in certain well favoured areas even three crops of rice might be produced, but this needs considerable farming skill. Most commonly in Asia, a single crop of rice is produced each year, with an alternative crop which can vary from groundnuts to sweet potatoes, vegetables or sugar cane for the remainder of the year. Where irrigation supplements rainfall, multiple cropping is easier than where there are two rainy seasons dictating the planting seasons. By raising plants in nurseries and transplanting into the field, often into a standing crop (relay cropping), a crop occupies the land for a minimum time. With this method the Chinese in Malaysia and Singapore raise as many as eight vegetable crops a year. The irrigated paddy areas of Djakarta and Soerakartra support over 1,100 people per km^2 (100 ha), and in the Red River delta of Vietnam this can rise to 1,500 people per km^2 compared with only 20 per km^2 on the surrounding hills where shifting cultivation is practised. Fortunately these flat areas can be cropped continuously with rice without yields declining or pests and diseases destroying the crops. A family can have a relatively high standard of living with 2 ha of irrigated land and many survive on much less. There is little land to use for tracks and in some parts a bare area to dry the rice crop is nearly a luxury. Arable irrigation requires a large amount of initial capital to build dams, water distributing channels and for pumping equipment. Where water is stored in tanks or dams good land has to be sacrificed as the water storage may occupy 25 per cent of the total irrigated scheme. It is essential to have good cooperation between the farmers in the area, a disciplined system of control as a farmer cannot organize an irrigation scheme alone, and a satisfactory system of land tenure - the government institution is frequently the landlord.

Wet paddy

Wet paddy lands producing one or two crops each year have supported dense populations in South-East Asia for many centuries. Production has not declined, which one might have expected as no fertilizers are used. Apart from its high nutritive value, rice has a big advantage that it stores well and needs little fuel for cooking, and can be carried out into the field cooked and packed for a meal. Production from these areas can be increased, particularly by growing varieties more responsive to fertilizers which will be needed and controlling the pests and diseases either by breeding or spraying. As the straw of many of these new rice varieties is short, the depth of water is too shallow for fish unless special trenches are dug for them. The area of irrigated rice can be extended by the construction of new dams and irrigation canals or by using water more economically. Since the war rice has become a more important part of the diet in much of West Africa, and irrigation schemes have been included in many development programmes. In the upper valley of the Niger for example many small irrigation schemes, ploughed with oxen, have been developed. West African rivers are less rich in silt than those of the Far East and the Nile. The Senegal river carries only 100 - 150 g of silt per m^3 of water, which is only one-tenth of that carried by the Nile. This is a disadvantage in maintaining the fertility of flooded areas. The protein in the diet of rice farmers tends to be supplemented by pork in Asia, ducks in China, and throughout Asia, fish, rather than cattle. Fish are often kept in the rice paddies themselves, as in Perak, Malaysia but this has not yet been developed in West Africa. Near the house fishponds, fertilized by human and animal excreta, are commonplace in irrigated areas in Asia. In Thailand, chickens may be kept over pigs which eat the excreta that drops down and the pig excreta drops into a fishpond. Chickens are found throughout the tropics.

Deep water and floating rice

In most of the important Asian river valleys and deltas including the Gangees, Mekong, Chao Phraya and Irrawaddy, the water level in the monsoon often rises over half a metre and the farmers can only grow low yielding *indica* varieties which can survive most floods. If the floods are severe, as in 1974 in Burma and Bangladesh, the crops may be killed. As these areas are very heavily populated and dependent upon rice, the consequences of prolonged submergence of the crop, particularly if it occurs in the early stages of growth, is widespread famine. This flood risk applies to about a quarter or even a half of the world's rice lands and the situation is deteriorating as the hills are deforested, run-off is increased, and the valleys and deltas silt up.

Where the water in the fields rises over a metre, even to 5 or 6 m, floating rice is grown. These varieties (*Oryza sativa* in Asia, *O. glaberrima* traditionally in West Africa) are well adapted to the deep water conditions of Asia and Africa. As the water rises, the internodes elongate, even growing as much as 30 cm a day, and this keeps the leaves floating on or above the water. Adventitious roots which form at the nodes absorb nutrients from the silty water. Tillers develop from the nodes just below the water level and the upper stems of these and the main stems curve upwards as a 'knee' keeping the panicles above the reach of fishes and out of the mud when the water recedes. It also keeps the first three leaves out of the water, and prevents them from rotting. If the plants are uprooted by the floodwater, they can fix themselves to the soil again when the floods recede. If they remain uprooted, however, they still produce some crop. In Thailand farmers grow floating rice on the acid sulphate soils (see p. 161) possibly because the adventitious roots supply the nutrients and lessen the dependence on the basal roots in the toxic soil conditions.

The rice seed is usually sown direct in the dry or moist soil. Fertilizers are rarely applied and then only in the very early stages to get the crop away. Though weeds, particularly water hyacinth, and pests cause crop loss, spraying is not practicable and the insecticides could kill the valuable fish and pollute the drinking water. The floating rice varieties are sensitive to day length, and flower after the peak of the floods but before the monsoon ends. Harvesting may start from boats but is completed on foot, collecting the seed heads from the inter-tangled straw.

Breeders are developing improved varieties of tall and floating rice which could increase rice production throughout Asia and West Africa including Gambia, Sierra Leone, Niger and Mali, and also help to bring 3.5 million ha of tidal swamp in Malaysia and similar areas in Indonesia into rice production (IRRI 1975).

Perennial crops

In the humid or semi-humid areas, where there is no prolonged dry season, perennial crops like coffee, cacao, oil palm, banana, rubber, coconut, tea, nutmeg, etc., thrive. They keep the ground well protected from beating rain and sun. These crops are normally grown as pure stands and separate areas are used to grow food crops. Some interplanting of tree crops occurs, coffee in bananas in East Africa and cacao in coconuts in Malaysia. Food crops are often grown in newly planted plots to help control the weeds, protect the soil, and give a return until the tree crop comes into bearing. Originally most of these crops were estate grown but now smallholder production is more important.

Perennial crops mean the farmer must have a permanent site for his house nearby. As the crops are mainly grown for cash, a local processing factory and a market are essential as are roads and trucks to move the crops. Perennial crops represent a big investment but give a high return. Perennial crop farming is less labour-demanding than fallow farming. In many countries the farmers growing perennial crops are the most prosperous. In areas of Central Africa - Bukoba (Tanzania), Buganda (Uganda) and in Rwanda and Burundi - where cooking bananas are the main food, about a third of a hectare will feed a family, and fertility is well maintained. Steep and rocky land unsuitable for cultivation can be planted to tree crops. The labour requirements for perennial crops are spread through the year and the variations in crop yield are generally small. By the application of improved methods yields can be appreciably increased. This is a different form of farming. Cultivation is only necessary before planting, but pruning and spraying are often important skills, and mechanization of much of the work, including harvesting, is not possible. Once the crop is planted and established, harvesting, which may be daily in the case of rubber, but more often seasonal, is the main workload of the year.

Total nomadism

Total nomadism is the farming system practised in the dry areas of Africa. The climatic conditions in these areas are unfavourable to reliable crop production, indeed water and grazing are insufficient to allow the animals to stay in one place. So these animal owners have no permanent home, the whole family moving around with the herd, constantly in search of water and grazing. Having no permanent home they have nowhere to store fodder or grain for the bad times. Cultivation of food crops on a permanent basis is not possible, though they may sow a crop like millet to which they return at harvest.

The wealth of nomads is their livestock and animal breeding is for survival, not production, as improved feeding is impossible. The animals must have fat reserves to survive long dry seasons, and long treks through areas of sparse grazing. Severe drought causes high losses, as in the Sahel disaster of 1972/74, and a large cattle owner may see his wealth dying by the day. Their only protection is to collect large herds, and the very large owners divide their animals into a number of herds which are looked after by hired nomadic herdsmen while the owner leads a more settled existence.

It is a high risk system in regions unusable even for semi-nomadism. Increased production is nearly impossible as the scant rainfall is the dominating factor.

Semi-nomadism

Semi-nomadism is more widespread in Africa than total nomadism. These stock owners have a permanent home base where some land is cultivated, though animal products, a mixture of milk and blood, are the major part of their diet. For long periods of the year, especially in the dry season, they travel long distances with their cattle in search of grazing areas; particularly forage species they believe have special value.

As grazing is communal and unrestricted the pastures are overgrazed and often eroded. At the end of the dry season pastures are burnt to destroy ticks and stimulate a new growth of grass. As stock raiding is a part of life and predatory animals are often around, the cattle are kraaled at night. As there is no means of conserving pasture, as hay or silage, the cattle lose weight in the dry season, and this is often made worse by the long distances they have to trek to water. Calving intervals are long and the death rate of the calves is high. To have survived, the older animals must be adapted to the environment and have some disease resistance, so are kept beyond their economically valuable life. Among these semi-nomadic people, which include some of the most respected tribes such as the Masai, Karamajong and Turkana, cattle have a very important social role. The Karamajong of north-east Uganda keep some milking cattle near the home compound for the women who cultivate mainly sorghum but also groundnuts, beans and millet, while the men take the main herd to the distant dry season grazing grounds, where milk mixed with blood is their main food, hence the disastrous famine of 1980 when armed raiders stole their cattle. Good grain storage is vital as crop failures are common. The whole system is very vulnerable to drought as shown in 1972–74 in the Sahelian zone of Africa where so many cattle and people died.

Ranching

Ranching involves commercial enterprises with large herds on large areas of land. It is generally practised in marginal zones for agriculture

Plate 2.2 The auto-feed jab planter designed at IITA. (Wijewardene 1978)

in Africa and Australia, but also in more humid areas of South America. Grazing may be open or fenced and watering points are vital. This is a high-capital system with a high risk due to droughts and disease; specialization is in a single final product, which is vulnerable to market fluctuations. Increased production depends upon more capital investment for fencing, irrigation, forage conservation and improved disease control.

No-Till system for the humid tropics

This system developed in Nigeria at the International Institute of Tropical Agriculture (Wijewardene 1978) eliminates soil cultivation. The weed cover is destroyed by paraquat or other weedkillers applied with a controlled droplet applicator (CDA) (see Ch. 4). The crop, such as maize or cowpeas is then planted in the mulch of dead weeds using an auto-feed jab planter (Plate 2.2). This is a hand-held tool of simple construction and easy to operate, which gives a precise spacing at the required depth into a slot cut only where the seed is planted. The crop germinates through the mulch and is protected by one or two insecticide sprays drifted on to the crop also from a CDA sprayer.

This system fulfils many of the requirements outlined in Chapter 1. The soil is continuously covered to protect it from the sun and pounding rain. This reduces soil capping, run-off and erosion in comparison with the traditional systems where the land is hoed or ploughed. (see Table 2.1).

Table 2.1 Effect of interaction between nature of crop and soil management techniques on run-off and soil loss at Ibadan, Nigeria

Slope (%)	**Run-off** (% of rainfall)		**Soil loss** (t/ha)	
	Maize (No-Till)	**Cowpeas** (Conventional till)	**Maize** (No-Till)	**Cowpeas** (Conventional till)
1	1.2	15.8	0.91	0.24
5	1.8	31.1	0.03	0.65
10	3.1	40.0	0.00	1.71
15	3.5	17.2	0.10	1.22

Note: Rainfall 781 mm.
Source: Lal (1976).

Wijewardene (1978) gives a comparison of the labour requirements of the two system in Table 2.2.

By this system it is suggested that a farmer could cultivate 4 ha a season without the need for a tractor and with minimal risk of erosion, instead of a 0.5–1 ha. A comparison of the two systems over six seasons showed no decline in yield from continuously cropped 'No-Till' areas and a dramatic increase in earthworm activity whereas yields declined as normally expected on the areas continuously cropped in the conventional way.

Though designed for the humid tropics this system has a much wider application. Where there is a marked dry season, cultivation of the seedbed must often wait for the early rains to soften the ground, and consequently planting cannot take full advantage of the nitrogen flush which comes with the rain (p. 26). This 'No-Till' system eliminates waiting as the ground cover can be destroyed before the rains. The speedier seedbed preparation also means that a second crop can be planted earlier and in some areas even three crops might be grown in the year where only two were possible before. The growth of the weeds is hindered by the mulch and crop competition.

Other tools and mechanized systems have been designed to reduce the time and energy input of the farmer by an increased input of an alternative source of energy, either animal or mechanical, as when the hoe is replaced by the plough. This 'No-Till' system reduces the farmers' input of both time and energy without replacing it by mechanized energy.

Table 2.2 Comparison of man-power requirements, using No-Till[(a)] versus conventional crop establishment systems, for maize and cowpea on savannah (*Imperata*) grass covered land in Nigeria.

Operation	Man-hours per hectare No-Till	Man-hours per hectare Conventional
A. *Field preparation*:		
a. Slash, burn and till manually.		180
b. CDA spraying with contact herbicide.	4	
B. *Seeding*:		
a. Manual planting into tilled soil with machete (low plant population).		20
b. Auto-Jab planting (maize-cowpea 75 × 25) through the mulch cover.	35	
C. *Weed control*:		
a. Manual weeding twice.		280
b. CDA spraying with pre-emergent herbicide.	4	
D. *Fertilizer application*:		
a. Banding by hand along rows.		25
b. Using hand propelled band applicator.	6	
E. *Plant protection*:		
a. Knapsack spraying of insecticide (on cowpea).		10
b. CDA spraying of insecticide, *twice* (over entire crop).	2	
Total man-hours spent to establish the crops each on one hectare.	51	515
Comparison of yields.	2,400	1,255

[(a)] No-Till tools used were CDA sprayers (herbicide and insecticide) and IITA automatic 'jab' planter.
Source: Wijewardene (1978)

The tractor on the farm is often justified by its alternative role in transportation, but by eliminating the high power need for tillage other simpler and cheaper solutions can be developed to help the tropical farmer with transportation and other farm operations.

FOOD CROPS

The characteristics of the main tropical cereals are given in Table 2.3, and those of root crops and plantains in Table 2.4. These crops provide

Table 2.3 Characteristics of main tropical food crops (a) cereals

Common name	Latin name	Ecological requirements
Rice	*Oryza sativa* L.	Swamp or good rainfall. Ample sunlight. Prefers irrigation. Certain salinity tolerance.
Barnyard millet	*Echinochloa frumentacea* (Roxb.) Link	Swamp.
Maize	*Zea mays* L.	Widened by breeding. More suited to high grass than short grass regions. Not drought resistant. Needs reasonably fertile soil. Dislikes dry heat. Vulnerable to water shortage around tasseling.
Finger millet	*Eleusine coracana* (L.) Gaertn.	Short grass regions with fairly high rainfall. Yields well on poor shallow soils. An irrigated type exists. Seldom attacked by birds.
Sorghum, dura, guinea corn	*Sorghum bicolor* (L.) Moench	Enjoys heat and little rain. Low transpiration rate and high drought resistance. Clear skies and not too humid. Suffers badly from bird damage and *Striga*.
Bulrush millet Pearl millet	*Pennisetum americanum* (L.) K. Schum [Syn. P. *typhoides*] (Burm. f Stapf & Hubbard)]	Similar to sorghum but requires less rain and grows on poor lighter soils but not as drought resistant.
Foxtail millet	*Setaria italica* (L.) Beauv.	Needs little rain and can stand high temperature but not swamp conditions. Grown on inferior soils.
Hungry rice	*Digitaria exilis* Stapf	Yields on poor thin soils with light rainfall. Grown in savana zone of W. Africa.
Panicum millet Common millet	*Panicum miliaceum* L.	Hot dry regions with poor soil. Lowest water requirement of any grain crop.
Panicum millet Little millet	*Panicum sumatrense* Roth ex Roem. & Schult, (syn *P. miliare*) Lamk.	Can stand both drought and swamp. Will mature some crop even when rains fail. Grown on inferior soils.

For further details see Purseglove (1972) Vol 1.

Table 2.3

Growing period	Grain (Rylha) Yield	Storage	Breeding
120 days	1,000-4,000	Good-unmilled.	Self-fertilized. Very many varieties. $2n = 24$.
45 days. Shortest of all cereals.	300-700	Good.	Generally self-fertilized. $2n = 36$ or 56.
90-200 days.	800-3,000. Green cobs also eaten, valuable when food short.	Poor. Dry cobs well before storage.	Crosses freely, can be selfed. Main races intercoss. Hybrid seed increased yield. Usually $2n = 20$.
4 months. Shorter than most maize varieties.	800-1,800. High calcium content	Very good in head.	Self-fertilization common. $2n = 36$.
Some short 110 days' maturation. Others up to 180 days.	750-1,500 Double if irrigated.	Poor. Hard grained varieties better than soft grained.	$2n = 20$. Cross-fertilizes easily. Numerous varieties. 1-4.5 m in height. Open or compact heads.
Shorter than sorghum.	400-900	Very good. Immune to weevils.	Cross-fertilizes readily, including wild varieties. Awned varieties prevent severe bird attacks. $2n = 14$.
90-120 days	600-1,000	Good.	Self- and cross-fertilization. Both common. $2n = 18$.
90-120 days	400-600		$2n = 54$
Quick maturing. 75-90 days	500-900	Good.	Both self- and cross-fertilized. $2n = 36$.
3 months	500-600	Good.	Almost entirely selfed. Can be crossed artificially. Emasculation a delicate operation. $2n = 36$.

Table 2.4 Characteristics of main tropical food crops (b) roots and plantains

Common name	Latin name	Ecological requirements	Growing Period	Yield t/ha
'Old' Coco-yams, Dasheen, eddoes, Taro	*Colocasia esculenta* var *antiquorum* (Schott) Hubbard & Rehder and *C. esculenta* (L.) Schott	Damp shaded habitats. Long wet season or poorly drained soil. Important in Pacific. Needs ample water	6-9 months	5 — 12
'New' Coco-yams, Tannia.	*Xanthosoma sagittifolium* (L.) Schott	As above. Slightly drier habitats and drained soils.	As above.	7 — 20
Bananas (Plantains)	*Musa* L. cultivars	Fertile soil and high moisture requirement. Well-distributed rainfall. Protects and maintains soil fertility.	Bears the second year after planting for 4 to 20 years or more, even up to 50 years.	7 — 20
Yams.	*Dioscorea* L. cultivars	Cannot tolerate frost. Poor growth below 20 °C 30 °C prefered.		7.5 average wide variations
White; winged; water, greater yam.	*D. alata* L.	1,000 - 1,500 mm minimum rainfall well distributed. Deep freely drained friable soil. Fertile loamy soil rich in organic matter.	8 - 10 months	Highest yielder Up to 30
White guinea yam.	*D. rotundata* Poir.	Good rainfall, sunshine and humidity. Can tolerate dry conditions but not so well as cassava.	8 - 10 months	7 - 15 or more 2.5 - 5 are needed for seed
Yellow guinea yam.	*D. cayenensis* Lam.		7 - 12 months	
Chinese yam.	*D. esculenta* (hour). Burk.	Differences between species in respect of drought resistance. Growth on mounds up to 1 m high, mulched, vine-trained up a long pole.	12 months	
Potato yam.	*D. bulbifera* L.	Wide distribution Asia and Africa	Up to 24 months	Up to 19 aerial and 22 underground
Cush-cush yam.	*D. trifida* L.	Only food yam originating in New World		
Cassava	*Manihot esculenta* Crantz	Prefers ample rain and moderate humidity but can withstand severe drought. Prefers deep fertile soils but can yield on leached tropical soils of low fertility.	9 - 24 months	6 - 30 One of the highest yielding of all plants
Sweet Potato.	*Ipomoea batatas* (L.) Lam.	Well-drained sandy loam with long, warm growing season. Needs less rain than other roots. 750 - 1,250 mm Too high a fertility level gives too much top. Drought resistant	3 - 6 months shortest growth period of main root crops.	5 - 10

Food value	Storage	Comments
Small grained easily digested starch. Protein content high for a root crop. Source of calcium, phosphate, and Vitamins A, B and C.	Stores reasonably well if fully mature but not in ground.	Only food crop grown under mature cocoa. Upper section of the corm planted. Poi of Hawaii is semi-processed Taro
More nutritious than Dasheen but less digestable.	As above.	Distinguised from Dasheen by leaf. Sagittate leaf and no purple coloration in foliage. Young leaves used as vegetables.
Preferred food of many areas. 20% the energy value per kg. compared with rice or maize. Easily digested. Good source of vitamins.	Must be dried for storage.	Leaves used for wrapping peeled fruit in cooking, but not eaten. Leaf base fibre used as rope. Some varieties suitable for beer.
Carbohydrate major dry matter content of tubers. Low protein but better than cassava. Vitamin C	Remove all soil before storage	Dioscorea tubers contain dioscorine, a poisonous alkaloid destroyed in cooking. Yam beetle only serious pest.
Not suitable for pounding into 'fufu', boiled or roasted. Poor quality. Very commonly grown.	Improves with storage.	Tubers large up to 2 m long and 45 kg in wt., large, coarse, spherical. Grown on poorer drier areas. Widely distributed Main sp. of W. Indies
The most palatable yam - white fleshed. Excellent flavour and cooking qualities.	Stores well.	Suited to long dry season of savanna zone. Hardy, frequently cultivated in W. Indies. Most important species in W. Africa. Cylindrical tubers 2-5 kg up to 20 kg.
Yellow fleshed. Not very palatable.	Short dormancy. Does not store well. Can be harvested all year.	Plants hardy. Tubers large. Yam of W. African forest zone. Flavour inferior *D. rotundata*
Low fibre, very palatable, free from toxicity, sweet flavour. Unsuited for pounding.	Easily bruised. Short dormancy. Does not store well.	Produces a large number (15-20) of very small (or 5-9 kg) yams. Thrives in drier areas. Most common in Pacific and Asia. Adapted to mechanical harvesting
Ground tubers not very palatable - bitter - particularly in Africa. Aerial tubers in Asia sometime eaten raw.	Can be stored considerable time	Tubers on aerial stems and at base. Small tubers about 0.5 kg Similar to Irish potato.
Delicious flavour. Low yield.		Common in Caribbean area. Tubers small, produced in groups. Flesh white, purple or pink.
The least popular staple food. Low protein content. Old tubers lignified. Starch of quality - tapioca. Important famine reserve. Used for gari and fufu in W. Africa and farinha in S. America.	Stores well in ground but not after lifting. Must be processed to flour, or sliced and dried.	Propagated by stem cuttings. Contains cyanogenetic glucoside which liberates prussic acid. Must be cooked or processed. Immune to locust attack. Mosaic virus serious.
Mainly a starch provider. Supplies vitamins A, B and C. Leaves used as vegetables.	Does not store well in ground or after lifting. Harvested over a period. Sliced and dried for storage.	Plant leafy stems of mature plants (slips). Subject to pest and disease attacks, including locusts. On cooking some cultivars are dry and mealy others moist soft gelatinous.

Plate 2.4 Three types of yams (Courtesy Sue Wing)

the bulk of the food intake of the people living in the tropics. To balance the diet and add flavour and variation a large number of vegetables and grain legumes or pulses are grown. (see Plates 2.3 and 2.4).

Pulse crops (Edible seeds of leguminous plants)

Table 2.5 covers the most important pulse crops.

Plate 2.3 Yams trained up wires. University Farm, Trinidad. Note shifting cultivation on Northern Range in the background.

Table 2.5 Characteristics of main tropical food crops (c) legumes

Latin name	Common names	Regions of importance	Ecological conditions	
			Soil	**Temperature**
Arachis hypogaea L.	Groundnut, Peanut.	All tropical and sub-tropical countries. India largest producer. Nigeria largest exporter.	Sandy loams with adequate calcium. Friable, free drained. No surface crusting.	Killed by frost
Cajanus cajan (L.) Millsp. syn. *C. indicus* Spreng.	Pigeon pea, Red gram.	Pan-tropical. Majority grown on Indian continent.	Tolerates poor soil and salinity but not waterlogging or lime deficiency.	18-29 °C. Tolerates heat but not frost.
Canavalia ensiformis (L.) DC.	Jack bean, Horse bean.	Minor grain legume with potential. Native of W. Indies and C. America.	Wide range. pH 5-6. Tolerant waterlogging and salinity.	Long growing season frost free, moderate temperature. Suitable for low altitudes high temperature.
Canavalia gladiata (Jacq.) DC.	Sword bean	Limited scale pan-tropical particularly East and India.	Fertile. No waterlogging.	High temperatures. 15-30 °C.
Cicer arietinum L.	Chick pea, Gram.	Represents over a third of pulses grown in India particularly in N.W. States. Most important food legume in S.E. Asian sub-continent.	Wide range with good drainage. Clay loams. Black cotton soils of Deccan. pH 6-9. Tolerates alkalinity and salinity.	Sub-tropical. Cool dry. Tolerates high temperature near maturity. 18-20 °C - variation between cultivars.

Ecological conditions		Growth period	Average yield	
Annual rainfall	**Husbandry**		(kg/ha)	**Comments**
1,000 mm/ annum or more. 500 mm growing season. Seasonally arid areas.	Grown alone or inter-cropped on flat or ridges. Close spacing. Seeds formed in ground. 100% inbred.	3½ months bunch types. 5 months spreading types.	400 - 1,600 unshelled. Shelling percentage 80 - bunch, 60 - 75 spreading.	Important source of vegetable oil. Oil residue for cattle cake. Peanut butter in USA. Green haulms for hay or fodder. Dry well after harvest. Store in shell.
600 - 1000 mm. Rain first 2 months important. Drought resistant. Deep rooting. Needs bright sunshine.	Often intercropped with cereals and left after cereal harvest. Weed control essential first 2 months. Plant on flat or ridge.	Pods mature 6 - 12 months according to variety	400 - 1,200. Higher with *Rhizobium* inoculation.	Short lived perennial - 3 - 4 years but yields fall after first year. Grown for canning in some countries.
900 - 1,200 mm. Down to 650 mm with subsoil moisture.	Principally broadcast as cover crop or green manure.	6 - 10 months for seed.	1,300 dry beans. Up to 20 t/ha green manure.	Require soaking and boiling to remove toxic constituents. Valuable green manure crop. Care needed for animal feeding.
Moderate. 900 - 1,500 mm/ annum. Some drought resistant Intolerant waterlogging.	Vigorous perennial climber needing support, but often grown as annual.	6 - 10 months As vegetable 3 - 5 months Perennial 2 years.	700 - 1,500	Toxic constituents must be removed by soaking and boiling. Tough seed coat, poor texture and strong flavour.
650 - 750 mm. Cannot tolerate heavy rain. Often grown as winter crop on residual soil moisture. Dull cloudy weather and high humidity reduce yield.	Rough tillage. Grown as pure stand or intercropped with cereals.	3 - 6 months.	1,600 good conditions. 600 average but experi-mentally up to 4,000.	Seeds split to make '*dhal*'. Green pods for vegetables. Very susceptible to storage pests particularly bruchids. Bland insipid flavour.

Latin name	Common names	Regions of importance	Ecological conditions	
			Soil	Temperature
Cyamopsis tetra gonoloba (L.) Taub.	Cluster bean, Guar.	India and Pakistan. Widely cultivated in S.E.Asia as a vegetable.	Wide range well drained non-acidic. Often sandy loams. pH 7.5-8.0. Tolerates high salinity and alkalinity.	Sun-loving. 25-30 °C. Intolerant forest and shade.
Glycine max (L.) Merr. syn. *G. soya* (L.) Sieb. Zucc.	Soya bean, Soybean.	USA largest producer and exporter E.Asia particularly China and Indonesia. Brazil as export crop. Not popular as food outside E. Asia as difficult to prepare and unusual taste.	More suited to sub-tropics as growing conditions required similar to maize. Needs specific strain of *Rhizobium japonicum*.	Hot damp summers, but no excessive heat. Dislikes shade particularly after flowering.
Kerstingiella geocarpa Harms syn. *Voandzeia geocarpa*	Kersting's groundnut, Hausa groundnut.	Savanna areas of W. Africa.	Sandy loam rich in lime but grows on poor sandy soil.	Tropical legume. 18-34 °C. Bright sunshine.
Lablab purpureus (L.) Sweet syn. *L. niger* Medik. *Dolichos lablab* L. *L. vulgaris* Savi.	Hyacinth bean, Dolichos lablab, Bovanist bean.	Grown throughout tropics and sub-tropics. Popular vegetable in Asia. Important supplier protein in S. India diet.	Tolerates poor soils if well drained. Sandy or heavy clays. pH 5-6.5.	Warm equable climate 18-30 °C. Tolerates high temperature. Withstands limited frost.

Ecological conditions		Growth period	Average yield	
Annual rainfall	Husbandry		(kg/ha)	Comments
500-700 mm. Hardy and drought resistant. Optimum yields under irrigation.	*Rhizobium* inoculation often essential. Scarified seed germinates more uniformly. Rarely grown as a pure stand.	2 months Green pods, 4 months seeds.	600-900 dry seed, 6-8 t green pods.	Green immature pods traditional Asian vegetable. Toxic constituents in seed. Source of vegetable gum.
Needs plenty of rain for germination. Not v. tolerant water stress but some tolerance waterlogging.	Rotated with maize, sorghum, millet or rice. Firm seedbed.	2½-7 months	600-1,300	One of world's most important sources of oil and protein. Unripe seeds used as vegetables, dry seeds eaten whole, split, sprouted, processed, fermented. Source of edible oil and residue, soya meal for livestock. Also hay, silage, fodder, cover, green manure. Easy to mechanize.
500-600 mm. Suitable for semi-arid conditions.	Often first crop in rotation or intercropped. As some seeds remain in ground and germinate in successive crops.	3-5 months	400-500	Seed-bearing pods develop in soil like groundnut. Highly nutritious. Pleasant tasting speciality food.
600 mm. Dry-land crop. Arid climates.	Dense foliage smothers companion crops so grown alone. Last crop in Sudan Gezira rotation. Climbing forms need support.	2½-10, months, 5-7 for seed.	450-1450, beans. 2,500-4,500 green pods.	Young pods and tender beans popular vegetables in India. Dried beans make split pulse or soaked and sprouted. Haulms for hay or silage. Remains green in dry season. Fodder, green manure or cover crop.

Latin name	Common names	Regions of importance	Ecological conditions	
			Soil	Temperature
Lathyrus sativus L.	Grass pea, Chickling pea.	Mainly India. Cheapest pulse in India, eaten by poor. Parts of S. America. Not widely distributed.	Wide range including very poor and heavy clays. Rice fields in India.	10–25 °C. Cold season crop in India.
Lens culinaris Medik. syn. *L. esculenta* Moench	Lentil, Split pea.	Mainly N. India. One of oldest grain legumes. Parts of S.America, Mexico and Ethiopia.	Wide range from sandy to heavy clay. Light loams, black cotton, and alluvial soils preferred with high fertility, vegetative growth at expense of seed.	About 24 °C. Varies with cultivar. Cool season crop in India. 15 °C for germination.
Macrotyloma uniflorum (Lam) Verdc. syn. *Dolichos uniflorus* Lam.	Horse gram.	Poor man's pulse of S. India Particularly Madras and Hyderabad.	Wide range, well drained not highly alkaline. Intolerant waterlogging.	Dry tropics. 20–30 °C.
Phaseolus acutifolius Gray var. *latifolius* Freem.	Tepary bean.	Cultivated by Aztecs in Mexico. Not widely grown. Quick crop in dry farming.	Light well drained not heavy clays. Yields on poor soil. Tolerates salinity and alkalinity.	Hot dry conditions with bright sunshine where other beans do not form seed.

Ecological conditions Annual rainfall	Husbandry	Growth period	Average yield (kg/ha)	Comments
375-650 mm. Very drought tolerant, often planted after rains have failed. Tolerates excess rain and land subject to flooding.	Often broadcast in standing rice crop. Grown with other crops including chick pea. Clean seedbed.	5-6 months. Few early maturing varieties 4 months	300-450 seed. Up to 1,000 as pure stand and equal amount of hay.	Continued consumption can cause Lathyrism - paralysis of lower limbs often irreversible. Efforts to ban sale, but important crop in adverse conditions. Used to adulterate other grain legumes. Steeping and drying reduces toxin. Hope also in breeding.
750 mm. Dry harvest. Likes sub-soil moisture. Enjoy heavy dews and moderately drought tolerant.	Pure stand or mixed cultivations. Sometimes sown in standing rice crop.	3-4 months	Pure stands up to 1,100, mixed crops 350-650.	Most easily digested of grain legumes. Split seed (*dhal*) used in soup. Ground with cereals. Young pods used as vegetable. Dried haulms, husk and bran for cattle fodder. Wide variation in protein content. Traded internationally.
Drought resistant. Under 900 mm.	Pure stand before cereals. Catch crop or drilled into standing castor. Crop receives little attention.	4-6 months.	700 down to 200	Seed boiled and fried but not split. May be grown for forage, or green manure.
500-600 mm. Arid conditions. Too much rain causes excessive vegetative growth.	Seed rapidly takes up soil moisture. Pods shatter. Harvest by hand.	2 months. Other types 3-4 months.	400-700	Laborious to harvest and cook. Strong flavour and odour. Never popular.

Latin name	Common names	Regions of importance	Ecological	
			Soil	Temperature
Phaseolus lunatus L. syn. *P. limensis Macf. P. inamoenus* L.	Lima bean, Sieva bean, Madagascar bean.	Originated in C. America. Spread throughout tropics. Important in humid regions of Africa, Burma and other parts of Asia.	Wide range, well drained. pH 6-6.8.	16-27 °C. Below 13 °C growth retarded. Over 21 °C quicker maturing, fewer smaller beans. Over 32 °C poor flower set.
Phaseolus vulgaris L.	Haricot bean, French bean, Kidney bean, String bean, Black beans.	Best know sp. of *Phaseolus*. Most important pulse of tropical America. Important in E. Africa and temperate countries. Local pulses preferred in India. Widely cultivated. Not suited to continuously wet areas.	Most light sands - heavy clays. Friable deep well drained. pH 6.0-6.8. Sensitive to trace element deficiencies and excesses.	16-24 °C. Killed by frost. 30 °C upper limit.
Psophocarpus tetragonolobus L.	Goa bean, Winged bean, Manila bean, Asparagus pea. Perennial herb grown as annual.	Main tropical Asia. Not grown in Africa.	Non-waterlogged loam.	Resistant high temperatures.
Vigna aconitifolia (Jacq.) Maréchal syn. *Phaseolus aconitifolius* Jacq.	Moth bean, Mat bean.	Native to Indian continent. Important semi-arid areas near S.E.Asian tropical deserts, India and Thailand. Not in humid and sub-humid tropics.	Many types. In India, light sands and red gravels.	Hot season crop in India. Requires high uniform temp. 27 °C soil temp. Very frost susceptible.

Ecological conditions		Growth period	Average yield	
Annual rainfall	Husbandry		(kg/ha)	Comments
900 - 1,500 mm. or over. Humidity important for flower set. Once established - drought tolerant.	Seeds brittle, need careful threshing.	3 - 3½ months. Small seeded. Up to 9 months large seeded.	Wide range 200 - 300 up to 3,000.	Large and small seeded types. Mature beans contain cyanogenic glucoside which can liberate HCN. Needs lot of cooking. Green or dry beans cooked.
500 - 1,500 mm	Seldom grown alone in Africa. In Latin America grown in maize - bean rotation. Often intercropped. Production limited by pests and diseases. Small growers often uproot whole plant dry and beat out beans.	2 - 5 months, 3 months common.	500 dry beans with wide variations. Max. yield 3 - 4 with irrigation.	Edible pods, green shelled beans and dry beans eaten. Important protein source in Africa. If pre-storage drying inadequate, toxicogenic moulds may develop. Insect damage in storage serious without precautions.
Hot, wet climate	Wide spacing with support for vegetables. Closer spacing and allowed to trail for roots.	Tubers 7 - 8 months		Grown mainly for immature pods cooked as vegetable. Ripe seeds eaten in Burma. Exceptional capacity for root nodulation. Pods 4-winged.
500 - 750 mm. Needs moisture during month after planting then drought tolerant. Often planted end of rains, and grows on soil moisture.	Usually brodcast in millet or cotton. Sometimes as pure stand. Needs well prepared seedbed.	3 months	200 - 600	Few insect pests. Sometimes grown as forage. Beans eaten whole or split for dhal. Low spreading habit makes harvesting difficult.

Latin name	Common names	Regions of importance	Ecological conditions	
			Soil	Temperature
Vigna mungo (L.) Hepper *Phaseolus mungo* L.	Urd, Black gram, Woolly pyrol.	Originally S.E. Asia now spread through tropics. Highly popular pulse of high caste Hindus. Green manure in W. Indies.	Heavy water retentive clays and black cotton soils in India. Also deep loams and paddy soils.	Tropical. Tolerates high temperatures. 25-35 °C. Prolonged sunshine, no frost, dislikes cloudy conditions.
Vigna radiata (L.) Wilczek syn. *Phaseolus aureus* Roxb.	Mung, green or golden gram	Important throughout S.E. Asia, lesser extent Africa and Americas.	Wide range, if well drained. Deep loam, adapted to clay and black cotton. Tolerant alkalinity and salinity.	30-36 °C. Tolerates heat but not frost.
Vigna umbellata (Thunb.) Ohwi & Ohashi *Phaseolus calcaratus* Roxb.	Rice bean, red bean.	India, Burma, Malaysia, E.Africa, W. Indies, Mauritius, Queensland.	Wide range. Fertile loam preferred.	18-30 °C. Tropical. Frost susceptible. Can tolerate high temperature.

Ecological conditions Annual rainfall	Husbandry	Growth period	Average yield (kg/ha)	Comments
900 mm. More drought resistant than *Vigna radiata*. Dislikes rain at flowering.	Quick growing. Intercropped with maize, sorghum, cotton or in pure stands. Grown on ridges in rainy season. Rough tilth.	2½ - 4 months	350 - 600	Attacked by many insects and diseases. Low habit makes harvesting difficult. Whole plant often harvested. Susceptible insect damage in storage. Used for *dhal* and papadams.
750 - 900 mm or less. Drought tolerant. Dislikes excessive rain particularly at flowering.	Good tilth. Frequently after rice or in cereal crop. Susceptible to many pests and diseases.	2½ - 4 or 5 months	200 - 700 or less when inter-cropped.	Very similar to *V.mungo*. Two main types yellow and green, also black and brown, seeds. Much studied in recent years. Pods shatter easily. Essential to separate green leaves and immature pods. Valuable grain legume. Popular, easily digested. Widely used for bean sprouts.
1,000 - 1,500 mm. Some drought tolerance.	Broadcast. Sowing date critical in W. Bengal. Free from most pests and diseases.	2 - 4 months	200 - 800	Short day. Pods easily shatter. Not normally attacked by storage pests. Nutritious pulse. Not popular in India as cannot be processed to *dhal*.

Latin name	Common names	Regions of importance	Ecological conditions	
			Soil	Temperature
Vigna unguiculata (L.) Walp. syn. *V. sesquipedalis Fruhw* *V. sinensis* (L.) Savier Hassk	Cowpea, Black-eye bean or pea, China pea.	Ancient food crop. Widely distributed. Very important W. African savanna regions. Second pulse of importance in Africa after *Phaseolus vulgaris*.	pH 5.5-6.5. Wide range if well drained. Heavy clays and fertile soils. Nematodes serious on sandy soil. Tolerant acidity but not salinity.	20-35 °C. Hot weather crop. Semi-arid. Over 35 °C flower and pod shedding. Cannot stand cold or frost.
Voandzeia subterranea (L.) Thou.	Bambara groundnut	Drier regions of tropical Africa on poorer soils; Zambia, Nigeria, Upper Volta, Madagascar.	Wide range, pH 5.0-6.5. Soils too poor for groundnuts.	20-28 °C. Very adaptable. Sunny, hot, frost-free.

Sources: Kay, D. E. (1979); Purseglove, J. W. (1968).

Note: The important legume crops *Pisum sativum* (pea), *Vicia faba, Lupinus* spp. (lupin) and *Phaseolus coccineus* though grown in the high tropics are regarded more as temperate crops and omitted from this table. *Mucuna pruriens* velvet bean is mainly for cattle feed.

Legumes

Among the vegetarian population of India, legumes are an important part of the diet providing much of the essential protein. Roughly a seventh of the arable land in India is planted with pulses, frequently as a cereal intercrop. *Dhal*, a cooked flour made from pulses, is served whenever possible with Indian meals. The large area of legumes grown is important in maintaining the nitrogen levels of these soils which are continuously cropped. In Central America, outside the towns, the brown bean (*Phaseolus vulgaris*) seems to appear at every meal. It is usually grown intercropped with maize. In other parts of the tropics

Ecological conditions Annual rainfall	Husbandry	Growth period	Average yield (kg/ha)	Comments
600-1,500 mm. Excess rain or humidity reduces yield. Tolerant heat and drought, but not waterlogging or shade particularly after flowering.	Firm moist seedbed. Follows finger millet in rotation, or interplanted in cereal crop. Susceptible to wide range of pests and diseases but tolerant cultivars. Protection against insects can increase yield 15-30 times.	2-8 months	100-900, 100-200 common. Raised to 1,000-2,000 with insecticides.	Wide range of forms. Widely used pot-herb in Africa as rapidly replaces harvested shoots and leaves. Susceptible to storage pests. Bruchids serious in Nigeria. Store in pods. *Striga* causes serious losses in Africa. Leaves and young shoots dried for dry season use as a spinach.
600-1,200 mm. Grows in dry savanna regions, also tolerates excess rain.	Intercropped with millet or pure stand as first crop. Fine tilth. Often ridged to encourage pod development.	3-4 or 5 months	650-850	Grown for edible seed not as an oil seed. Few serious diseases or pests in field but number of storage pests. Pods easily detached before harvest in some cvs.

legumes provide not only the beans as a major part of a meal but also green vegetables. In addition legumes are important for feeding animals, providing both grazing and hay. Groundnuts are an important cash crop in the savanna regions of Africa. Of the thousands of known leguminous species only 20 are used extensively. Research has been concentrated on a few species such as the soya bean and the groundnut. However there is a great need for research on the other species to overcome major weaknesses. For example many grain legumes are deficient in cystine and methionine, some contain toxic constituents, some species are difficult to harvest, others shatter easily when ripe, there is a wide variation in temperature tolerance, and tolerance to pests and diseases, both in the field and in storage, in some cases the cultivars grown are short-day responsive and the yield is influenced by the sowing date. Among the different cultivars of each species there is generally sufficient variation in these characteristics to suggest the correction of these deficiencies would respond to breeding.

Some of these legumes are of great value in areas marginal for cultivation, where temperatures are high or the rains short or unreliable. As a legume crop may add the nitrogen equivalent of 200 kg/ha of ammonium sulphate, fixed by its root nodules, the importance of legume crops in maintaining fertility without the use of expensive fertilizer is quite considerable.

Fruits

Apart from the well known tropical fruits, orange and its relatives, pineapple, mango, banana, pawpaw and avocado, there are at least a hundred others grown in tropical Asia, and another hundred in the subtropics which flourish in certain parts of the tropics. (For brief descriptions of the most commonly grown fruits, see Macmillan 1956; Purseglove 1968).

Many of the tropical fruits very popular in specific areas have not been studied, nor developed, nor successfully introduced elsewhere. The durian (*Durio zibethinus*), said to have a unique flavour, is widely cultivated in Malaysia and the adjacent countries, as is rambutan (*Nephelium lappaceum*) and mangosteen (*Garcinia mangostana*), arguably the most delicious of all tropical fruits, all these thrive in hot humid climates but are rarely found outside the Malaysia region.

Sapodilla (*Manilkara achras*) is a native of the New World tropics. Besides providing a fruit, in many parts of Central America a latex is tapped from the stem which is used to make chicle, the basis of chewing gum.

Many of these fruits could, with strict quarantine, modern techniques and air transport be introduced into Africa and South America and even produced for export. Though the West African farmers do not grow fruit trees to the extent the Malaysians do, they have the kola nuts which are chewed as a stimulant, particularly in the drier areas. The betel nut (*Areca catechu*) sometimes wrapped in betel pepper leaves (*Piper betle*) serves a similar function for the East Indians, Malaysians and Indonesians, but these cannot be strictly regarded as food.

Vegetables

Herklots (1972) describes around 200 vegetables grown in South-East

Asia, including 25 different beans, 6 forms of cabbage, 6 peppers, a wide range of spinach-type plants, as well as cucurbits, tomatoes, lettuce, onions and others common in the West. These vegetables form an important part of the diet, both in Asia and Africa. The permanent garden plot near to the house, so common in Africa, may have 20 or more different plants used as vegetables. Most tropical people have their local beer made from bananas, maize, sorghum, palm juice, etc., which is an important source of vitamin B complex.

Staple foods

The amount of rainfall, and more particularly its distribution, is the major factor determining which food crops are grown. The local people have their preferences, plantain, rice, palm oil, and yam for example. Where the rainfall is less well distributed or more erratic, the choice of food crops is reduced. In the short grass areas of Africa for example, sorghum and millet are the staple food.

Sorghums and millets are hardier crops than maize but do not yield very heavily and require a lot of labour for threshing, pounding or milling, and winnowing. A woman can thresh and winnow 6 kg/hour which is then ground between two stones. A kilogram is sufficient for two adults for a day with other foods such as beans, groundnuts or meat. Rice and maize mills have been set up in many areas where these crops are grown. Often the owners of such mills will process a peasant's crop for a small payment or part of the crop. Using a pestle and mortar, a woman can pound only 1 kg of maize in an hour, thus there is a large economy of time and effort by mechanization; a trip to the mill also has the attractions of a market. As the grains can be stripped from the maize cobs in a simple hand stripper, the bulk of the labour between the dried cobs and the flour can be mechanized. Mills have not been set up to deal with the smaller grains; this in itself has encouraged the change from millet and sorghum to maize and in regions marginal for maize, soil exhaustion and erosion have followed. Heavy storms are characteristic of the millet areas. *Eleusine* in particular offers much better resistance to soil wash than the maize crop which is replacing it. Hill rice would be better than maize. All should be planted across the slope of the land.

As the fertility of an area declines, there is an increasing reliance on cassava which can yield heavily on poor soil. Peeled cassava roots usually have only about 1 per cent protein, about half that of yams, but there are varieties with more, some as much as 5 per cent and these varieties could either be introduced or used for breeding. In East Africa surplus cassava, sweet potatoes, and plantains are peeled, dried and stored, and when needed ground into a flour. In West Africa much of the cassava is converted into gari before consumption. The roots are peeled, washed and grated, then placed in a bag, where some fermentation occurs and the juice extracted. It is then roasted, sieved and dried. (See Onwueme 1978 for details of the process.) About 100 kg of cassava roots are needed to make 40 to 50 kg of gari, and over 90 per cent of the prussic acid is lost in the process. Gari can be stored for a long time, and is bought by the town dwellers and even exported to the UK. It is an instant food which can be eaten dry or mixed into a paste with water and eaten with stew or soup like 'fufu', the sticky balls of starchy material made from boiled and pounded cassava or other starchy roots, or plantains. In South America cassava is mainly eaten in a form similar to gari. The preparation of gari takes a lot of hand labour, but most of the processes can be mechanized as they have been in South America. A processing plant however requires a regular supply of freshly harvested cassava roots, as they deteriorate rapidly after being dug out of the ground. With the growing demand for food for the big Nigerian towns such as Ibadan and Lagos there is interest in developing gari production on a commercial basis. In Thailand cassava is an important cash crop, over 3½ million t being exported each year mainly as factory produced pellets for feeding cattle in continental Europe. This trade is greater in volume but not value, than rice exports.

Improving food crops

Plant breeders are now paying more attention to improving food crops. The scope for improving cassava is shown by some of the current aims where progress is being made; mosaic and bacterial blight resistance, reduced fibre and prussic acid in the tubers, higher protein, and better storage after harvest. The improvement of yams by breeding has been made easier by the observation at IITA in Nigeria that the seeds with the capacity to germinate could be selected by their plumper feel, and seeds that did germinate had a 3 month dormancy period. Planted yam tubers from seed show a wide diversity of plant type and leaf forms, including vigorously growing plants and dwarfs.

The local varieties of sorghum in Northern Nigeria, where the crop is planted on a quarter of the arable area, grow to 4 m high and lodge easily. Though they yield only 700-900 kg/ha the grain quality is good and the protein content higher than that of maize. New hybrids have been developed which only grow half as high, have a much better grain : straw ratio and yield 2,000-4,500 kg/ha (Andrews 1970). For top yield however they must be planted much closer than normal and supplied with phosphate and nitrogen. Some of these new lines are more resistant to *Striga* than the local varieties. How to maintain the supply of seed of the new hybrids; how the shorter strawed varieties will fit into the mixed cropping system; and whether the need for tall stems for surrounding the house compounds will outweigh the extra yield of the dwarf varieties, are important questions still to be answered.

Pearl millet varieties with a high resistance to both ergot, a major disease, and downy mildew which caused devastation of pearl millet crops in India in the 1970s, have been selected at the International Crops Research Institute for the Semi-Arid Tropics (ICRISAT).

In all food crops there is a greater interest in improving nutritional value, by selecting for higher protein content and a better amino acid balance, as in the case of high lysine maize (see p. 236) which has been paralleled in sorghum.

INCREASING AGRICULTURAL PRODUCTION

Enough research has been done in the tropics to show that production can be increased by improved crop culture techniques including:

1. Growing crops, or crop varieties suited to the climate and farming system. This includes quality seed and clones of improved varieties.
2. Practising non-exhausting rotations which include a legume crop and or using cattle manure and mulch.
3. Preparing the land in time and in a way that conserves moisture.
4. Planting at the correct time. Peasant farmers often plant too late. This may be due to the need to ensure food supplies by waiting until the rains have started before sowing precious seed, and not planting cash crops until food crops are established.
5. Correct spacing and early thinning.
6. Maintaining soil cover all the year.
7. Growing the crops free from weeds, particularly in the early stages.
8. Harvesting at the correct time.
9. Sowing indigenous legumes when land is rested.
10. Discarding all inferior planting material.

Most of these can be applied without any cash expenditure. Where extra yields have a cash value a farmer can use more advanced techniques including:

1. Use of fertilizers for more intensive cropping.
2. Controlling pests and diseases by seed dressings and spraying.
3. Irrigating.
4. Mechanization of cultivation, harvesting or crop processing.

All these factors which lead to improved yields are integrated. A higher yielding variety may need closer spacing and an increased fer-change in one factor should be accompanied by a consideration of all the others to determine where the limitation to yield occurs.

There is little point in increasing yields if, due to lack of care, the extra yield is lost in storage, a loss which a little care could reduce or eliminate.

In 1960 Richardson calculated that on the basis of field trials and existing knowledge, rice production in India could be doubled. The average yield was 1,280 kg of paddy per hectare, produced without any fertilizer. With a small application of fertilizer, 34 kg of N and 34 kg of P_2O_5 yields rose by 52 per cent. Most growers plant a mixed seed, and if this were to be replaced by the pure seed available, yields could be increased a further 20 per cent, better cultivation would give a further 20 per cent raising the average yield to 2,580 kg. Newer varieties with double the rate of fertilizer would raise yields to 3,020 kg. The introduction of these simple measures throughout the rice areas of India would produce 50 million extra tonnes of paddy (30 million t rice), but it would need 17 million tonnes of sulphate of ammonia. India plants more rice than any other country but its annual shortage is around 15 million tonnes and is rising with the population increase.

Perennial crops respond to care in their early years. Studies at River Estate in Trinidad showed that 64 per cent of the yield variation in a cocoa experiment was essentially due to variations in early growth of the trees, those trees which had made the greatest growth at 3 years of age providing the greatest yield of cocoa at 6–8 years of age. The high yields were around 2,240 kg/ha of dry cocoa, irrespective of fertilizer addition (*Colonial Research* 1958–59, p. 66). This probably applies to most tree crops and it is vitally important to give tree crops a good start, free from weed competition.

CROP ROTATION

The succession of crops on a peasant farm is far from fixed: indeed plot boundaries are themselves not fixed. A newly opened area is planted to food crops such as yams, maize, groundnuts or millet or a cash crop such as cotton or tobacco. Rarely are these crops grown as a pure stand but are mostly interplanted. In areas with two rainy seasons this is followed by another crop the same year. The crops planted are influenced by the level of food reserves, the price and market for cash crops and surplus food, pressure from local officials, the earliness of the rains, indebtedness, and the state of health and social commitments of the farmer. The optimum planting time for each crop is determined by the distribution and reliability of the rainy seasons.

The rotation is adjusted to the productivity of the soil; a sandy area may grow two groundnut crops in a rotation, another area may produce a cotton crop every year it is not under fallow, with short season crops between. Sweet potatoes are seldom the first crop of a rotation, except on special food plots, and are generally planted in the less reliable rainy season where two occur. Sim-sim is rarely the first crop of a rotation, except where it is a cash crop. Cassava yields well as the first crop on new land but seldom gets the privilege. Often it is the last crop before the fallow, and is left in the ground until needed though work by the Belgians in Zaire suggested that cassava is detrimental to the rapid establishment of the bush fallow. Sorghum is often a crop of the second year.

Most rotations include leguminous crops such as groundnuts, beans, grams, cowpeas or pigeon peas which leave behind nitrogen in the roots and foliage. These are generally followed by cereals which can make good use of this nitrogen. One crop follows another very quickly where the rain permits. The ground is not left bare, being protected either by a growing crop, or a mixture of crops, or by the remains of the previous crop.

At Yangambi, in Zaire, an attempt was made at continuous cultivation, growing dry land rice, cassava and groundnuts and a leguminous cover crop. The soil in this rotation was ploughed deep and left exposed for more months than it was protected. Aeration and insolation destroyed the organic matter and the soil structure, reducing the area to an infertile waste. The cover crops did not prevent this deterioration. In none of these experiments, however, was there any attempt to return nutrients.

The Magadi system of Malawi is a 5 to 10 year rotation based on maize-beans-maize-groundnuts. The maize is often spaced at 1.5 m and the groundnuts interplanted on mounds. The maize and bean crop that follows may be interplanted with cowpeas or bambarra groundnuts. The weeds are buried in the mounds. This system combines mixed cropping, the use of legumes, and the conservation of organic matter.

In rotation experiments in eastern Nigeria with yams, maize and cassava on deep permeable acid soils, the highest yields were obtained when the crop followed a bush fallow cut and burnt in the year of cropping. A delay between clearing and cropping allowed leaching of nutrients and yields were lower. When the area was replanted with the same crop, the yield of yams was 67.5 per cent, maize 82.7 per cent and cassava 59.8 per cent of the first crop (Obi 1967).

Wet paddy land which is used for a single rice crop each year is often ridged and planted to crops such as tobacco and sugar cane in Indonesia or groundnuts, soya beans, maize or sweet potatoes often mixed. This rotation besides opening up the soil has a beneficial effect on the subsequent rice crop, and reduces the carry-over of pests, diseases and nematodes (probably *Ditylenchus langustus*) where they occur.

In the wetter parts of the tropics a monoculture of tree crops, rice, or sugar cane is common. The evils predicted to befall farmers not rotating crops rarely arise in these cases, though suggestions are being made that nematodes may eventually become a limiting factor in yields of perennial crops. For this reason a break of at least 6 months is recommended between uprooting and replanting a perennial crop such as bananas or tea. The Indicative World Plan for Agricultural Development says that ratooning sorghum gives promise of very high output of grain per annum (FAO 1970, Vol. III, p. 29). This has been done in Western Australia and Swaziland under irrigation. Ratooning would not appear to be practical where *Striga* and stem borers are serious.

CULTIVATION

Cultivation is the first step in preparing the land for planting a crop, and it is done with the object of removing the existing vegetation, opening up the soil to aid aeration and water penetration, to help root growth, and preparing a clean seedbed for planting. Seedbeds in the tropics are considerably rougher than in temperate climates.

Tillage should be kept at a minimum to achieve these objectives. Excess cultivation may pulverize the topsoil, reduce rainfall acceptance, increase run-off and cause sheet erosion. After forest or bush is burnt the crop is planted in the undisturbed ash.

In preparing a seedbed, all that is required is to reduce the resistance of the soil to root penetration and shoot emergence, and allow the seed to contact the soil water.

Clearing a piece of resting land for cropping is much more difficult than preparing land that has already been cropped. In the forest and tall grass areas, burning is the traditional and easiest method. A good burn leaves the land reasonably clean and fertilized with ash. Large trees are usually left, due to the difficulty of dealing with them, or their protection by law or superstition. Tall grass, even elephant grass (*Pennisetum purpureum*), can be dug into the land with hand tools, but it is heavy labour, and must be done at least 2 months before planting to prevent nitrogen starvation of the young crop. Burning tall grass before cultivation is general. In the 'Ladang' system of the Far East, material for burning is cut from resting areas to add to the ash, but this is unusual in Africa, though the 'Chitemene' system of Zambia is similar. In short grass areas, most of the non-cultivated land is fired during the dry season, either by accident or intention, to remove the dry 'foggage', to control animal parasites and give an early flush of grazing when the rains start. This burn leaves the land in a condition easily opened up with a hoe. Land already under cultivation is easier to prepare. A minimum of cultivation is generally done between crops; the wisdom of this has already been explained. Cultivation is not too deep as this would accelerate mineralization of organic matter and increase the loss of nutrients by leaching. Traditionally, only the topsoil is stirred and opened up during cultivation and so the active, fertilized soil is kept at the top.

The hoe, in a variety of shapes and sizes, with a short or long handle, and the cutlass or machete, are the main implements used for cultivating. The spade and fork are rare. During this century, animal-drawn iron ploughs have increased in importance in the short grass areas where cattle are kept and the land is reasonably free from tree stumps. The introduction of the plough in Africa led to increased erosion and the production of soil pans, until the farming system was modified to accommodate the change. Wooden ploughs have been used for centuries in the East to prepare the paddy fields.

In Botswana the land is ploughed after the rains start in November. As the oxen are in poor condition after the long hard winter the land is not ready for planting sorghum and millet until December. It would be better to plough immediately after harvest while the oxen are in good condition. The early rain which is lost in run-off could then penetrate the soil, crops would be planted earlier, and give higher yields. However, such a change would be socially difficult as the farmers live in large villages 48 - 64 km (30 - 40 miles) from their farms and hurry back to their villages after harvest.

The preparation of wet paddies is a long process of flooding, ploughing, harrowing and puddling with implements often made of wood, which have been developed over centuries to meet the needs of local conditions (Plates 2.5 and 2.6). The consequences of flooding and puddling are dealt with by Sanchez (1976). Puddling, which is a vital part of wet rice production, breaks down the soil aggregates of the flooded soil to a uniform mud into which rice seedlings are transplanted. The structure of the topsoil is destroyed, water losses are cut and air is kept out maintaining the reduced state of the soil. This cultivation pulls up the weeds and buries them under the water where they decompose anaerobically to form ammonia which the crop can use (p. 19).

Where animals are used for cultivation they are frequently in poor condition at the end of the dry season when the heavy cultivation work is required. Similarly malaria, which has had a resurgence since 1965 due to the *Anopheles* vectors having acquired resistance to the previously effective insecticides and the failure of some eradication programmes, is most severe in the rainy season and the warm season in irrigated lands, when the labour demand is highest (Pant and Gratz 1979).

Plate 2.5 Traditional method of rice cultivation. Palmyra palms in the background. Kerala, South India. (Courtesy Cadbury Bros. Ltd.)

Mechanization of cultivation

Throughout the tropics during the past 30 years, governments and farmers have been trying to introduce tractors and mechanical equipment into the local farming systems. There are a number of good technical reasons for this. Where land preparation is done with tractors, the soil can be prepared before the rains start and crops planted at the best time. There is also the sociological need to take much of the hard and tedious labour out of agriculture, and give it a modern image to attract young educated people and stop the drift to the towns.

The Indicative World Plan (IWP) (FAO 1970, Vol. 1, p. 226) prepared before the oil crises and subsequent inflation, proposed that initially power mechanization should in general be concentrated on the following tasks:

1. Land clearance and preparation, including levelling; also the con-

Plate 2.6 Harrowing rice. Philippines. (Courtesy Philippines Tourist and Travel)

struction of ditches and bunds where these cannot be done properly by hand.

2. Ploughing and seedbed preparation to allow early planting for increased yields. Where double cropping is feasible, mechanization removes power input bottlenecks and thus increases the intensity of farming and consequently also the demand for labour.
3. Doing other cultivation work that cannot be done effectively by human or animal power, such as subsoiling, chisel ploughing or stubble mulching. Power for such operations is particularly essential in dry areas where draft animals cannot cope with hard soil conditions.
4. Threshing.
5. Transport.
6. Pumping water for irrigation.
7. Spraying of certain crops, which cannot be done effectively by hand equipment, for example fruit trees.
8. Increasing the value of crops by drying, partial processing, grading, etc.

Unfortunately there are many difficulties in introducing tractors and implements into a peasant system of farming. Mechanization in India in the guise of the Green Revolution has created a few larger and more prosperous farmers and sent a larger number to join the workless in the cities. Agriculture in Latin America is more mechanized than most of the developing countries, and it is estimated that in Chile, Colombia and Guatemala each tractor replaces three to four workers, even taking into account the new employment opportunities created by mechanization. Tractors are profitable for the large farmers but not the national economy (Abercrombie 1972). Mechanization makes a great demand on foreign exchange and less use of labour but in most tropical countries labour is abundant and foreign exchange scarce.

Among the technical as opposed to sociological disadvantages of introducing tractors are:

1. Mechanization in the field is mainly applicable to annual crops of low market value.
2. The cost of a tractor, equipment and fuel is very high, in relation to farmers' income from annual crops.
3. The farmers' holdings are small and often difficult to reach with tractors over the poor roads.
4. The land is cleared for three or four years' cropping and the tree stumps and roots are left in the ground as the manual effort needed to remove them is too great. These cause delays and breakages of equipment, but help the regrowth of the fallows.
5. The people are generally unfamiliar with machines, hence there is a low standard of driving, maintenance and repair in regions where soil and its unfamiliarity with the plough, the dust, and temperature, all shorten the effective life of equipment and cause numerous breakdowns, requiring spare parts which are often unavailable.
6. The tractor and equipment only work for a minor part of the year making an inefficient use of the capital investment.

In the long sunny days of favoured cotton growing areas of Greece and Turkey with irrigation, fertilizers and insecticides, yields of 3,000-4,000 kg/ha of seed cotton make mechanization an economic proposition, but in most of the cotton growing areas in the tropics yields are less than half this and the costs of mechanized cultivation and harvesting higher.

To encourage mechanization governments have been prepared to pay heavy subsidies, and encourage the consolidation of holdings into blocks which can be worked as single units. The minimum for a mechanized operation is 10 ha in a contiguous block. In Northern Nigeria a 65 hp tractor can cultivate 120 ha in a block (twice the estimate for Indian farms, FAO 1970, p. 230) which is equivalent to 12 pairs of oxen but at a much higher investment (Haynes 1966). Out of the cultivation season tractors are often used for threshing, pumping irrigation water, crop transport, and even wood sawing. Most of these jobs, however, can be done just as efficiently at less cost using a much smaller engine.

If a farming system is modified towards a greater utilization and dependence on mechanical equipment, a ready availability of spares and servicing is essential in case a breakdown occurs at a critical work period. Governments encouraging such developments must work closely with the major tractor and equipment manufacturers who maintain this essential service, and also make available experience from other countries.

While experience with mechanization in parts of Africa has not been very encouraging, in Thailand and Sri Lanka individuals have set up as contractors, offering farmers a variety of services throughout the year. In the maize growing area of Thailand, 'lalang' grass is controlled by tractor ploughing with big discs which bury the grass about 25 cm deep. Without such mechanical help the land would be abandoned for the bush fallow to crowd out the 'lalang'. This is an example of mechanical cultivation being able to provide a service which ox-drawn equipment or hand tools cannot match.

Following the native practice of retaining the topsoil *in situ* and using the hoe to open it up, disc implements and digger-type ploughs, rather than mould-board ploughs have been used in mechanization schemes. These do not bury the topsoil deeply and break up the turned soil better. Also they are less liable to breakage on recently opened land full of stumps and roots over which they are able to ride. Deep cultivation with tines could be more effective than ploughing, as tines open up the soil to a good depth and the fertile topsoil is left on top. Cultivation, both hand and mechanical, has been shown (Pereira & Jones 1954b) to reduce the rate of entry and percolation of water into coffee soils in Kenya and leads to sheet erosion. The forked hand hoe causes less damage and gives greater yields than weeding with a disc harrow or mould-board plough. With the 'No-Till' system previously described, the topsoil is left in position and the crop drilled into the mulch of dying weeds.

On larger agricultural units such as estates, the position is rather different and cultivation has been mechanized as much as practicable. The sugar estates are very advanced in their mechanization, but maize, sorghum, groundnuts and tobacco are the only annual crops grown to any extent in the tropics where large-scale farming occurs and cultivation is mechanized. In the case of perennial crops, mechanical means are now in use for clearing, ploughing and cultivating the land prior to planting and cultivators or sprayers may be used in the early years to keep the plantations weed free. On the whole the less disturbance of the soil that occurs, the better. Most of the tree crops are surface rooting and tea, cocoa, rubber, banana, coffee, citrus, oil palms and coconuts suffer when their surface roots are damaged by excessive cultivation and this may help the entry of root diseases as does damage to the bark of tree crops.

The application of fertilizers and crop protection chemicals by aircraft, possible on large areas under the same crop, is often both cheaper and more effective than ground application, as in the Gezira cotton scheme.

Mechanization of wet paddy*

The introduction of tractors into rice cultivation is especially complicated as the smallholdings are divided into small separate paddies surrounded by water retaining bunds which make it difficult for tractors to get on to the land and operate. Cultivation of the dry hard clays with animal drawn ploughs is difficult and is not normally started until the rains commence. The weeds decompose more rapidly in wet soil to provide available nutrients for the new crop. Hard clay does not deter tractors, but often they must work in a few centimetres of standing water. Growers are often reluctant to release any water which collects on the paddies before planting, in case the subsequent monsoon rain is delayed or light. In Surinam dry tillage has given increased rice yields, and the beneficial effects were not reversed by puddling later.

Manufacturers have developed wheels which help the tractor to work in mud, and the various traditional land preparations can be reproduced by rotary hoes and puddling equipment.

Mechanized rice production is a major part of the Mwea irrigation settlement in Kenya and this crop in the early 1960s brought the average family with 1.6 ha of land a net return of £135 each year, compared with an average income of £12 15s. 0d. for the majority of African farmers (Giglioli 1966). Average yields in this mechanized scheme were 5 to 7 t/ha.

As much of the cultivation in rice paddies is to control weeds by drowning them, it is not only necessary to have close cooperation between paddy farmers, but also to examine all the functions of the traditional cultivation practices, particularly puddling and transplanting, to see if these can be carried out or replaced to better effect by a combination of mechanical equipment and modern herbicides. Reorganization of land holdings assists not only cultivation, but could make it possible to harvest the crop with the modern rice combine harvesters and to apply fertilizers, herbicides and insecticides from the air.

In Thailand, the Division of Engineering have developed the 'Iron Buffalo', a low cost tractor for rice cultivation. Many cheap petrol or diesel motors have been bought to turn a dragon pump or Archimedes screw pump for the rice fields, propel a boat to transport the rice sheaves after harvest, drive the thresher or winnower, and turn a grindstone.

* See 'Mechanization and the World's Rice', a conference organized by Massey-Ferguson with FAO, 1966.

The use of single axle tractors, some 2 million of which are used in Japan, has not been very successful in other rice growing areas as they are difficult to handle in thick mud. Over half the rice in Japan is transplanted mechanically. The machines that can handle washed seedlings are preferred as they are more precise and save labour.

Efferson (1952) pointed out that the wholesale introduction of mechanization of rice growing would cause widespread unemployment and starvation. Such a change is neither practical nor possible, but the use of machines should help to produce heavier crops and take much of the physical labour out of rice production including water raising and threshing.

Even with the trend to greater mechanization water buffaloes and oxen will remain the most important source of power for rice production. They are well suited to working small muddy paddies, are cheap to feed, easily moved about the country and have a slaughter value when too old to work. They can however become sick or die at a crucial part of the season and they cannot work continuously day after day without rest.

Not only could the selection of work animals be better, but the equipment designed for them improved. The method of hitching bullocks is often unsatisfactory. The Nagpuri yoke, which is the best available in India, while strong and simple only utilizes power from the hump and neck muscles of the animals, and concentrates the force on a small part of the neck which often causes sores. The angle of hitch is high, which reduces draft capacity of the bullocks. Swamy Rao (1964) has suggested the essentials of a good yoke. The training, management and efficient use of oxen is explained by Howard (1980).

Ox-drawn implements for dry land cultivation

The high cost of tractors and tractor equipment has made mechanization uneconomic except for high value crops, large farms, and work that cannot be done without tractors. Consequently there is increased interest in animal drawn equipment in areas where draft animals can be kept.

The cost of tractor fuels has risen considerably in recent years but a tractor only needs fuel when it is working, whereas oxen must be fed all the year whether they are at work or not, and this may be difficult when grazing is short. From the food the draft animals produce manure which is valuable if it can be used.

The ox-plough is usually the first implement the farmer purchases, and with it he often increases the area of land he cultivates. Unless he has the tools or extra labour to help at critical times, such as the first weeding, the result is a larger area producing a lower total yield from ill-tended fields. Boyd (1980) catalogues a wide range of low cost ox-drawn equipment now available including harrows, hoes, ridgers, planters, groundnut harvesters, as well as equipment for dealing with harvested crops. Small engines to pump irrigation water and drive threshers are usually more economic than tractors or animal power and equipment.

Greater use of ox-drawn equipment has been encouraged in many countries, for example in Northern Cameroun there has been a good demand for ploughs and ox-carts, available on 3 year credit. Ox-drawn hoes and harrows have been much less popular, possibly because the design is not sufficiently good to help the farmer as much as he requires. The number of ploughs used in Mali increased from 700 in 1961 to 20,000 in 1970, and also the use of other ox-drawn tools mainly for cotton cultivation. This has meant a greatly increased demand for blacksmiths (Chantran & Grimal 1971).

In the Korhogo area of the Ivory Coast, the production of food crops, including rice, and cash crops has been increased by the use of ox-drawn equipment, as this has enabled farmers to cultivate a larger area and obtain much higher yields. Experiments in India have shown that by using a steel plough in place of the traditional 'desi' plough, ploughings can be reduced from six to two (FAO 1970, p. 228).

The use of small horticultural equipment for smallholdings is not an economic proposition as these specialized machines are generally more expensive to buy and operate than larger machines, when compared on an area basis.

The use of push hoes and drills is difficult in anything but the lightest soils due to the roughness of the seed bed.

It has been suggested that the hand cultivation tools should be improved, but few practical suggestions have been put forward. The peasant has a wide range of similar hoes to choose from and is no doubt using one adjusted to himself, his needs and his land, and man's inventiveness can be put to better use on many other manual tasks, particularly the processing of crops.

PLANTING

Whereas the farmer of temperate climates largely drills seeds, the tropical farmer has not only a wide range of planting material to use but often a choice in the way he uses it, grouped as follows:

1. Seed planted direct into the field - Cereals (maize, sorghum, millet, rice, etc.); legumes (groundnuts, beans, cowpeas, cover crops, etc.); and cotton, sunflower, jute, sim-sim.
2. Seed too small for planting in the field, sown in nursery and transplanted - tobacco, pyrethrum.

3. Seed planted in a nursery for further selection or grafting before planting out - tea, coffee, cocoa, rubber, oil palms, citrus, coconut, rice, tung.
4. Vegetative planting material, planted direct in the field - sugar cane, banana, plantain, pineapple, sisal (suckers) and root crops - cassava sweet potato, yams, arrowroot, eddoe, tannias, pangola grass.
5. Vegetative material planted in a nursery before planting out - tea, cocoa, sisal (bulbils), coffee, rubber, manila hemp.

When planting stem cuttings of sugar cane (Plate 2.7) and elephant grass, the size of the sett is restricted to two or three buds. With larger setts the influence of apical dominance in the buds results in an uneven pattern of germination as the buds sprout in sequence. Perennial crops occupy the land for many years and only the best material should be planted. It is better to lose a year than use poor quality seedlings.

Nurseries

The extensive use of vegetative planting material often simplifies the way in which the farmer can derive full benefit from the work of plant breeders and the introduction of the nursery stage allows a second selection of planting material before a field is committed to a crop for a number of years. Full advantage has not been taken of these opportunities, and 'planting of seed at stake', still occurs with crops that would benefit from an early life in a nursery and a rigorous selection before planting out (Plate 2.8). Often a number of seeds are sown at stake so the eventual plant has to compete not only against the natural hazards of a poor soil tilth, weeds and drought, but also fellow plants which will be later discarded.

It is far easier to supervise new plants when they are grouped together in a nursery than when they are spread over the fields at their final spacing distance. Fertilizers, shade, water and protection against pests and diseases can be applied more readily and economically in the nursery. The use of nurseries for rice enables two or even three crops to be produced annually on irrigated areas, and the seedbeds can be heavily fertilized and sprayed with a selective weedkiller to give the plants a good start. Intensive relay cropping is dependent upon full use of nurseries.

The use of soil filled polythene bags perforated in the lower half for drainage, produces faster nursery stock of tree crops, which have a high rate of survival after transplanting, at a reduced cost per plant. New soil can be used for each planting and it is not necessary to treat the soil against nematodes. More attention has to be paid to watering bags than when the plants are in conventional nurseries. Green nylon netting

Plate 2.7 Planting sugar cane. Jamaica. (Courtesy Association, Inc.) West Indies Sugar Company)

Plate 2.8 An oil palm nursery seedling ready for field planting. Nigeria. (Courtesy WAIFOR)

makes an excellent shade for nurseries; it is rot proof and the nursery can be watered through the nylon by overhead sprinklers.

Provided that planting out is done at the start of the main rains or is accompanied by irrigation, a good stand should be achieved. Nursery raised tree crop planting material is hardened off or stumped before transplanting, and shaded after planting out. Poorly developed and diseased plants should be burnt.

Planting methods

Though certain crops are suited to mechanized planting, very little advance has been made in mechanization, largely due to the factors mentioned under cultivation and the rough nature of the seedbeds which are unsuitable for the small hand planters except the jab planter (see p. 107). Small seeded crops like sorghum and millets are broadcast, and the larger seeded crops including maize, cotton, beans and groundnuts, are thrown into a hole made with a hoe, stick or cutlass. The peasant farmers are adept at this and plant regularly, evenly and quickly with their apparently crude method. There is a general tendency to plant too many seeds in each hole, and unless thinning is carried out early this reduces the crop yield. Extensive trials in Bangalore (India) using a simple automatic seed drill, made locally, to plant finger millet (*Eleusine coracana*), showed not only a saving of seed and labour, but yields of grain and straw were 25 per cent higher (Patil 1963). The jab planter (Plate 2.2) described in the 'No-Till' system is a simple tool to improve the planting of large seeded crops.

The optimum plant distance for a crop is that which will give sufficient foliage canopy to utilize the maximum amount of solar radiation without excess self-shading, while the roots have adequate room to forage.

Consequently small insignificant yield differences are recorded in spacing experiments with crops which tiller or branch freely or have a spreading habit so that the foliage canopy is soon complete. Conversely a crop such as maize with a more erect habit shows significant yield variations in spacing trials.

The newly developed shorter, erect varieties of rice or sorghum should be planted closer than the older tall varieties, particularly if they are early maturing.

If a perennial crop such as bananas is planted at a wide spacing, the first crop may be produced before the leaf canopy closes and the yield will be lower in the first years than with closer spacing but evens out as the foliage canopy meets. Broadleaved cocoyams which use the surplus solar energy are often interplanted in cocoa and bananas to protect the soil from sun and rain. Perennial crops planted too closely together may show a declining yield with age due to mutual shading and a lack of root room. Some tree crops have often been planted to a dense stand initially, then thinned out by removing alternate rows as the trees matured. With the high cost of quality planting material and the need to pass through with spraying and other equipment, this approach has become rare except in forestry. Too often thinning out is delayed and the crop suffers from overcrowding. Thinning out may involve removing some vigorous high yielding trees and leaving less promising trees, which growers dislike doing.

The Rubber Research Institute in Malaya has shown that rubber trees planted closer together than normal give an increased yield, but it is not an economic gain due to the increased time necessary for tapping. Browne in Cameroon planted fewer trees per hectare on a diamond pattern produced trees with a flatter close canopy less subject to wind damage and a yield comparable to more closely planted trees. In the Sudan Gezira closer spacing was considered necessary to obtain the maximum yield from late cotton sowings (Lambert & Crowther 1935), the effect of the closer spacing being mainly to reduce pest damage. With the control of pests by modern insecticides this may no longer apply as the protected foliage may develop a full leaf canopy early enough to give the full yield. Table 2.6 gives some typical crop spacings.

Cocoa illustrates the varying views on spacing. In Trinidad it was considered that a more fertile soil could support a closer spacing, whereas in the Congo the reverse view was held. Spacing experiments have proved little, except that higher yields are obtained with closer spacing in the earlier years but there is little difference later (Wood 1975). In West Africa the initial close spacing is thinned out by intention or the death of plants, and in Nigeria there appears to be no 'best' planting distance even in one locality on any one soil type (Russell 1953).

In a spacing trial with arabica coffee at Zuezon in the Philippines (Handog & Bartolome 1966), the yield of cherry for the first five crops was:

1 × 3 m - 34.2 t
1.5 × 3 m - 33.2 t
2 × 3 m - 18.3 t
2.5 × 3 m - 14.8 t
3 × 3 m - 13.6 t

It is possible that the wider spacing, apart from reducing the amount of crop bearing wood, allowed more light to reach the ground, not only wasting solar energy but also encouraging weed competition.

With large areas of tree crops such as cocoa, arabica coffee, and bananas that are sprayed regularly, a spacing between the rows which will allow tractor mounted spraying equipment to pass, will make a big

Table 2.6 Crop spacing

The plant population per hectare is easily calculated from the spacing

$$\text{No. of plants per hectare} = \frac{10{,}000}{\text{metres between plants} \times \text{metres between rows}}$$

Table of sample spacings and populations

Spacing in metres	Crop example	Approximate no. of plants per hectare
0.3 × 0.3	Groundnuts	110,000
0.3 × 0.6	Soya beans	55,000
0.9 × 0.3	Maize	37,500
0.9 × 0.45	Cotton	25,000
1.20 × 0.6	Pineapple	13,750
0.9 × 0.9	Sweet potatoes	12,500
1.2 × 0.9	Tobacco	9,000
1.2 × 1.2	Tea	6,750
1.5 × 1.5	Areca	4,250
2.1 × 2.1	Arabica coffee	2,250
2.4 × 1.8	Sisal	1,700
2.7 × 2.7	Robusta coffee	1,350
3.0 × 3.0	Bananas	1,100
3.7 × 3.7	Plantains	750
4.6 × 4.6	Cacao	500
6.1 × 3.05	Rubber	550
6.1 × 6.1	Clove	275
7.6 × 7.6	Coconut	175
8.5 × 8.5	Grapefruit	138
9.2 × 9.2	Oil palm	125
12.2 × 12.2	Nutmeg	68

saving on spraying costs. Close planting along the rows will compensate for the reduced plant density and provide a closer target for the spray.

Time of planting

This is probably more important than spacing. Peasant cultivators generally plant after the optimum date. If the land is carrying a crop being harvested over a period, growers are reluctant to destroy the remains before it is completely harvested. Once the rains start, planting should normally commence while nitrate production is at a maximum (Birch 1958), but the land often cannot be prepared until some rain has fallen, and preparation is a slow process. Food crops obtain priority, and if seed is scarce planting is delayed until the rains can be relied upon.

The preceding crop may delay planting, and changes in the cropping system may aggravate the situation, as in the case of maize replacing finger millet in East Africa, as the food crop preceding cotton. The maize harvest is about 6 weeks later than the millet harvest and is well after the optimum planting date of cotton. The optimum planting date for cotton in Uganda was determined as being about 6 weeks prior to the actual planting date (Manning 1949). Advancing the planting date and contracting both ends of the season offers a more immediate return in increased production than breeding and fertilizer usage, at no extra cost.

Ruston (1962) summarizing a number of years' observations on the effect of delayed sowing on the yield of rain-grown cotton gives the loss in yield from a 2 week delay beyond the optimum as 14 per cent, a delay of 4 weeks 40 per cent, and 6 weeks 50 per cent of the potential crop. Even in irrigated cotton in the Sudan an experiment gave a loss of 23 per cent with 4 weeks delay in sowing and 31 per cent with 6 weeks. A further benefit from early sowing is that the early picked cotton is of higher quality than the later pickings.

Groundnut yields at Gezira (Sudan) were reduced by 40 per cent by delaying planting from early August to September. Early planting particularly benefits later maturing varieties of groundnuts. Unfortunately early plantings are more liable to be attacked by *Cercospora* leaf spot (Ishag 1965).

Delay in transplanting rice from the nursery is a frequent cause of reduced paddy yields.

Yields are further reduced by planting too many seeds per hole, particularly with cotton and maize, and allowing the seedlings to grow too big before thinning.

Intercropping

Intercropping (Plate 2.9) is an important characteristic of farming in the tropics. Apart from rice it is doubtful if any other food crop was planted as a pure stand until cash crops were introduced. The farmer and his family need a small but regular supply of fresh food, particularly vegetables and root crops, throughout the year. They have no freezers in which to store surplus vegetables, so the practical solution is to sow little and often in the other crops. Crops do not need all the space, light, moisture and nutrients which are available at all times during their life. Maximum requirement is often at flowering and grain filling. Intercropping is a system worked out by experience to make more efficient use of the environment. The total yield from intercropping depends upon the competition between the crops interplanted and their ability to compensate for reduced populations (Andrews 1972 a,b).

An association of food crops is a system carried out by the native

Plate 2.9 Mixed cropping in Antigua, West Indies. This is a typical mixed stand of cotton, maize and sweet potatoes. (Courtesy Cotton Growing Corporation)

cultivators of South Asia, India, East and West Africa, the Sudan and Central America. It has stood the test of time and has a number of advantages:

1. Intercropping makes more efficient use of solar radiation.
2. The soil is better and more continuously covered from the beating rain and sunshine, and weeds are suppressed.
3. Land planted with an association of crops is better protected from erosion.
4. It is an insurance against a crop failure and gives more stability to food supplies and income.
5. The harvest is spread, which is important where the harvested crop does not store well. By planting early and late maturing varieties of a crop such as sweet potatoes, supplies are available over an extended period.
6. A site is found for crops only required in small quantities, such as maize for eating green, vegetables, plants for sauces, medicinal herbs and gourds.
7. Where legumes are mixed with other crops they appear to supply nitrogen to the companion crop.

8. The roots of the various crops feed at different levels in the soil and withdraw a different ratio of nutrients.
9. An association is less susceptible to an attack from a pest or disease, which is generally confined to one crop of the association.
10. Interplanting young tree crops with food crops not only protects the soil and controls the weeds, but also produces a return from the land while the tree crop comes into bearing.
11. There is up to 60 per cent total yield increase from growing two crops together.

The disadvantages of intercropping come when the farmer wishes to spray one of the crops, and when he wishes to harvest mechanically. There is less chance for, and often less need for, tillage. In Java this is being overcome by interplanting in rows so each row or even each plant can be weeded and fertilized.

Farmers around Zaria in the north of Nigeria have developed a rotation of annual crops which also involves intercropping and changes in the position of the ridges. The Hausa word *hura* means 'Putting a leguminous or other crop in the field in order to derive a better Guinea corn (sorghum) or millet crop the following year'. The *hura* crop which may be cotton, peppers, sweet potatoes, yams, cassava, okra or a legume is planted on the ridges and next year on the ridges will be Guinea corn. With Guinea corn on the ridges, millet, cowpeas, sweet potatoes, okra or other crops may be interplanted in the furrow with the rows running across the ridges. If cotton or pepper is the *hura* crop, Guinea corn is not interplanted as the secretion of the aphids on the Guinea corn has an adverse effect on the cotton or pepper (Buntjer 1971).

The Azande of the Sudan live in a tsetse infested area and must maintain the productivity of their soil without the advantage of animal manures. Intercropping for food production follows a definite series of systems which have been developed over the centuries (Schlippe 1956). Groundnuts are planted as early as possible even at the risk of losing the seed, and the harvest comes when food is becoming short. Finger millet (*Eleusine*) is sown through the groundnuts and occupies the land after the groundnut harvest. Other crop associations of the Azande include groundnuts/maize, *Eleusine*/sorghum, cassava/*Hibiscus cannabinus*, also sesame/*Eleusine* - an association found in Northern Uganda and other short grass areas but shown in experiments in the Sudan to give a poor yield of sim-sim and a reduced yield of *Eleusine* (Tothill, 1948).

With the introduction of cotton as a cash crop in Azande the possibility of growing a food (groundnuts or beans) and a cash crop in association needed investigation. Unfortunately the cotton suffered from intercropping (Anthony & Willimott 1957) to an extent varying from year to year. Good results depend upon good early rains and heavy late rains. To ensure reasonable yields of both cotton and legumes, the legume should be planted in early April, and the cotton in June the established optimum planting time in the area.

After 1945, when maize became more extensively grown as a cash crop in Uganda wider spacing 1.8 m × 0.3 m intercropped with groundnuts was tried in comparison with the 0.9 m × 0.3 m pure stand of maize. The object was to reduce soil erosion, but in many trials a worthwhile crop of groundnuts was harvested without an appreciable reduction of the maize crop. After harvesting the groundnuts, cotton could be interplanted at a sowing date earlier and nearer the optimum than if the cotton followed maize. The maize stalks eventually served as a mulch between the cotton rows.

In Tanzania in 1958-59, a year of unfavourable rainfall, intercropping groundnuts with maize or sorghum generally gave appreciably higher production than pure stands. The production of 1 ha intercropped was equivalent to 1.5 ha cultivated in a pure stand (*Colonial Research* 1958/59, p. 171).

The majority of the beans (*Phaseolus vulgaris*) produced in Latin America, where they are an important part of the diet, are grown in association with maize. Studies at Centro International de Agricultura Tropical (CIAT) in Colombia suggest that intercropping has little effect on the maize yield and in addition about 2 t/ha of beans are produced. Beans planted alone can yield 4 t/ha.

Intercropping is not confined to annual crops. On the fertile soils of East Africa bananas are interplanted with coffee. The bananas shade the coffee and provide material for mulching.

Where coconuts are grown on good soil as on the West Coast of peninsular Malaysia, there is a good case for interplanting with cocoa (Leach *et al*. 1971). The coconuts are normally planted in avenues 8 or 9 m apart, and cocoa planted in two rows 3 m apart and equidistant from the palms. Within the rows, the cocoa is spaced at 2 m intervals giving just over 1,100 to the hectare. This high planting density is reduced by selective thinning to about 1,000 trees per hectare at the end of the second year in the field. No subsequent thinning is carried out until the cocoa yield is 900 kg/ha when large gaps will not be left in the canopy by thinning. Gaps in the coconut canopy caused usually by lightning strike are filled with a shade tree such as *Gliricidia* before the cocoa is planted. Interplanting of cocoa and coconuts has been practised extensively in New Guinea and the Philippines since the war.

Cassava intercropped in oil palms competes for light and nutrients. On a fertile coastal clay in Malaysia Chew and Khoo (1976) reported a 14 per cent reduction in yield over 3½ years with 2 years cassava cultivation.

One disadvantage of intercropping is that it is not possible to control weeds by spraying. Experience in the Ivory Coast is that when a cotton crop follows maize interplanted with cowpeas, not only is the weed population less than if the previous crop is maize alone, but also there are more broadleaved weeds than the more serious grass weeds.

Interplanting of cocoa in Trinidad with coffee during the 1930s depression, led to the coffee being pruned in the same way as cocoa, which is quite wrong.

Relay cropping

Relay cropping is where the second crop is interplanted while the first crop is maturing prior to harvest. This is commonly done in Central America where the maize crop is interplanted with beans. The corn cobs may be bent down on the stalks at that time, to protect them from moulds while ripening, and to allow the beans to climb up the maize stalks. In the Far East, vegetable growers are highly skilled at relay cropping and rice growers use this method to produce an extra crop of rice or vegetables in the year. Some crops are too sensitive to shading in their early growth stages, others are unaffected, and other crops like the Central American bean grown better with shade in the young stages.

Ruthenberg (1976) describes relay cropping as practised by the Kikuyu in Kenya. When the rainy season begins at the end of March or the beginning of April, the women plant maize and pigeon peas broadcast in the plot, which has been prepared with hoes by the men. When the seed has germinated the women go over the field again and plant an early-maturing and a late-maturing variety of bean among the growing maize and pigeon peas. When the beans have germinated a third planting is made, and the plot weeded. Sweet potatoes are then planted among the maize, peas and beans. The early-maturing beans are available from mid-June, and the maize and late-maturing beans are harvested in mid-August. Sweet potatoes and pigeon peas are taken from the plot as needed from September onwards.

SOIL AND WATER CONSERVATION*

'Soil erosion' brings to the mind of anyone who has visited the tropics vivid but different scenes. A striking appreciation is obtained from an aircraft over the Orinoco delta, seeing the dark-coloured river carrying soil from the Andes into the Caribbean and changing the sparkling blue seas to a dark green. To others it means the 'dongas' of Lesotho; the

* See also Reese, 1966; Constantinesco, 1976; Greenland and Lal, 1977.

Plate 2.10 Soil erosion in Buganda. Poor plantain growth on murram. (Courtesy Department of Information: Uganda. Photograph by Ron Ward).

rivulets through the maize rows that run down the steep hills of Central and South America; the small banana plant struggling for existence on the murram remains of a Buganda garden (Plate 2.10); the rapid rise of rivers after rain. The water cycle is shown in Fig. 2.1

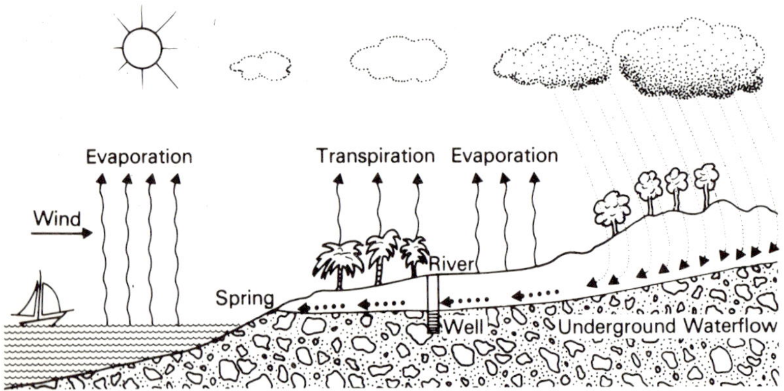

Fig. 2.1 The water cycle.

These different scenes have a common origin, the misuse of the land. The soil, exposed to the elements and exhausted by overcropping, sacrifices its fertility to the wind and the rain. The legendary fertility of Egypt originated in erosion of the Ethiopian highlands.

The traditional farming systems of the tropics, the ancient terraced cultivation of the East and of South America, mixed cropping, shifting cultivation, the mulched plantain gardens which the explorer Stanley found in Africa, all maintained a balance between human population and soil fertility, and restricted soil erosion. The introduction of new cash crops, and increase in cattle and human populations have led to greater demands on the soil. No traditional methods were available to guide the planters of Sri Lanka, or the peasant cultivator with his new crops of maize or cotton. Large fields, planted in rows that ignored the contours, and clean weeded, proved disastrous. The Dutch in Java saw the damage done in Sri Lanka and modified their methods, in particular avoiding clean weeding.

Erosion can be controlled by reducing the impact of raindrops and slowing down run-off water. 'Run-off' is the vehicle of erosion and is determined by:

1. The amount and intensity of the rainfall.
2. The soil cover from crops and mulch.
3. The receptivity and erodability of the soil.
4. The slope of the land.
5. The land usage.
6. The barriers to run-off.

The intensity of the rainstorms of the tropics, often accompanied by strong winds, and the size and speed of the raindrops, beating on the exposed soil, are the origins of erosion. The large amount of kinetic energy is dissipated in moving the soil or in breaking down the soil crumbs and sealing the surface. If the impact is broken either by natural cover, a crop, or mulch, the kinetic energy is reduced and the effect on the soil surface minimized.

If the soil structure is poor or the surface capped, due to overcropping, trampling of cattle feet, erosion, beating by the rain, then the receptivity of the soil to rain will be reduced and run-off will increase. Receptivity is better where worms, ants and some termites are active. This applies particularly to forest soils as Shown in Table 2.7. As in many parts of the tropics, particularly Africa and India, rainfall is marginal for crops, the loss of rainfall by run-off could be critical.

The velocity of the run-off is determined by the slope of the land, the velocity being doubled if the slope increases four times (i.e. the velocity of run-off is doubled increasing the slope from 1 in 100 to 1 in 25). The velocity is reduced if the surface is covered with crops, vegetation, mulch or litter, or soil erosion barriers. If the rate of flow is doubled, it has 4 times the scouring capacity, 32 times the carrying capacity and can carry particles 64 times as large.

Erosion is lessened by reducing the influence of these factors which increase run-off. The soil must be protected from the beating rain by plant cover from either crops, cover crops or mulch, and left bare as short a time as possible during the year. This is one benefit from intercropping and relay cropping.

The receptivity of the soil should be increased by breaking up any cap which forms either in cultivated or grazed land, by improving the soil structure by crop rotations, adequate resting fallows, and maintaining the organic matter by manuring or mulching.

Land usage

The hilltops which receive most rain and are steep should be kept as forest reserves. Much of the flooding and serious erosion caused by the cyclones in Malagasy would have been avoided if the eastern mountain slopes had not been deforested. The steeper cultivated slopes should be used for grazing or planted to permanent crops with cover crops, or mulched, but not clean weeded. Annual crops should be planted on the more gentle slopes in strips. Wherever possible all cultivations, ridging, planting, weeding, etc., should be carried out along the contour, and not up and down the slope. Planting crops on tied ridges, provided that the ridges are well maintained, prevents both erosion and the loss of available rainfall by run-off. Experiments in East and West Africa have given marked yield increases under certain rainfall and soil conditions, with both cotton and other annual crops by planting on tied ridges (Plate 2.11). Ridging however increases the soil surface from which rainfall is evaporated, and in some cases this extra evaporation loss may be greater than the extra rainfall conserved.

In Northern Nigeria the soils 'cap' with the impact of rainfall which reduces the infiltration rate, and standing water can be similarly detrimental by reducing aeration. The average yield of seed cotton in this area is 290 kg/ha, but cotton grown on tied ridges mulched in the

Table 2.7 Run-off soil erosion at four locations in West Africa.

Country	Site	% slope	Annual rainfall	Forest Run-off	Soil loss t/ha
Upper Volta	Ouagadougou	0.5	850	2	0.1
Senegal	Sefa	1.2	1,300	1	0.2
Senegal	Bouake	4.0	1,200	0.3	0.1
Ivory Coast	Abidjan	7.0	2,100	0.1	0.03

Notes: Run-off is % of annual rainfall.

Plate 2.11 Tie-ridges conserving water and soil. Lake Province, Tanzania. (Courtesy Cotton Growing Corporations.)

furrow will yield 2,240 kg/ha if protected by insecticide. Tying of alternate furrows without mulching is much easier than mulching, and provided that the water loss by run-off is unimportant, can give yields similar to mulched tied furrow cultivation (Lawes 1963).

In trials over 11 seasons in Sukumaland, Tanzania, cotton yields without fertilizer ranged from 450 to 1,465 kg seed cotton per hectare cotton yields with fertilizers ranged from 590 to 1,775 kg seed cotton per hectare. Response to the ridging varied from -15 to +300 kg seed cotton. Millet yield ranged from 225 to 1,200 kg grain per hectare. Response to ridging varied from -70 to +240 kg grain per hectare (Brown 1963). Equipment has been designed for mechanical tie ridging (Boa 1966).

Cultivated land		**Bare soil**	
Run-off	**Soil loss**	**Run-off**	**Soil loss t/ha**
2-32	0.6-8.0	40-60	10- 20
21	7.3	40	21
0.1-26	0.1-26	15-30	18- 30
0.5-20	0.1-90	38	108-170

Source: Charreau (1972).

Plate 2.12 Aerial view of young rubber planted in clean weeded rows along the contour, and interplanted with legume cover crops - the dark strips. Malaysia. (Courtesy The Rubber Research Institute of Malaysia)

Certain crops such as millets which are close planted, resist wash. Widely spaced crops such as maize should be intercropped, planted on contour tied ridges, or trash mulched. Young tree crops should be planted on the contour and the area between the tree rows sown to a cover crop such as *Centrosema* (Plate 2.12).

Catchment studies in the Mbeya range (Tanzania) comparing the run-off from the steep cultivated hillsides typical of the region with the run-off from a similar forested area with a gradient of 1 in 2.2 overall showed:

Cultivated area 0.9 - 76.0 cusecs per square mile
Forested area 0.03 - 12 cusecs per square mile

[One cusec = 0.02832 cubic metres]. (East African Agriculture and Forestry Research Organization (Anon, 1959)

Even in fair, clear, dry weather stream flow, the run-off from the cultivated area carried 12 times as much soil as from the forested area. The crops cultivated in this area are maize and vegetables, restricted in their growing season by a 6 month dry season.

Plate 2.13 A well-mulched banana garden, with anti-erosion bund, in Buganda (Courtesy Department of Information, Uganda. Photograph Ron Ward)

In Kericho (Kenya), forest has been replaced by tea. In the early stages run-off was increased but as the tea became established with deep roots and full ground cover, run-off was reduced to the same level as with the original forest cover (Dagg and Blackie 1965).

Physical barriers to soil erosion

Cultural measures are frequently not sufficient to prevent soil erosion in which case physical protection is necessary. These measures include:

1. *Terracing*. The vertical faces may be made with stones. Terracing involves considerable work. The slopes around Darjeeling are so steep that only a single row of tea may be planted in each terrace. In areas like the Kenya Highlands, broadbased terraces can be constructed mechanically.
2. *Bunding*. These are about 0.9 m wide and 0.6 m high and run along the contour. About every 20 m along on the upper side of the bund, a short bund is made at right angles. This tie bund collects the water if the main bund is off the contour. These bunds may be planted with a short grass to stabilize them (Plate 2.11 and 2.13).
3. *Grass strips*. This is a method favoured in parts of East Africa. A

grass strip about 0.9 m wide is left uncultivated along the contour every 15 to 30 m down the slope. The width of this cultivated strip depends on the slope. Unfortunately grass strips are a source of weeds such as *Digitaria scalarum* and of rats.

4. *Stop wash lines*. These are single lines of grass, such as vetiver grass (*Vetiveria zizanioides*), a useful thatching grass, which retain much of the soil and allow the water to pass on.

These last three methods, if properly maintained, slowly terrace the land.

Pits to collect silt are used in certain regions but are only temporary expedients and need frequent maintenance.

Cattle and other farm stock are often the cause of erosion. At the end of the dry season, areas are overgrazed and overstocked and the tracks to the watering places trodden bare. With the onset of the rains, run-off is high and the cattle tracks are the source of gulley erosion. An increase in the number of watering points for cattle is of great help, but limitations of cattle numbers may be the only solution. Overgrazed areas should be closed to grazing to regrow their cover, but if closed too long, bush encroachment may occur and exclude the cattle. Goats have been blamed for the poor state of many Kenya reserves where all the vegetation is destroyed.

The protection of gullies with stop wash lines and small dams may give a visible indication that anti-erosion work is in progress but the answer to the gulley lies in dealing with the source of run-off. This is frequently a non-receptive area such as a house compound (Plates 2.14 and 2.15), a road, a rocky outcrop or an unprotected area of cultivation. Soil conservation measures should start here. Compounds and road drains can be planted with short grass, such as *Cynodon dactylon* or *Paspalum notatum* (Plate 2.14).

In areas where rainfall is a limiting factor to crop yield, measures taken to prevent soil erosion will conserve the rainfall and increase the crop yield. A combination of tie bunding and manuring with cattle dung gives a very obvious increase in the robusta coffee yield in Buganda. The better growth of the trees also shades out the weeds and protects the soil.

Better water conservation improves the flow of natural springs. In the Far East, excess rainfall is common and it is necessary to remove the surplus through open drains as quickly as possible. Such drains should have a gentle slope or be grassed to prevent them eroding.

Wind erosion

Wind erosion is most serious in dry areas, or at the end of the dry season, particularly where the soil has a low organic matter. Soil blown

Plate 2.14 Soil conservation. Protection of a house compound from run-off, by planting lines of *Paspalum* grass, Sayi, Buganda, Uganda. A crop drying table, robusta coffee and banana are in the background. The thatched house in on the left. (Courtesy Uganda Department of Information; photograph by Ron Ward.)

by the wind is not only lost but has an abrasive action on crops and vegetation which is often lethal. Large sand dunes may form which can travel across country at a steady pace engulfing fences and houses. Where this is likely to occur, windbreaks at regular intervals are essential. The land must be farmed to encourage a good structure, and cultivations kept to a minimum. The stubble and remains of old crops must be left on the soil surface (stubble mulch) until the next crop is planted to reduce wind velocity and hold up soil particles.

Swamp reclamation

In many tropical countries including East and West Africa, Malaysia and the adjacent territories, there are large areas of swamp little used for cultivation. As these swamp soils have a high humus content and an abundance of water, an obvious development is to drain and reclaim these areas. This has been done very successfully in certain cases, but elsewhere trouble has occurred. In Uganda where there are an

Plate 2.15 Soil erosion of an unprotected house compound in Bugunda. (Courtesy Uganda Protectorate Department of Information; photograph by Rou Ward.)

estimated 0.5 million ha of papyrus (*Cyperus papyrus*) swamps, reclamation has been carried out by cutting a main channel near the centre of the swamp and lateral channels. There are checks in the main channel which control the water flow to maintain the water table within 60 cm of the surface. Three-quarters of these swamps can then be cultivated and yields of sweet potatoes and sorghum are very high.

Many coastal areas in Vietnam, Guyana and West Africa and other tropical areas, and deltas including the Mekong, have acid sulphate soils, where the sulphate in the sea water is reduced to hydrogen sulphide in the anaerobic conditions of the flooded soil high in organic matter. The hydrogen sulphide reacts with iron compounds in the soil to form iron pyrites (FeS_2). Pyrites can be formed in soils inland, as on the Bangkok plain; in some freshly dug soil pits in Uganda swamps there is a strong smell of hydrogen sulphide. Where the swamps contain considerable amounts of iron pyrites (FeS_2) they become quite sterile after draining as the iron sulphide is oxidized to sulphuric acid and iron is deposited. A pH as low as 2.4 and 2.7 was recorded in Kigezi (Uganda), 1,800 m above sea level, and pH 2 to 3 is usual on acid sulphate soils after draining. The deposited iron gives the soil a bright yellow mottled appearance, and the sulphuric acid releases toxic amounts of aluminium and also manganese and iron. Drainage, liming and leaching of acid sulphate soils would take 7 to 10 years to get rid of the iron pyrites. Work in Rothamsted showed that raising the water table of previously drained acid sulphate soils considerably increased the yields of the oil palm plantations in Malaysia (Kasasian *et al.* 1978).

Swamp reclamation should start on a pilot scale and be monitored carefully with frequent laboratory checks. The oxygen should not be let into the peat too rapidly and the water table kept about 60 cm below the surface. If possible the swamp should be flushed periodically to remove the sulphuric acid and toxic elements and any other undesirable decomposition products. Phosphate fertilizer may be needed as these swamps are often deficient in phosphate. Reclaimed swamps can yield heavy crops.

FERTILIZERS

The value of fertilizers in increasing the yields of tropical crops was clearly shown on growers own plots in the FAO fertilizer programme (Tables 2.9 and 2.12). Previously trials were mainly on experiment stations where the farming system maintained the soil fertility and responses to fertilizer though significant were seldom economic. On the growers' own plots in 19 different countries average yield increases were 50 per cent from fertilizer alone (Richardson 1966).

Each year the FAO issue a review of fertilizer production, use and trade, from which it is seen that more phosphate and potash are used at present in the United Kingdom than in the whole of the tropics, the bulk of the fertilizer used in the tropics is applied to plantation crops, and the current fertilizer use among the small tropical farmers is insignificant.

There are, however, important economic factors to recognize. Peasant farmers need a local supply of small packs of fertilizers and small packs of any product are expensive. In most tropical countries a village distribution system for small packs does not exist. Distribution through the village shopkeepers is costly as these tradesmen need a high margin particularly if they have to give credit until harvest. The true cost of fertilizer to the farmers in underdeveloped countries is much higher per unit than in developed countries, as there is the cost of shipping and inland transport, to add to the high packaging and distribution costs. However, this true cost is often obscured by government intervention either directly in the form of a cash subsidy or indirectly by providing free distribution.

Furthermore, the peasant farmer regards expenditure on fertilizer as a high risk venture and needs to anticipate a high return before it becomes worthwhile. A 10 per cent return on his investment is of no interest, as cash is a precious commodity. A value : cost ratio of at least 2, i.e. 100 per cent return, would seem to be necessary to make the use of fertilizers interesting, provided also that the crop has a ready local market. Such situations do occur in the tropics particularly for cash crops, and for food crops near towns. For example, in 398 rice demonstrations organized by FAO in Senegal, 45 kg/ha of nitrogen gave an average yield increase from 1,108 to 1,441 kg with a value : cost ratio of 2.9, an average return throughout West Africa. The value : cost ratio for yams, a high value crop, was often over 5. On the other hand frequently it did not pay to use fertilizers with maize in West Africa owing to the low price for the crop. In Thailand, where rice is an important export crop, fertilizers are used on about a quarter of the arable land, two-thirds of this fertilizer is applied to rice but even so two-thirds of the rice receives no fertilizer.

The FAO Indicative World Plan (FAO 1970) estimated that developing countries would require 16.5 million t in 1980/81 and 33 million t in 1985/86. To provide the foreign exchange to purchase this fertilizer or the feedstock for manufacture will be a major problem for these countries, assuming supplies are available.

Before the oil crises more fertilizer was being used because of:

1. Greater availability and lower prices of fertilizers, particularly nitrogen, but this position changed at the end of 1973.
2. Greater awareness of farmers of the value of fertilizers and how to use them. In certain countries this has been stimulated by the FAO project.
3. Better distribution and credit facilities in many countries which have brought fertilizers within reach of small farmers.
4. The encouragement of crop varieties which respond to fertilizer, in particular rice varieties which utilize added nitrogen.
5. Better marketing facilities and prices for increased production of certain crops in some countries.

Determining which fertilizers to use

Fertilizers feed the soil not the crop, and plants can only take up that which is not fixed and remains within the space foraged by the roots.

With an indication of the amount of nutrients in the soil and the amount taken up by the particular crop (Table 2.8) it might be thought that the fertilizer application for a crop could be made on the basis of soil analysis. While soil analysis indicates the amounts of the important elements in the soil, and may give an arbitrary indication of their availability, this is no guide to the amount the particular crop can take

Table 2.8 Major nutrients removed annually by crops

Crop		Yield per hectare (kg)	Remove the following amount of nutrient (kg/ha)		
			N	P_2O_5	K_2O
Cereals					
Maize	Grain	2,000	40	5	10
	Stover	3,000	15	2	14
Rice	Grain	2,000	24	6	6
	Straw	1,600	14	2	32
Sorghum	Grain	1,000	20	2	5
	Straw	1,200	6	1	2
Finger millet	Grain	1,000	16	10	65
Root crops					
Sweet potatoes	Roots	15,000	66	17	96
Cassava	Roots	30,000	28	6	36
Legumes					
Groundnuts (unshelled nuts)		1,000	45	11	30
Soya beans		1,000	45	15	22
Cash crops					
Bananas (plantains)		40,000	70	20	200
Cocoa		700 (Dry beans)	19	10	14
Cotton		800 (Seed cotton)	27	9	7
Coffee (Arabica)		750 (Dry beans)	124	28	175
Coconuts		1,200 (Dry copra)	55	15	50
Rubber		1,000 (Dry rubber)	8	3	6
Tea		1,000 (Dry leaf)	40	9	21
Oil Palm		4,500 (Fruit)	80	33	130
Tobacco		1,000 (Cured leaf)	105	30	210
Sugar Cane		80,000 (Millable cane)	40	50	130
Sisal		55,000 (Leaf)	80	18	100

Note: The above figures are only a rough indication. Figures from different sources show wide variation.

up from the soil. Phosphate is particularly difficult as it is often firmly held by the iron and aluminium compounds in tropical soils. Insufficient is known about the mechanism by which crops take up nutrients, or even the extent to which they can draw on different soil depths.

Leaf analysis is a guide to what the crop is getting from the soil and what must be added to balance the nutrients for maximum yield. Unfortunately for this purpose the amounts and ratios of the different elements in the leaf are not static and change as the leaves mature, also as light, climatic conditions and seasons vary. Nevertheless standard sampling techniques and nutrient levels have been established for most tropical crops (these are summarized by Chapman 1964). There is also the further difficulty that the difference between the level at which a nutrient is adequate, and the level at which it is deficient without causing visual deficiency symptoms is very small. With modern analytical techniques these differences can be reliably detected. Where important deficiencies occur on tropical soils, if they are not detected by soil analysis, they will be shown up by foliar analysis. As crops may accumulate elements above the level of their needs there is no strict relationship between mineral composition of leaves and crop yields, though for most crops arbitrary levels have been laid down for field management use. Sugar cane, rubber, tea and oil palms are estate crops where this has been done.

Pot trials are a further help, but there is no real substitute for field trials. Unfortunately field trials are slow and tedious and the results only apply to that particular soil, and that particular season. Laboratory analyses of soils and leaves, on the other hand, can be done very accurately and very rapidly. A small staff with the modern instruments such as the Technicon auto-analyser can provide quick quantitative determinations of calcium, potassium, magnesium and phosphorus simultaneously and automatically. Nitrogen and other elements can also be determined. The results are printed out for direct reading. Where these analyses serve a single crop on a restricted range of soils, they can be easily translated into fertilizer recommendations on the basis of field trials and experience. Many of the metal analyses can be done with an absorption spectrophotometer which is a cheaper instrument.

The limitation on the output of such a modern laboratory is the sampling. Provided that the samples are not mixed up, human error is eliminated. Results are available so quickly that even with a fast growing crop such as tobacco steps can be taken to counter adverse nutrient ratios in the plant due to the season. Such a laboratory would seem to be an essential for every government agricultural department, and large estate groups who plan to use fertilizers on a rational basis. Such equipment is especially valuable in the tropics where minor elements may be deficient due to heavy leaching and may be the limiting factor preventing response to additions of NPK.

There is frequently a residual effect for a number of subsequent years after fertilizer applications. Where this applies to nutrients such as potash, nitrogen and sulphur which are easily leached away, it illustrates the nutrient cycle and the way in which these nutrients are returned to the soil in an organic form as crop residues, which decay and become available to succeeding crops. This is the basis for building up soil fertility with fertilizers.

Perennial crops should be fertilized well in their early years to ensure rapid development and early bearing. Much of this fertilizer will be bound up in the wood of the tree, but a certain amount will continue to circulate through leaf fall.

It is only possible to discuss a few examples of the use of fertilizers in the tropics.

Nitrogen

Nitrogen is almost universally deficient in the soils of Africa and most of the tropics, and is the nutrient that most frequently limits yield. Lack of nitrogen results in poor plant growth and a uniform yellowing of the leaves. Application of nitrogen produces a larger leaf which is darker green in colour. If applied too near to harvest, the increased leaf area may not have time to make a contribution to the yield, and indeed may reduce yield by diverting the plant into greater growth rather than production. Cotton is a crop where this can easily occur. A big leafy plant develops which is attractive to insect pests, but the crop ripens few bolls.

Hardy (1946) showed that the nitrogen level in the soil builds up very rapidly at the start of the rains but is then rapidly leached out. The most effective time for application could be early in the rains when the crop has developed sufficient roots to take up the nitrogen before it is leached away rather than at planting.

Grass crops, sugar cane, maize, rice, millet and sorghum all respond well to nitrogen, provided that the crop variety can respond to nitrogen, and rainfall is not limiting. Breeding work in rice has recently concentrated on developing varieties with a greater response to nitrogen fertilizer than the commonly grown varieties. This is illustrated clearly in Fig. 3.6. The graph based on a dry season experiment, illustrates how response to nitrogen differs between varieties. Dwarf varieties are typically more responsive to nitrogen than tall ones, Binato & Peta. It also illustrates the high yield potential of improved plant types when carefully managed and protected against pests.

The soil and water factors influencing nitrogen application to rice have been explained on p. 19.

There are two periods in the growth of a rice crop when nitrogen requirements are highest, at the tillering stage and later when the panicle primordia start to form. Yield is correlated with nitrogen availability at these growth stages. As tillering is not desirable in maize or sorghum, nitrogen need is highest when growth is at its maximum, and at flowering. Sugar cane on the other hand has its maximum nitrogen requirement at tillering.

Nitrogen applied to root crops (yams, sweet potatoes) will increase the area of foliage and so should be applied as the leaf area begins to increase rapidly.

Some typical responses of food crops to the application of 22.4 kg/ha of nitrogen in a series of trials and demonstrations carried out in the FAO Fertilizer Programme are shown in Table 2.9.

Table 2.9 Some typical yield responses to 22.4 kg/ha of nitrogen

Crop	Region	Zone	No. of demonst. or trials	Yield (kg/ha)			Value : Cost ratio
				Control	Increase	%	
Maize	W. Nigeria	Savannah	62T	970	187	19	1.3
	W. Nigeria	Savannah	99T	1,205	424	35	2.6
	Mid-West Nigeria	Forest	32T	1,431	381	27	2.7
	Ghana	Savannah	147D	1,005	352	35	3.2
	Ghana	Forest	759D	1,296	451	35	4.3
Rice	W. Nigeria	Savannah	106D	1,301	372	29	6.8
	W. Nigeria	Forest	64D	1,942	764	39	14.0
	Ghana	Forest	66D	1,484	322	22	4.7
	Ivory Coast	Forest	14T	2,298	364	16	2.5
Sorghum	Ghana	Savannah	25T	450	151	34	1.1
	Senegal	Fleuve	9T	256	74	29	1.3
Millet	Ghana	Savannah	3D	329	136	41	1.5
	Senegal	(Average country)	83T	626	287	46	5.1
Yams	W. Nigeria	Savannah	94T	9,088	2,370	26	10.5
	W. Nigeria	Forest	35T	12,196	1,073	9	4.9
	Mid-West Nigeria	Forest	57T	7,522	1,704	23	13.7
	Ghana	Savannah	141T	7,644	1,397	18	4.2
	Ghana	Forest	29T	8,408	4,222	50	12.7
	Ivory Coast	Savannah	24T	6,889	310	4	1.2
Cassava	W. Nigeria	Forest	5T	8,447	2,386	28	2.6
	Mid-West Nigeria	Forest	8T	7,059	3,842	54	11.9

Source: F FHC Fertilizer Programme. Consolidated review of trial and demonstration results. FAO (1969).

Tea, which depends for its yield on new growth, responds well to nitrogen; indeed, few crops respond as well, provided that the density of shade does not limit the response. Evidence from Tocklai (Carpenter 1938), confirmed by others, showed that the increase in crop has a direct relationship to the amount of nitrogen applied, up to a rate of about 125 kg of nitrogen per hectare.

Nitrogen applied annually per hectare	Average annual gain in crop of tea per hectare
45 kg	287 kg
90 kg	600 kg
135 kg	808 kg

Over a 4 year period, 67 kg of nitrogen per annum as sulphate of ammonia gave twice the increase in yield as an annual dressing of cattle manure containing the same amount of nitrogen. Ready availability to the tea plant appears to be all important.

The response to nitrogen varies with the soil and other conditions, the higher the cropping capacity, the lower is the response. As shown below, at different sites there is a variation in the increase of made tea for each kilogram of nitrogen applied per hectare where 45 kg nitrogen were added annually.

Tocklai (Assam)	7.2 kg)	
Tulsipara (Bengal, Dooars)	5.7 kg)	(Carpenter 1938)
St Coombs (Sri Lanka)	4.9 kg	
Kericho (Kenya)	6.5 kg	(Goodchild 1959)

From the introduction of a standard annual application of nitrogen, there is an annual rise in the yield increase as was shown at Tocklai.

	Year				
	1st	**2nd**	**3rd**	**4th**	**5th**
Kilograms crop increase per kilogram nitrogen	3.7	5.6	7.2	7.2	9.0

(Carpenter 1938)

No differences were obtained between split and single applications, and broadcast as compared with digging in. The regular addition of nitrogen increased the weight of prunings and the soil organic matter.

Heavy nitrogen does not affect tea quality. With heavy nitrogen dressings, estate tea yields over 2,240 kg/ha are quite common. Blocks of clonal tea are giving over 4,480 kg/ha, some over 5,600 kg/ha and the limit has not been reached. Only a few years ago the average yield in Sri Lanka and Assam was under 1,000 kg/ha. Heavy rates of sulphate of ammonia have reduced the soil pH in some areas to 3.8 or 4.0 and yields of clonal tea in these areas have fallen to 1,112 kg/ha. Non-acidifying nitrogen fertilizers, calcium ammonium nitrate (CAN), urea, or ammonium nitrate are advised in such cases to bring the pH back to 4.5 or 5.0. CAN is no longer recommended for tea in East Africa as yields with CAN have been lower than with equivalent nitrogen applications in other forms.

When the annual application of nitrogen ceases, the yield drops rapidly in the following year and to a lesser extent in the subsequent years (Gokhale 1955).

Legumes, as might be expected, give a poor response to nitrogen fertilizer, though a little available at planting gives the crop a good start before the nitrogen from the root nodules becomes available, except where planting takes place at the start of the rains when there is adequate nitrogen.

Eroded soils tend to be deficient in nitrogen due to the loss of the more active and fertile topsoil. Land cleared from a grass fallow shows a better response to nitrogen than land which has been resting under bush, as the grass roots seem to take up the soil nitrogen and make it unavailable. Sown pastures in the humid tropics respond well to nitrogen fertilizer.

With the rapidly increasing oil price, the cost of nitrogen fertilizer is becoming too expensive for many small growers to use and so there is a greater need to replace the soil nitrogen by growing legume crops. There is hope that the nitrogen-fixing ability of legumes might be bred into other crops.

Phosphate

After nitrogen, phosphorus is the most common nutrient limiting production in the tropics. Phosphate stimulates root growth, which spreads the roots through a greater volume of soil to forage and improve drought resistance; and assists nodulation in leguminous plants, hence its importance when establishing legumes in pastures. Phosphate also encourages earlier maturity which partly counters late planting and the effect of nitrogen. Tillering of cereals, particularly rice, is encouraged which can be important when the crop is recovering from an early attack of borers. Phosphate accelerates the ripening process of tobacco leaf, and produces a leaf with a high proportion of monosaccharides. This is desirable in good cigarette tobacco but not for cigars. Excess phosphate causes the leaf to mature too early and the yield is reduced.

Lack of phosphate is often associated with a purple leaf colour, particularly at the edges, but in certain crops such as cotton and tobacco, the leaves become dark green. Growth is stunted, especially the roots.

A moderate deficiency of phosphate is widespread throughout tropical Africa, and in certain areas this is so serious that phosphate fertilizer must be added to grow reasonable crops. Phosphate deficiency is very common in the Far East, including more than 800,000 ha of Java and large areas of Thailand, especially the heavy clays around

Bangkok. In Malaysia, the soils from sedimentary rocks are especially phosphate deficient, and many of the rubber soils have insufficient phosphate for optimum growth.

Many tropical soils immobilize phosphates. Under acid conditions iron and aluminium phosphates are formed and in calcareous soils calcium and magnesium phosphates, all of which have a low solubility. By placing pellets or granules of phosphate fertilizer a centimetre from the seed, a low application rate will give the maximum effect on soils that readily fix phosphate. The response to phosphate is associated more with the soil than the crop as in Indonesia where about 400,000 ha of the 5½ million ha of rice respond to phosphate. Quite a lot of the soil phosphate is present in an organic form, on some soils this may be half or more, and the phosphate level is maintained by maintaining the amount of organic matter in the soil.

In Northern Nigeria and Ghana (Hartley 1937; Greenwood 1951; Nye 1953 and 1954) and the adjacent areas of West Africa, groundnuts show a spectacular response to phosphate. The yield of shelled groundnuts on average soil can be raised from 785 kg/ha to 1,000 kg from a single application of 14 kg/ha phosphate as single superphosphate. An application of 224 kg/ha of superphosphate had a residual effect on groundnuts planted 4 years later (Greenwood 1951). The same amount of phosphate applied as cattle manure did not have the same residual effect. A very light dressing, 12 kg P_2O_5 per hectare gives a very good increase in the yield of groundnut kernels. Single superphosphate, contains 45 per cent calcium sulphate and as both the calcium and sulphur are deficient in Northern Nigeria (Goldsworthy & Heathcote 1963) both contribute to the crop response and residual effect.

In the FAO Fertilizer Programme, the yield of food crops in West Africa was increased similarly with 22.5 kg/ha of P_2O_5 as with the same rate of application of nitrogen. The responses of groundnuts and yams to phosphate were particularly good.

Coffee rarely shows symptoms of phosphate deficiency even though phosphate is lacking in many coffee soils and response to P_2O_5 has seldom been shown in fertilizer trials. Coffee appears to have remarkable ability to take up phosphate when the level in the soil is low.

Rubber, both in Sri Lanka (Constable and Hodnett 1953) and Malaysia (Haines and Crowther 1940), responds well to phosphate. Phosphate, to a greater extent than nitrogen, increased the girth of the rubber trees. When fertilizer was applied from planting out, the trees in Malaysia reached a girth of 50 cm at which tapping starts, 6 months to 2 years earlier than the unfertilized trees. Mainstone (1963) reporting a long-term fertilizer trial on rubber in Malaysia recorded double the yield from trees which had received phosphate in comparison with the trees without phosphate.

In addition to the direct effect of phosphate on rice it has been suggested that the phosphate stimulates the algae associated with the fixation of atmospheric nitrogen in paddy soils.

At Urambo in Tanzania (Hagenzieker 1956), the phosphate and calcium are concentrated in the top 15 or 20 cm of soil, particularly in the gley soils. The roots of soya beans, groundnuts and sunflower are similarly restricted in their distribution, though the roots of maize, sorghum, millet and cotton, provided that there is no hard pan or permanent water table, penetrate deeply, and are well distributed. The nodulation of the soya bean and groundnut roots is confined to the surface soil. By drilling superphosphate at a depth of 30 cm there were spectacular yield increases, 76 kg of P_2O_5 increasing the yield of soya beans by 80 per cent and maize 45 per cent. Broadcasting or placing the phosphate in the top 15 cm had no effect on yield.

A similar concentration of phosphate in the surface soil occurs in some of the short grass regions of Uganda, and may be fairly common in the drier tropics.

Cotton in East and West Africa shows a response to phosphate (King 1960). At Samaru (Northern Nigeria), a yield of 2,516 kg of seed cotton per hectare was obtained by a combined dressing of superphosphate, sulphate of ammonia and compost where water was conserved and insects controlled. This is 10 times the yield of unmanured plots on the same farm.

Response to phosphate does not always occur. In a trial on a Buganda clay loam applying a high rate of triple superphosphate (190 kg/ha) alone or with a fertilizer mixture of N, K, Mg, farmyard manure and lime to maize and cotton over 3 successive years, reduced rather than increased the yield, though the fertilizer mixture increased yields (Stephens 1966). Possibly sulphur was limiting. In an experiment on groundnuts carried out by Institut de Recherches pour les Huiles et Oleagineux (IRHO) in Senegal, triple superphosphate raised the yield from 1,750 to 2,175 kg/ha, but when sulphur as well as triple superphosphate was applied the yield was 3,015 kg/ha.

Potash

Potash deficiency occurs in more localized areas than the other two major nutrients. Many forest soils are low in potash, which may be accentuated by growing crops such as yams with a high potash requirement.

After burning the forest for shifting cultivation, there is a big increase in the amount of potash and other bases in the topsoil, but this is often barely adequate after the second crop.

As potassium is very mobile in plants, deficiency symptoms appear first on the older leaves initially as chlorotic spots near the leaf tip and margin developing into a necrosis between the leaf veins.

Fruit crops in particular respond to potash, including oil palms, coconuts, coffee and bananas.

Confluent orange spotting, a leaf symptom of oil palms found throughout West Africa, is associated with potash deficiency (Hale 1947). Oil palms on a worn-out soil show a marked response to wood ash, this had been observed before the importance of potash was demonstrated.

In an experiment in the Eastern Region of Nigeria, a single application of 628 kg/ha, of potassium chloride gave nearly a 100 per cent increase in yield from 30-year-old oil palms the second year after application, and maintained the yield increase over the control palms for a further 6 years (Chapas and Bull 1956) (Fig. 2.2). The incidence of *Fusarium oxysporum* wilt has been reduced considerably by the use of potash (Prendergast 1957).

Research work on coconuts in Jamaica has shown that nitrogen fertilizer encourages the coconut to produce more flowers and potash helps these flowers to set and produce more nuts. This also means a long delay between application of fertilizer and increased yield.

The effect of potash on sugar cane varies with the soil and rainfall, ratoons giving a greater response. The sugar content of the cane is frequently increased by the use of potash.

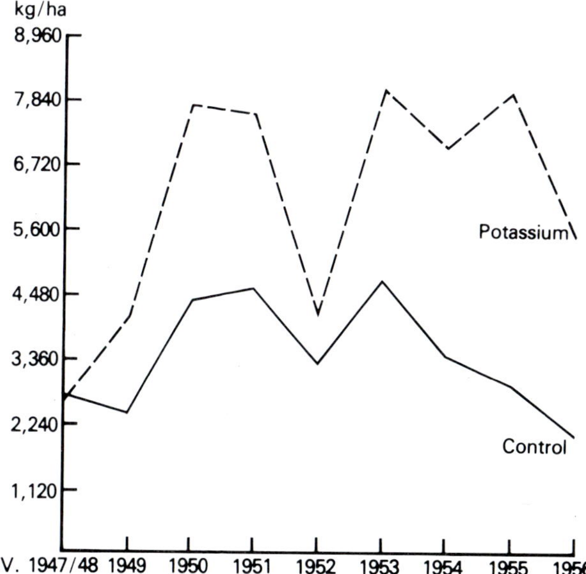

Fig. 2.2 The 8-years' effect of a single application of potassium on the yield of oil palm fruit bunches in Eastern Nigeria. (Hartley 1958)

There is a balance between potash and magnesium, and where mulch with a high potash content is applied regularly in Kenya to coffee, magnesium deficiency symptoms appear. This is corrected by adding magnesium sulphate to the fungicide sprays. The same effect occurs on soils low in magnesium liberally fertilized with potash. Shortage of potash causes 'banding disease' of sisal in East Africa, reducing both the quality and yield of fibre.

As rice is largely grown in heavy soils, in most cases adequate supplies of potash are available. In Sri Lanka and Malaysia response to potash fertilizer only occurs on some lighter soils.

The possible long-term effects of fertilizers have been referred to and are well demonstrated in experiments which have been carried out on tea. The effect of potash was not seen in an experiment at the Tea Research Institute in Sri Lanka until the treatments had been continued for 12 years (Portsmouth 1949). In the absence of potash, there was a marked reduction in the number of bushes which was even more marked with the application of 45 kg/ha of nitrogen. [See Table 2.10]

Table 2.10 Average crop response to individual nutrients in kilograms per hectare of made tea. S. India

Treatments	**Cycle 1** 1940–43	**Cycle 2** 1943–46	**Cycle 3** 1946–49	**Cycle 4** 1949–52	**% Decrease in bush population**
N_0	972	1,016	1,010	1,102	18.6
$N_{44·8}$	1,079	1,197	1,072	1,173	31.25
$N_{89·6}$	1,261	1,519	1,426	1,651	21.7
P_0	1,049	1,166	1,119	1,296	—
$P_{22·4}$	1,141	1,273	1,193	1,299	—
$P_{44·8}$	1,120	1,292	1,241	1,332	—
K_0	1,133	1,184	926	846	46.1
$K_{22·4}$	1,042	1,202	1,183	1,403	21.2
$K_{22·8}$	1,134	1,345	1,404	1,667	3.4
Significant difference					
5%	124	221	295	374	
1%	188	335	447	567	

Source: Jayaraman and de Jong, (1955).

The traditionally high import of potash into Sri Lanka was partly explained by its critical effect on the survival of tea bushes. Potash is important in improving the burning quality of tobacco leaf, and a high potash status is required throughout the growing season.

The response of many crops to potassium often increases in magnitude with successive applications.

Sulphur

It is not always realized that certain crops such as cotton, legumes and crucifers may take more sulphur out of the soil than phosphate, and many other crops have a high sulphur requirement. Sulphur appears to be held in the soil in an organic form but taken up by plants as inorganic compounds. Where plants rely upon the mineralization of organic matter for their nitrogen, this process should also liberate an adequate amount of sulphur. This possibly explains why sulphur deficiency is most common on sandy soils, particularly in savanna country and heavily leached areas and becomes increasingly apparent when non-sulphur-containing nitrogen fertilizers are used. Where the land is cleared by burning much of the sulphur is volatilized; this is particularly important on grassland which is fired annually.

In the early 1930s, sulphur deficiency in tea was found to be widespread in Malawi (Storey & Leach 1933). Attention has been drawn to the low level of sulphur in many African rivers and lakes (Beauchamp 1953), suggesting that sulphur deficiency may be more widespread than thought.

At Maboka in Malawi, Matthews (1972b) found a depression in yield where calcium ammonium nitrate was applied without sulphur, but a much improved yield with sulphur, as shown in Table 2.11.

In a trial carried out on cotton in the Upper Volta, the addition of sulphur to a basic fertilizer treatment increased the yield of rain grown cotton from 2,185 to 3,085 kg of seed cotton per hectare. The crop was protected from insects.

Table 2.11 Yield of seed cotton (kg/ha)

Nitrogen (kg/ha)	**Sulphur** (kg/ha)			
	Nil	**22.4**	**44.8**	**Mean**
Nil	1,029	1,334	1,286	1,217
44.8	1,059	1,568	1,671	1,433
89.6	809	2,254	2,355	1,806
Mean	966	1,719	1,771	

Note: The cotton was planted on ridges.

Legumes suffer from sulphur deficiency in parts of West Africa (Greenwood 1951) and in consequence there is a better response to phosphate in the form of single super than triple super due to the difference in gypsum and calcium sulphate content. As groundnuts contain a high calcium content, the gypsum may have a double effect. Sulphur has also been found to be deficient in Java, the Punjab, Malaysia, Central America, Brazil and Australia.

Sulphur deficiency was found in oil palms in the Ivory Coast and in coconuts in Madagascar where the condition can be very serious on young plantations. In both situations the deficiency occurred where the palms were planted after the mechanical destruction of *Imperata cylindrica* (Ollagnier & Ochs 1972).

Chenery (1960) has suggested that the carrying capacity of Trans-Nzoia pasture in Kenya, and many areas in Uganda, will be more than doubled by dressings of gypsum, provided the soils are not also deficient in phosphorus. These conditions occur in Eastern and Northern Uganda.

Any fertilizer trial carried out in Africa, or any other country where sulphur could be deficient, should include sulphur on the same basis as NPK, as in many cases failure to respond to NPK has been due to sulphur being limiting.

Other deficiencies

Many other mineral deficiencies occur in the well leached tropical soils, and as crop production is increased by the use of more fertilizers, more deficiency symptoms will appear. Iron, manganese and zinc may be deficient on alkaline soils. Iron deficiency is not confined to alkaline soils, but may occur in soils with a pH of 5 due to antagonism by other elements including Zn, Cu, Ni and Mn. Leaf chlorosis is often associated with a low iron/phosphate ratio. A high organic content in the soil may cause manganese deficiency. In Malaysia however, manganese is sometimes lacking in the soil, not just unavailable, and this may be the case with iron on some well leached sands, as in Florida.

Magnesium deficiency, which is fortunately easily corrected by spraying with Epsom salts, is common on many tree crops including coffee in East Africa, tea in Sri Lanka, rubber in Malaysia, oil palms in the Ivory Coast and Zaire, and coconuts in Sri Lanka.

Zinc deficiency of rice which occurs on continuously submerged or badly drained soils causes rusty brown spots on the older leaves 2 to 4 weeks after transplanting. In severe cases it may look like potassium deficiency as the leaf margins dry up. Growth and tillering are restricted. Zinc deficiency in swamp rice is more common than was previously thought. In some areas, as on the Agusan plain in the Philippines, the land will not produce a crop, and in other areas only low yielding varieties tolerant to zinc deficiency can be grown. This can be corrected by dipping the seedling roots in a 2–4 per cent suspension of zinc oxide before transplanting. With this simple, cheap treatment, better yielding varieties can be grown and large areas can be used to produce rice on soil that otherwise would not grow this crop.

Zinc deficiency in cacao in Dutch New Guinea has occurred on acid silty loam as well as on alkaline soil. Coffee and cotton are particularly susceptible to a low level of zinc. Lucerne is sensitive to boron and copper deficiency, and in Malawi improper fermentation of plucked tea has occurred through a lack of copper in the leaf. Part of the benefit from Bordeaux spraying for banana leaf spot control in Central America may be due to the correction of a copper deficiency. Copper is deficient in the peat soils of Malaysia. Iron and copper deficiency has occurred in the sugar cane fields of Natal, most of which are short of phosphate. Groundnuts in Sierra Leone form empty shells due to a lack of calcium.

Deficiencies are generally best corrected by a foliage spray of the sulphate or other soluble salt in water, or borax in the case of boron deficiency. Soil application is less efficient. Nitramolybdenum dusted on cowpea seed at a very low rate increased the vigour, colour and yield of that crop grown on an acid soil in Sierra Leone, but it had no effect on soya bean growth or yield even though it increased the root nodules per plant (Rhodes and Nangju 1979). A basal fertilizer dressing was applied. Incorporation of trace elements in the seed dressings could be a simple and cheap method of overcoming minor element deficiencies in annual crops. Chelating agents are used, particularly where iron deficiency occurs on alkaline soils planted to tree crops or other high value crops.

Results of the FAO fertilizer programme in West Africa (Richardson, H.L. 1965)

This programme has already been referred to, and it is worth examining the results for one area in greater detail to show the effect of the three major nutrients on growers' land not research stations.

From 1961 to 1965, 19,569 demonstrations and 5,160 trials were carried out in Nigeria, Togo, Ghana, Senegal, Gambia and the Ivory Coast. Using only 22 - 44 kg/ha of N, P_2O_5 and K_2O the overall average response was about 50 per cent of the unfertilized yield.

The results for 3 years in three countries are summarized in Table 2.12 where the response was not so great. Three-fourths of the results in West Africa showed a cash profit on fertilizer use from at least one treatment, and fertilizer used efficiently gave an average profit of 270 per cent on local 1960's prices. Cereals and cassava were the least profitable, due to the low crop prices. It is understandable why the average rate of increase in fertilizer consumption in the countries that had taken part in the scheme was double that of other developing countries without a fertilizer scheme.

Table 2.12 FAO Fertilizer programme
Response of some crops to NPK in forest and savanna zones of West Africa

Crop	**Zone**	**No. of trials**	**Control** (nil) **yield** (kg/ha)	**Main effects** (kg/ha)		
				Nitrogen	**Phosphate**	**Potash**
Maize	Savanna	159	1,190	+ 181	+ 205	+132
	Forest	343	1,223	+ 176	+ 154	+156
	All	502	1,212	+ 178	+ 170	+148
Sorghum	Savanna (only)	21	402	+ 93	+ 26	+ 28
Millet	Savanna (only)	91	570	+ 132	+ 128	+ 65
Rice	Savanna	199	1,265	+ 215	+ 208	+174
	Forest	153	1,311	+ 219	+ 211	+ 80
	All	352	1,285	+ 217	+ 209	+133
Yams	Savanna	160	8,405	+1,041	+1,061	+767
	Forest	60	9,030	+1,018	+1,051	+616
	All	220	8,575	+1,035	+1,058	+726
Onions	Forest (only)	5	801	+ 125	+ 27	+ 96
Forage	Savanna (only)	18	1,774 (Dry wt.)	+1,188	+ 526	+124

Note: The averages are weighed by the number of trials in each group.

Source of nutrients

All the major nutrients are available in a number of forms and concentrations as shown in Table 2.13. Many of the fertilizers contain certain of the minor elements, such as basic slag which has a wide range of elements: superphosphate which incorporates calcium sulphate; and kainite which contains both magnesium chloride and sulphate, and a lot of common salt besides potash.

Certain materials have associated disadvantages. Sulphate of ammonia increases the acidity of the soil. Calcium ammonium nitrate is hygroscopic and will scorch damp foliage. Ammonium nitrate under certain conditions can be explosive, while nitrate of lime is very deliquescent and hence difficult to handle and store in the wet tropics. Calcium cyanamide is poisonous and acts by liberating ammonia in wet soils; this ammonia can damage a newly germinated crop. Anhydrous ammonia requires special injection equipment for its application. Rock phosphate and certain other phosphatic fertilizers are of low solubility and hence not as effective as superphosphate. The phosphate deposit in Uganda is of little value as a fertilizer in its natural form.

Many fertilizer companies manufacture special blends suited to specific crops, which are frequently granulated for speedier and easier

Table 2.13 Sources of major fertilizer nutrients

	Analysis (%)			
	N	P_2O_5	K_2O	**S**
Nitrogen				
Sulphate of ammonia	21	—	—	24
Aqueous ammonia	25	—	—	—
Anhydrous ammonia	82	—	—	—
Ammonium nitrate	34	—	—	—
Nitrate of soda	16	—	—	—
Urea	46	—	—	—
Calcium cyanamide	20.6	—	—	—
Calcium ammonium nitrate ('Nitro-chalk')	23-26	—	—	—
Ammonium sulphate nitrate	26	—	—	15
Potassium				
Muriate of potash (chloride)	—	—	50	—
Sulphate of potash	—	—	48-52	17-18
Sulphate of potash magnesia	—	—	28	11
Kainit	—	—	14	1-12
Phosphate				
Rock phosphate	—	27-40	—	—
Single superphosphate	—	18-22 (soluble)	—	12
Double superphosphate		32-40		3.5
Triple superphosphate	—	48 (soluble)	—	1.5
Ground rock phosphate	—	27	—	—
Basic slag	—	9-18	—	0.5
—	—	2-16 (soluble)	—	—
Rhenian phosphate	—	25-30	—	0.5
Mixed fertilizers				
Potassium nitrate	13	—	44	—
Mono-ammonium phosphate	12	52	—	—
Di-ammonium phosphate	18	48	—	—
Nitro phosphate	11-20	11-20	—	—
Ammonium superphosphate	2-5	14-20	—	11-13

Note: Compound fertilizers with N, P and K in varying ratios are becoming more commonly used in the tropics.

handling and better distribution. For example a 10 : 10 : 15 (NPK) mixture is suggested for young cocoa in Ghana, with a higher potassium ratio for more mature cocoa. Minor elements can also be incorporated for ease of application. Developments in fertilizer production are reflected by an increased concentration of nutrients in these granulated mixtures. More concentrated fertilizers are important in tropical agriculture where haulage distances may be long and freight charges high, and in addition, there is a saving in bagging, handling and storage charges. Triple superphosphate has however been shown to be inferior to superphosphate on a basis of equal phosphate content, in parts of West Africa where the calcium sulphate in single superphosphate is also needed.

Plants vary in their ability to utilize various forms of nitrogen though often this is more associated with the soil and cultivation methods than with a physiological preference. For rice, the relative efficiencies of various forms of nitrogen were rated as shown in Table 2.14 (de Geus 1954).

Table 2.14 Efficiency of various forms of nitrogen as a fertilizer for rice

Fertilizer	**Rating**
Sulphate of ammonia	100
Ammonium nitrate	92
Ammonium phosphate	86
Urea	82
Calcium cyanamide	64
Potassium nitrate	44
Sodium nitrate	40

Apparent differences in the ability of certain crops to utilize various forms of nitrogen may result from other factors. Crowther (1941), comparing the effect of ammonium sulphate and calcium nitrate on cotton growth in the heavy alkaline Gezira soil, found the nitrate more efficient and showed a loss of about 22 kg/ha of the nitrogen applied as sulphate of ammonia at rates up to 130 kg/ha of nitrogen. This loss and difference in efficiency did not appear to occur if the fertilizer was placed 15 cm deep. There is considerable loss of ammonia when ammonium sulphate is broadcast over this soil (Jewitt 1945). Urea is now used on the cotton in the Gezira where it can be moved into the soil by the irrigation water before it is hydrolysed and lost by volatilization of the ammonia.

Urea is the most concentrated of nitrogen fertilizers with 45 per cent N, and has a considerable advantage over other fertilizers when transport and storage per tonne are expensive. Urea manufactured in modern plants should not have a toxic amount of biuret as an impurity. Parish and Feillafé (1960) after comparing urea with sulphate of ammonia concluded that 'in view of the intense biological activity occurring in tropical soils, it would seem reasonable to think of urea merely as a convenient way of storing and applying the unstable ammonium carbonate'.

The free ammonia released from urea by hydrolysis can damage germinating seedlings. Nitrogen from urea may be lost by the ammonia volatilizing or nitrite accumulating. Urea should be cultivated into the top 5 cm of topsoil, or washed in by irrigation, or applied with an acid fertilizer such as superphosphate, to prevent the ammonia being volatilized. Urea does not acidify the soil as much as ammonium sulphate even when nitrite accumulates, as it does not leave a residual sulphate anion. Neither will it have the beneficial effect of sulphate of ammonia on sulphur-deficient soils.

Little ammonia is lost from light dressings of urea but with urea providing 112 kg N/ha the loss can be high on soils with a low base exchange capacity (Gasser, 1964).

Field experiments in India showed NH_3 losses under no rainfall conditions increased from 0 to 66.4 per cent when the rate of urea application was increased from 2.8 to 1,107 kg of N per hectare. When the urea was applied at a depth of 1.25 cm before irrigation the loss was 30 times greater than when applied after irrigation. Under both conditions there was no loss of NH_3 when the urea was applied at a depth of 7.5 cm (Shankaracharya & Mehta 1971).

In experiments in Uganda (Griffith & Mills 1952) no differences were found between sodium nitrate and ammonium sulphate when fertilizing finger millet with 112 kg of nitrogen per hectare, but at Yambio in the Sudan, Willimott and Anthony (1958) obtained the responses shown in Table 2.15.

Table 2.15 Response of finger millet to two forms of nitrogen (Sudan)

Treatments	**Yield**	
	kg/ha of grain	**% control**
Control	1,212	100
Sodium nitrate at 47 kg/ha N	1,362	112
Sodium nitrate at 94 kg/ha N	1,342	111
Ammonium sulphate at 47 kg/ha N	1,478	122
Ammonium sulphate at 94 kg/ha N	1,907	157
Significant difference $P = 0.05$	441	

Unfortunately the role of sulphur was not investigated and this, rather than the nitrogen form, may account for the differences. Obviously some factor unexplained was limiting the response to sodium nitrate at a level of 47 kg nitrogen per hectare or less. On tea at Kericho (Kenya), sulphate of ammonia was more effective than urea at a rate of 45 kg/ha nitrogen. Urea only gave a reasonable response in the third year after pruning. This amounted to about 60 per cent of that given by ammonium sulphate. To eliminate sulphur as a possible cause of this difference, it was added to the urea (Goodchild 1959).

Organic manure

While the emphasis is placed on fertilizers for increasing production, wherever possible organic manures such as cattle and pig manure, chicken manure from intensive poultry farms, coffee husks, and other organic waste materials should be applied to the soil, rather than used as fuel or wasted as is so common in India. The fertilizer values of some organic wastes are shown in Table 2.16. Farmyard manure (FYM) provides trace elements, or alternatively holds elements such as phosphorus and sulphur in an organic form for slow liberation, and prevents the phosphate being made unavailable. Besides the main nutrients very striking crop responses and residual effects have been obtained with cattle manure in the tropics. Peat and Brown (1962) observed the effect of a 7 application of FYM, 13 years after application. Various trials in Africa suggest that 1 t of FYM per year applied yearly, alternate years, or even 5 t every 5 years will maintain soil fertility under continuous cultivation once the fertility of the soil has been built up. The effect of the organic matter in increasing the rainfall receptivity of the soil may be an important factor in maintaining soil fertility. Higher rates will be needed on poorer soils and are desirable for crops such as cereals and

Table 2.16 Analysis of some widely varying organic fertilizers

Material		**Analysis** (%)			**Nutrient** (kg) **in a 1 t application**		
		N	**P_2O_5**	**K_2O**	**N**	**P_2O_5**	**K_2O**
Pen or kraal manure							
Trinidad	75% moisture	0.37	0.57	0.22	8.3	12.8	4.9
Nigeria	63% moisture	1.00	0.35	1.58	24	8	35.4
Mauritius	59% moisture	0.7	0.24	0.22	15.7	5.4	4.9
India	31% moisture	1.0	0.6	1.8	22.4	13.4	40.3
Guano (Peruvian)		13	12	2.5	291	269	56
Sugar cane residues							
Molasses		0.8	0.3	3	17.9	6.7	67
Bagasse		0.2	0.1	0.2	4.5	2.2	4.5
Filter Press Mud		2	4.5	0.3	45	101	6.7
Sugar Cane trash		0.9	—	—	20	—	—
Crop by-products							
Cocoa husk	84% moisture	0.2	0.1	0.4	4.5	2.2	9
Coconut meal	10% moisture	3.0	1.5	1.0	67	34	22
Cottonseed meal	10% moisture	4.5	1.2	1.2	101	27	27
Rice straw	71% moisture	0.2	0.1	0.4	4.5	2.2	9
Groundnut cake	10% moisture	8.0	1.4	1.2	179	31	27
Soya bean meal	10% moisture	7.0	1.5	2.5	157	34	56

root crops (yams, sweet potatoes), bananas and coffee that benefit particularly from FYM rather than cotton, groundnuts and sim-sim.

Fertilizer sprays

The low percentages of the nutrients applied to the soil which are recovered in the crop suggests that this method of application is inefficient. Striking results are obtained with the minor elements when sprayed on to plant leaves to correct deficiencies and the amount needed as a spray is only a fraction of the dosage required for soil application. Phosphate, which is rapidly fixed by many tropical soils, is an element that could possibly be applied more efficiently by a foliar spray. Elements required in larger amounts at various times in the crop development are difficult to supply in a single spray, but in certain crops, such as tea, bananas and coffee, regular fungicide sprays are applied to which nutrients can be added. Urea is a useful form of nitrogen to apply as a foliar spray as it is very soluble in water, non-corrosive, non-ionic and therefore less liable to cause leaf scorch, compatible with many standard insecticides and fungicides and their added wetting agents, and is easily absorbed by leaves. In Trinidad foliar applications of urea with copper fungicide, and when necessary an insecticide, are applied with mist-blowers to cocoa. The leaf size is considerably increased and the yield response very profitable.

Fertilizer sprays incorporating the major elements were tried on tea and while some planters were convinced of their value, others obtained no visible response. Urea is largely absorbed through the under surface of the tea leaf.

IRRIGATION*

As long ago as 4,000 B.C. efficient systems of irrigation existed on the Tigris and Euphrates, and today 160 million ha of the world's agricultural land is irrigated, including all the cultivated land of Egypt, and half the agricultural land in Pakistan and China. Irrigated paddies along the main rivers and deltas of Asia have maintained the highest population densities in the world, and along many of these rivers new dams are under construction to expand the area irrigated (Plates 2.16 and 2.17).

* For details, particularly engineering aspects, see Withers and Vipond (1974).

Plate 2.16 Irrigation in Thailand. (Courtesy Aerofilms Ltd)

Farmers of Africa south of the Sahara, however, have not this tradition of irrigation. The system found on Kilimanjaro is a remarkable exception to this generalization. Until recently there was no population pressure on the land in Africa to make irrigation essential, and there is the risk of bilharzia which is carried by the snails infesting damp and wet places.

In recent times, faced with a growing population wanting a higher standard of living, and more rice, many of the African governments including Kenya, Nigeria, Senegal and Ghana have developed irrigation projects, and more are planned. Where rainfall is short the skies are sunny and crops can use this sunshine if water is provided. Many facets of irrigation in Africa at the farmer level are covered in 'Africa and Irrigation' a symposium sponsored by Wright Rain in 1961.

Irrigation is the controlled application of water to crops, to supplement the available soil moisture where this is limiting yield. It is not a means of making the deserts fertile. Water should be applied according to the needs of the particular crop at that particular growth phase. It must be applied in such a way that it will neither scour nor silt up the ir-

Plate 2.17 The 3,000 year old Ifugao rice terraces may be rightfully counted among the wonders of the ancient world. Carved out of mountain sides and employing an ingenious irrigation system, this particular group of terraces covers an area of 400 km^2 (154 square miles) If placed end to end, the terrace would cover an estimated distance of 22,500 kilometres (14,000 miles), or more than half way round the world. This group is the oldest, highest and most extensive of the numerous rice terraces in the Mountain Province, Luzon, Philippines. (Courtesy Philippines Tourist and Travel Association, Inc.)

rigation channels, will not cause waterlogging and will not build up the salinity of the soil. 'When a farmer has to work hard to raise irrigation water he uses it economically, but when it is supplied by flow irrigation he uses it excessively and salinity results'. (Sir Joseph Hutchinson) A simplified plan of an irrigated area is shown in Fig. 2.3.

Infiltration of irrigation water

The measurement of soil infiltration, usually done with a ring

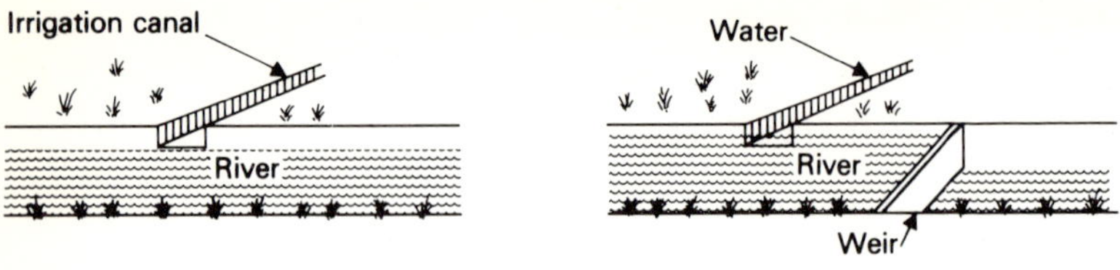

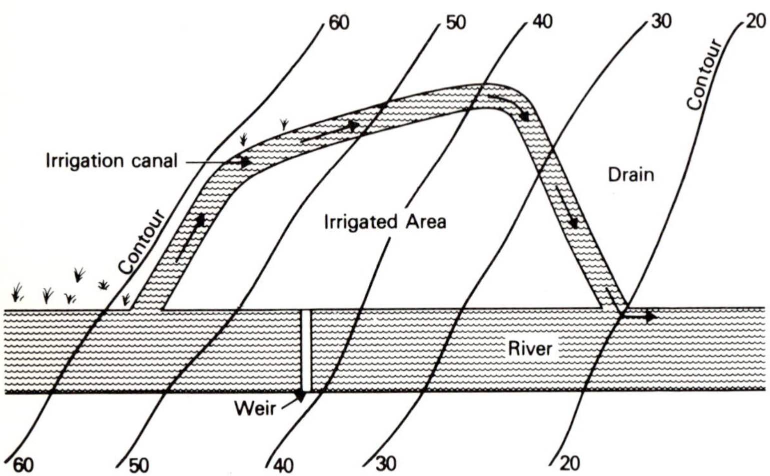

Fig. 2.3 Plan of irrigated area.

infiltrometer, is an essential part of planning an irrigation system and in determining which method of irrigation to use. The figures given in Table 2.17 give a general indication.

Methods of irrigation

The earliest form of irrigation was the impounding of natural flood water on the flat land bordering rivers, as in Egypt, in the same way that rainfall can be impounded by bunding. Flood water impounding is a relatively simple method where the land is reasonably flat. A development of this system is to supplement the impounded flood and rain with water drawn from the river which overflowed. This system is often used in tree crops where a basin is constructed around each tree and flooded at intervals. The bunded areas may be fed in turn from laterals of the main irrigation canal or from one enclosed area to the next. On heavy

Table 2.17 Suitability of soils for irrigation

Infiltration rate (cm/hour)	**FAO infiltration category**	**Suitability for irrigation**
< 0.1	Very slow	Only suitable for rice.
0.1-0.5	Slow	Hardly suitable. Deep percolation difficult. Roots waterlogged.
0.5-2.0	Moderately slow	Most suitable.
2.0-6.0	Moderate	Most suitable.
6.0-12.5	Moderately rapid	Marginal. Small basin irrigation.
12.5-25.0	Rapid	Excessive leaching. Too high for surface irrigation.
> 25	Very rapid	Overhead irrigation only.

land where complete flooding may cause damage to the soil tilth, the field may be left in small ridges or corrugations and flooded in such a way that all the soil is not covered. This system is difficult to manage efficiently and there is a constant risk of soil erosion. Where complete flooding is not desirable or possible, the land may be ridged and the crops planted in the furrows or on the ridges, which may be up to 1.8 m high and subsequently the furrows are flooded. The furrows must not be too big or deep percolation will cause unnecessary water loss. This is a useful method for row crops but care must be taken to avoid erosion down the furrows, or the water collecting at a low spot and breaking through the ridges to form a gulley. Where a single row of a crop is grown on the ridge, alternate furrows may be irrigated, the other furrows being used at alternate irrigations (Fig. 2.4).

Plastic siphon tubes are a cheap and simple way of taking water from an irrigation canal to an area at a lower level without breaking the canal bank. The canal lining can then be more permanent and less permeable to reduce water loss.

Sprinkler irrigation*

For the production of more valuable crops such as fruit and vegetables, water can be supplied by overhead irrigation through a rotating sprinkler head, or through a series of holes in a pipe (Plate 2.18). This system is of no practical value for paddy.

* See Pillsbury & Degan, 1968

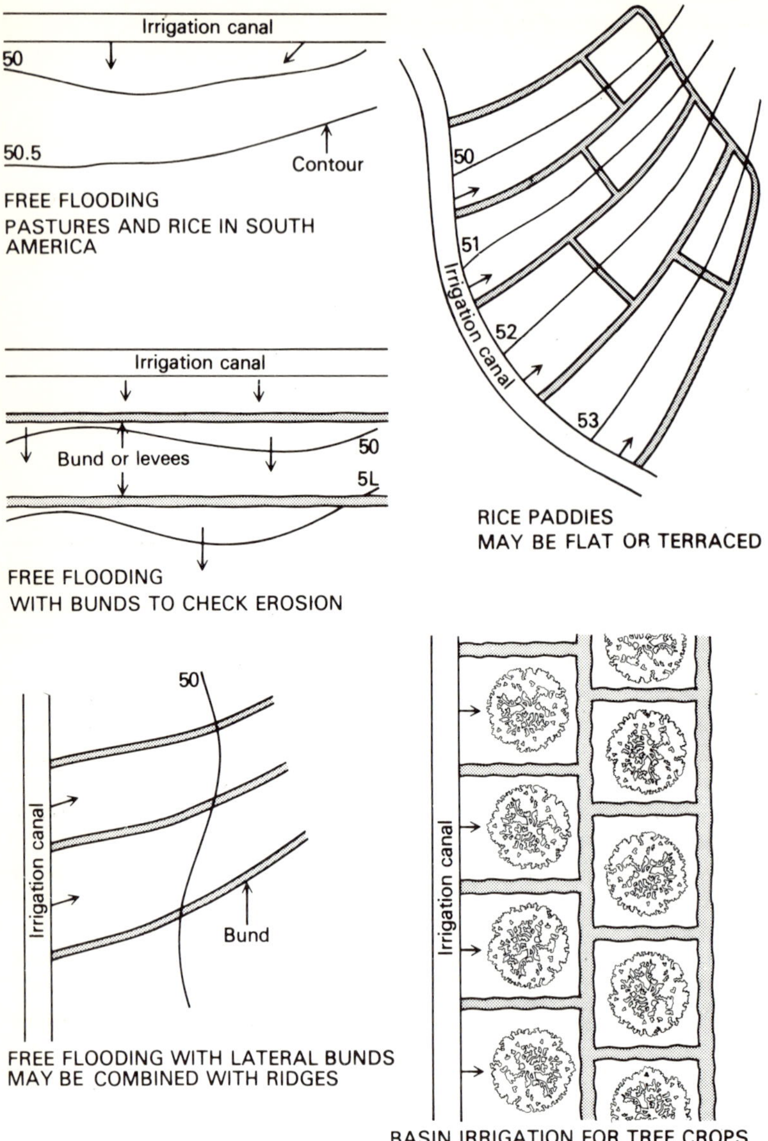

Fig. 2.4 Some methods of irrigation.

With a sprinkler system a specified amount of water is applied to the crop at predetermined intervals through slowly rotating low precipitation sprinklers spaced along a lightweight aluminium pipeline made in

Plate 2.18 Sprinkler irrigation of tobacco in North Queensland.

easily carried lengths which can be quickly coupled and uncoupled. This sprinkler line is moved at intervals along the main water supply line into which water is pumped under pressure. The areas wetted by individual sprinklers are planned to overlap so that the overall water distribution is even. (See Fig. 2.5) The 'rain' produced by the sprinklers can be delivered at a rate which will not cap the soil, cause waterlogging or leaching, or damage the crop. It can be used on sloping or rolling land. To achieve this the distance between the sprinklers, the rate at which the water is pumped and the type of sprinkler have to be chosen to suit the crop, the soil type, the prevailing winds and their strength, and the terrain.

The capital cost of a sprinkler system is quite high for the buried mains, booster pumps and spray equipment, apart from the cost of making the dam and supplying to the main. The pipes and nozzles can be moved around the farm. Such equipment has a life of at least 20 years. If used efficiently it may not be much more expensive over the life of the system than surface irrigation which is less flexible and uses part of the land for canals and may cause soil wash.

It is advisable to have a scheme planned by specialists, as a well planned scheme operates efficiently without any skilled labour, but a badly

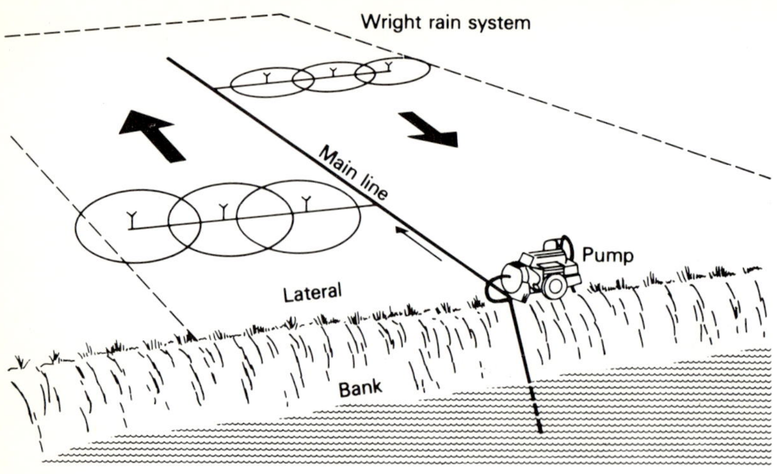

Fig. 2.5 Wright rain system for sprinkler irrigation (Wright Rain Ltd.)

planned scheme will produce areas of waterlogging and dry areas resulting in uneven crops (Plate 2.19).

Sprinkler irrigation is being used on a variety of tropical crops including Pangola grass in Barbados, tea in Central Africa and India, coffee in India, sugar in Central Africa and Mauritius, tobacco and oranges in Australia. In Central America a high pressure system with nozzles to throw water 60 m has been used on bananas.

A sprinkler system makes more efficient use of valuable water supplies, and may allow two or three times more land to be efficiently irrigated than if surface irrigation were used. Foliar nutrients such as urea can be metered into the sprinkler system to allow more critical crop feeding, and the rapid correction of any deficiency symptoms which might occur.

In areas subject to high wind, results will not be very satisfactory due to the uneven distribution pattern the winds create. In very dry areas the loss of water by evaporation of water from the sprinklers is high.

Sprinkler irrigation creates a humid atmosphere which can encourage the spread of diseases, particularly leaf diseases. This can be serious when a low application rate is applied over a long period. Provided that soil conditions can accept a higher rate of application, when there is a disease risk it is better to irrigate at a heavier rate for a shorter time.

Tube wells

In India an increasing area is irrigated with water from tube wells. These are an investment made on the larger holdings. Deepak Lal (1972) studied the relationship of water availability and agricultural in-

Plate 2.19 Sugar cane unevenly irrigated. Pale circles result from over-watering around sprinkler. (Courtesy Wright Rain Ltd.)

comes in the famine belt of the Deccan which lies in the rain shadow of the Western Ghats. The availability of water reduced the variability of crop yields and made it practical to use fertilizers and plant improved crop varieties. The cropping pattern changed from sorghum and millet which are drought resistant but of low cash value, to sugar cane which is profitable if adequate water is available. Ultimately the new wheats may be grown. Socially it would be preferable to use the water to irrigate a larger area of sorghum than the smaller area of the less vital crop, sugar.

With tube wells the profit per hectare rose from Rs 7.9 to Rs 100.6. The cost of making a well varied from around Rs 1,000 for a 2.2 m well to Rs 5,000 for a well 6.7 m deep. These costs only made the use of wells profitable if a surplus of labour was available and the land not planted to sugar cane could be double cropped. The chances of siting a well successfully are only 30 per cent, but this chance is increased if a hydro-geological survey is carried out first and used to choose the sites.

Underground irrigation

Of the water applied by surface irrigation 20–30 per cent is lost by evaporation. This loss is eliminated with the underground system where the water is placed directly in the root zone, using perforated plastic tubes through which water is pumped under pressure. Care must be taken in the choice of plastic where termites occur. The water spreads from the pipes in all directions. With a special plough, the pipe can be laid about 50 cm deep at 1.5 m spacing in crops. The cost is well over £300 per hectare. Wider spacing could be used in some tree crops and pasture. Such a system could be useful in the arid Middle East where irrigation water is valuable and evaporation losses in summer very high. It may also have a place where it is necessary to control the water table, using the pipes as drains in the wet season, and for irrigation in the dry season. It can only have a place where crops and water have a high value.

Trickle irrigation

Trickle or drip irrigation is a system developed in the South of England for glasshouse crops, which is applicable to high value crops where water is scarce. The system moves the water in plastic piping through a main to laterals, sub-laterals, and eventually through a very fine tube to each individual plant where it releases a continuous drip when irrigation is needed. The system must be carefully designed and the piping is left permanently in place. It can be used on steep terrain.

As the final outlets are very small they can be easily blocked, so all the water and fertilizer solutions must be filtered before they are fed into the system. There are a number of ways of unblocking the outlets; in America they use a double walled tubing which can be back-flushed. If the plastic piping is above ground, algae may grow and block the flow of water.

Trickle irrigation is designed to place a precise quantity of water where and when it is most useful as dictated by the needs of the crop. Besides fertilizers, certain crop protection chemicals such as nematocides can be fed into the irrigation water. Ammonia applied in this system is not nitrified in the saturated soil below the drip, but is released slowly as it percolates away. Trickle irrigation has been used successfully with avocados, bananas, citrus, cocoa, coffee, mangoes, pawpaws, sugar cane and nut trees, as well as fruit, flowers and vegetable crops.

There is a considerable saving in water use with this system which is less likely to encourage leaf diseases than sprinkler irrigation as the humidity around the foliage is not increased, and it is less likely to damage the roots by waterlogging.

The capital cost of this system is very high and its use is largely confined to the USA and particularly dry countries like Israel and the Middle East.

The water requirement of a crop

Profitable irrigation of a crop requires full knowledge of the moisture requirements of the crop from planting to harvest. This means not only the total amount required during the year, but the critical periods during the growth cycle when more or less water is needed.

Seedlings with their poorly developed root systems have a low water requirement. As the crop develops this rises to a peak at flowering and fruit setting. The land is usually dried out as crops such as rice and sugar cane ripen. Moderate water stress at certain stages of growth may increase the yield or quality of some crops. With the increased evaporation and transpiration in hot dry weather, the calculated crop requirements often need to be increased 50 per cent under arid conditions. In an irrigation scheme, the water available is sufficient to supply the needs of a given crop area and is referred to as the 'duty of water'. This relationship is expressed differently in different countries. In Egypt it is the number of cubic metres of water flowing continuously during the life of the crop which is required per hectare. In India it is the area of a crop which the continuous supply of 1 cusec. (1 cubic foot of water per second or 0.0283 cubic metres (cumecs)/second, will bring to maturity. For example, 1 cusec. will irrigate 16 ha of rice in Egypt, 16-18 ha in Bombay, and 24-28 ha in Behar.

The American system refers to the application of a certain depth of water at stated regular intervals, e.g. 2 in every 12 days. To flood an acre to the depth of 1 ft requires 272,000 gallons of water. To cover a hectare with 1 mm of water requires 10,000 litres.

In estimating the crop area to be irrigated by a given volume of water, allowance must be made for the loss from percolation, evaporation and run-off, between the water source and the crop. This loss may account for a quarter to half the water take-off, and in certain cases it is taken into account is assessing the duty which then applies to the head of the main irrigation canal. If the irrigation water has a high salt content 'flushing' may be necessary and the water thus used to remove the salts will not be available for crop growth. Flushing should wherever possible be carried out at the time of the year when water is freely available and crop needs at a minimum. Care must be taken not to apply excess water, which does not drain away until the crop roots have been damaged by lack of oxygen.

Slope of the channels

The flow of water in an irrigation channel is roughly according to the formula

$$V \quad = 0.92 \quad (2RS)$$

where V = Velocity in ft per second. The mean velocity is about 80 per cent of the surface velocity.

R = Hydraulic mean depth in ft (i.e. when studying the water-flow in cross-section it is the area divided by the circumference in contact with the channel, expressed in ft).

S = The fall in the water channel in ft per mile. This must be steep enough to carry silt and not too steep to scour the channel, thus the velocity must be greater than 3 ft per second if silt is a problem, which will scour a clay or sand channel.

Factors related to an irrigation project

When a large-scale irrigation or land development project is planned the services of experienced consultants and engineers are essential. Today large new schemes are often of a multipurpose nature; generating hydro-electric power, formation of navigable waterways, eradication of public health pests, and the resettlement of the occupants of areas to be flooded may also be necessary. This should not deter an enthusiastic Agricultural Officer from investigating the possibilities of irrigation by setting up a small pilot irrigation area, or where irrigation is a traditional practice, considering the extension of irrigated areas by better water control or a more economical use of the water which is available.

The project could be approached as follows:

1. Survey the area. If it is large, aerial surveys may be the most economical method. The resultant maps or plans showing the area, the vegetation, soil type and contours are vital to planning the layout.

2. Study the soil. Heavy clays are usually only suitable for irrigated rice crops due to the problem of aeration. Sandy soil and shallow soils respond to irrigation but are wasteful of water as frequent application is necessary. Deep alluvial soils with good drainage respond best to irrigation for crops other than rice.

3. Study the water available. How much is available and at what time of the year. Measurement over 3 years is generally considered a minimum and at least 30 years before the full limits of variation are covered, but it may not be planned to utilize the maximum available water at one site. The measurement of river flow is done either by special instruments or by constructing a weir.

4. Assess the quality of the water (Anon 1974b). This is determined by six major factors:

(a) The total concentration of soluble salts.

(b) The sodium ion concentration. Under 600 ppm usually safe, over 1,000 ppm dangerous.

(c) The ratio of Na to Ca and Mg ions. The higher the ratio of sodium to calcium, the less desirable is the water for irrigation.

(d) The concentration of bicarbonate ions.

(e) The absence of toxic elements, and minor elements at a toxic level. Boron, selenium and lithium should be determined.

(f) The crop to be grown. There are wide differences in the tolerance of crops to salinity.

Sodium ions become important when they exceed 10 per cent of the exchangeable ions. With increasing amounts of sodium ions the soils become less permeable and by the time they reach 50 per cent the plant has difficulty in taking up calcium. Magnesium ions act in a similar way but to a lesser extent. The effects are a result of base exchange which does not occur until the sodium ion concentration is at a high level, thus if the total salt content is low the ratio of sodium to calcium ions can be higher (Greene 1948). The increase of salt content of the soil increases the osmotic pressure of the soil solution and growth retardation varies directly with the increase in osmotic pressure. Certain crops are more tolerant of salt than others but eventually the salt accumulation may prevent the growth of any crop.

An analysis of Kano river water (Northern Nigeria) showed the following:

Na	3.9 ppm
K	2.7 ppm
Ca + Mg	7.2 ppm
Bicarbonate	36.6 ppm
Chloride	2.1 ppm

The calcium and magnesium is adequate for crop needs, and the potassium will meet a lot of the crops' needs (Stockinger 1971).

Saline water should not be used for sprinkler irrigation as leaf absorption of salts is more rapid than root absorption, and saline water may scorch the leaves.

5. *Measure the silt.* The presence of large quantities of silt in irrigation water can lead to a breakdown of the system. It has been suggested that the highly developed civilization of Babylon disintegrated due to the increasing problem of removing silt from the irrigation canals. Dams which silt up are of decreasing value for water storage. Erosion control on the watershed is an integral part of a well planned scheme but it may often be under different authority. The catchment areas around the dam should be planted with a grass such as *Paspalum notatum, Cynodon dactylon* or *Pennisetum clandestinum* or possibly a legume such as *Stylosanthes guianensis (= S. gracilis.)*

6. *Site the dams, weirs or pumps*. Care must be taken to ensure that the river flow is not changed and the banks damaged. Small dams are often constructed by non-engineers in a manner similar to Fig. 2.6. For

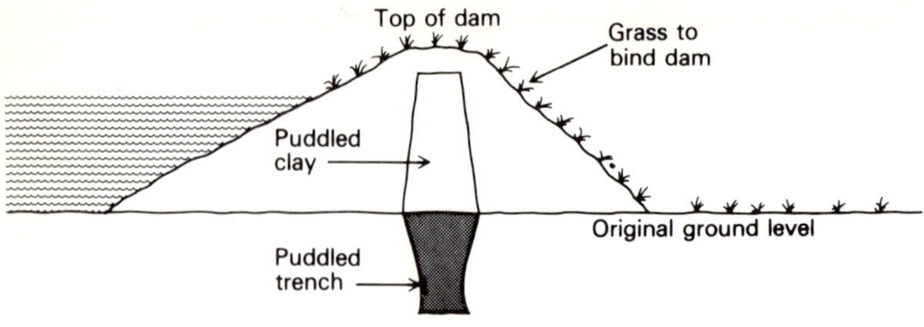

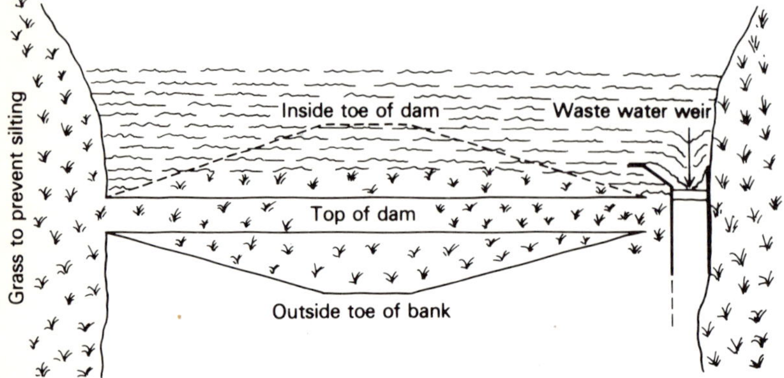

Fig. 2.6 Earthen dam.

details of construction of small earth dams see Cormack (1948). The dam wall can be protected from wave damage by planting a reed such as *Phragmites communis* along the water line.

7. Calculate the crop area. The crops to be grown are generally known. From the rainfall figures and the 'duty' the area of crops that can be irrigated is calculated, or alternatively the amount of water required to irrigate the planted area.

8. Plan of layout. The plan of the layout can be drawn, showing the area irrigated, the aquaducts and the drains. If the salt content of the drainage water is satisfactory, the drainage water may be used for further irrigation.

9. Economics. The economics of the scheme should now be studied. What is the capital cost, the cost of supplying the water and maintaining the scheme (running costs)? Is there a market for the crops to be grown? What is the estimated net return from sale of the produce?

In a new area a pilot scheme is essential before a large irrigation (or any other) project is initiated, as the soil may not be suitable for irrigation over a number of years.

Strict soil conservation at the head waters of the dam is very important if the dam is to have a long life. The forest must be protected and the land not overgrazed otherwise the dam will silt up.

Problems of an established irrigation scheme

Where alkaline or saline water is used for irrigation the soil may deteriorate by:

1. The accumulation of sodium salts on the soil surface.
2. The accumulation of alkali salts in the soil at root depth.
3. The formation on clay soils of a tilth which is plastic and sticky when wet and dries hard without ever having a crumbly structure.

Salt accumulation

Salt accumulation is the major threat to all irrigation projects, particularly in arid areas. Irrigation water adds salts continuously to the land and unless they are removed at an equal rate, either by the crop or in the drainage water, they will soon accumulate to a level toxic to the crop. There are many sources of irrigation water with a high salt content, particularly when water is pumped from wells in arid areas. If the irrigation water has a high salt content and the soil surface is not very flat, there is a tendency for the salt solution to accumulate in the hollows. Due to the lower vapour pressure of this salt solution these hollows dry out more slowly than the surrounding soil, and water is drawn to the area from the surrounding soil. Once a salt accumulation has started to occur there is a tendency for it to build up rapidly rendering the soil sterile. To avoid salt accumulations:

1. The soil surface should be flat if flood irrigation is carried out.
2. Water with a high salt content should be avoided if possible.
3. The irrigation water applied should at certain applications exceed the field capacity of the soil to such an extent that the surplus water will carry the unwanted salt out of the root area and into the drainage system. This may cause crop damage from waterlogging and lower the 'duty' of the water.
4. If salt accumulation is a risk the water table should be kept low, about 2 m below the soil surface, to prevent salt moving upwards.

Soil salinity is usually assessed by measuring the electrical conductivity of a saturated soil extract (EC_e) in milliSiemens per centimetre (mS/cm) (previously mmho/cm) at 25 °C. One mS/cm is roughly equivalent to 640 mg per litre or ppm salt in the extract. Bernstein (1966) gave the following approximate figures for the tolerance of certain tropical crops to salinity in EC_e, mS/cm at 25 °C.

Crop	Yield reduction		
	10%	25%	50%
Cotton	10 mS/cm	12	16
Sorghum	6	9	12
Soya bean	6	7	9
Paddy	5	6	8
Maize	5	6	7
Sweet potato	2.5	4	6
Cynodon dactylon (Bermuda grass)	13	16	18

Rice is especially sensitive during the seedling stage before flowering and during flowering. During these development phases the EC_e should not exceed 4. Citrus trees have a low salt tolerance.

Salinity may reduce the quality of the crop as well as reducing the yield as growth is generally stunted.

Good drainage and the use of sufficient water to leach out excess salts are both essential to maintain a productive irrigation system. The water channels should be kept free from weeds as these can cause waterlogging and salinity (p. 364).

Irrigation of specific crops

Traditionally, irrigation in the tropics was largely confined to rice cultivation. It spread to those crops such as cotton, sugar cane, vegetables and millets grown in rotation with rice, and it has also been extended to most perennial crops. For example, irrigated cocoa trees grow more quickly and yield more in their early years, particularly with drip irrigation (Jadin & Jacquemart 1978). In some countries irrigation might reduce the loss of young cherelles. To recover the high capital cost of introducing irrigation, and to pay for the water used, it is necessary to grow high value and high yielding crop varieties which benefit from irrigation, feed them with fertilizers and protect them against pests, diseases and weed competition. Mechanization, where practicable, becomes part of this higher standard of farming.

Bananas

As bananas need a minimum of 100 mm of rain each month throughout the year there is obvious scope for using irrigation to produce fruit the year round, and particularly at the time of year when prices are good. A variety of techniques have been used to supply the water, depending on circumstances. Simmonds (1966) summarizing the practices or recommendations for different territories showed a variation from 5 cm a month in French Guinea to 56 cm per month in part of India. In the

Canary Islands where the bananas are all irrigated there has been a tendency to over water, usually by flooding, without replacing the bases lost by leaching. In many countries including Australia and Jamaica water is often used more economically by applying it with a sprinkler system.

Coffee

Irrigation of coffee has been practised in the Yemen for centuries, but it was only with the advent of high post-war prices that serious attention was given by the main coffee-growing countries to irrigating coffee on a field scale. Soil moisture appears to be particularly important to stimulate flowering, to sustain bean growth about 3 months later, and to keep the crop in condition for the following season. Many of the important coffee areas in Brazil and East Africa have a rainfall which is marginal for coffee, or have long dry seasons. It is in these areas that most of the irrigation development has taken place, resulting in yield increases of 30–50 per cent, but in exceptional cases the irrigated crop has been more than twice as heavy as the rain grown crop, and the quality higher by one or two grades. With such yield increases coffee planters can recover their capital expenditure in about 4 years when prices are good.

Water is applied either by sprinklers attached to lightweight portable piping or by surface application to basins around each tree. A minimum of 50 mm per application per month or 75 mm to 100 mm every alternate month is suggested for Kenya (Littlehales 1960).

On estates in South India irrigation has increased the yield of both robusta and arabica coffee by two or three times the yield without irrigation. For optimum blossoming and setting 20 - 30 mm of rain is required and this must be available within a period of a few days. When the 'blossom showers' are insufficient to meet this requirement, as much as 75 per cent of the blossom may fail to set. The yield increase is achieved by supplementing the showers to meet this critical moisture requirement. The same estates use the equipment to carry new coffee plantings through the dry season, and to help the main crop through a severe dry season. Nitrogenous fertilizer is applied through the sprinkler system.

Irrigation does not eliminate biennial bearing, though this is concealed in the early years of irrigation due to a general increase in yield. No evidence has been found to support the idea that a resting period which coincides with the dry season is essential to the coffee tree, and no harm appears to result from maintaining the soil moisture at a level sufficiently high so that growth is continuous. As flowering is associated with the onset of the rains, if necessary flowering might be controlled by irrigation.

The discovery that rainfall is an important factor in the distribution of rust spores (Rayner 1958; Nutman 1959) suggests that sprinkler irrigation without efficient fungicide spraying could cause a rapid build up of rust, particularly if the leaves remained wet overnight.

Cotton

As cotton thrives under hot sunny conditions, water is often the limiting factor in producing economic yields and fibres of good length. Irrigation has introduced a lucrative cash crop into areas previously without a useful source of income, as is seen particularly along the Nile in the Sudan and Egypt. The Sudan Gezira scheme has converted a semi-nomadic people into settled prosperous agriculturists. There were 120,000 ha irrigated in the Sudan in 1926 when the Sennar Dam was inaugurated and the gross revenue of the Sudan was £6 million. By 1952 this had risen to 400,000 irrigated ha. When the Sudan Plantation Syndicate was dissolved in 1950 the revenue from the Gezira plantations from cotton sales was £54 million.

Furrow irrigation is mostly used though both basin and flat irrigation are also common. The water requirement varies according to the variety, the rainfall which occurs and its distribution, the soil characteristics and the humidity. Apart from the removal of excess rainfall, drainage is not necessary in the Gezira but is essential in Egypt due to the higher salt content of the soil and irrigation water. The Gezira soil is too impervious for subsoil drains. The roots of the cotton plants extend to a depth of 2 m or more, unless conditions are unfavourable through a lack of air or moisture or the occurrence of a hard pan or toxic salts. As the cotton crop can take up plenty of water from the top metre of the soil where the maximum lateral root spread occurs, there is no need to irrigate to a greater depth.

Irrigation gives a better germination of cotton seed but also of weed seeds. It was generally believed that the withholding of irrigation water in the early growth stages forced the cotton roots down, but work in California has cast a doubt on this. This idea was popular in the early twentieth century in Egypt (Brown 1953) when the irrigation water was withheld 30-50 days after planting, but this was reduced to 21 days in more recent years to avoid the growth check. For maximum yield, the cotton crop should never be subjected to water stress. In Egypt, cotton is watered every 18 days in the cooler part, increasing in frequency to every 12 to 18 or 20 days depending upon the weather. Irrigation should not be carried on too long otherwise there will be an excess of vegetative growth, nor must it be excessive as cotton does not like 'wet feet'.

Maize

In most of the maize growing areas of the tropics, where the crop is grown on a field scale rather than as odd plants intermingled with other crops, the market price is usually not considered sufficient to warrant irrigation and the use of fertilizers. Farmers in the corn lands of the USA have experience with irrigation, and this could become applicable as the crop value rises.

The plants are usually grown on ridges and the irrigation water runs down the furrows, or it is applied by a portable overhead sprinkler system; 400–600 mm of water are usually applied during the year. A high soil moisture level should be maintained throughout the flowering period (tasselling and silking). Water stress during early growth delays flowering. The water level in the soil should be maintained until the hard dough stage of the grain. Maize draws its water requirements from the surface soil initially, sending its roots deeper as necessary. In a soil offering no resistance to root penetration maize may draw water from the top 2 m. Irrigation lessens the need for such penetration and the bulk of the water is drawn from the top 60 cm if it is readily available. Maize rapidly shows signs of wilting and irrigation water should be applied before this occurs. Care should be taken to keep the top 30 cm of soil well watered during flowering. Irrigation should be combined with a denser plant stand than is optimum for a rain grown crop and a high level of fertilizer application is normally justified. Nitrogen may be applied in the irrigation water, the optimum time of application frequently occurring when the crop is knee-high, but phosphate should be applied early as it gives the greatest response on young plants.

Rice

Rice is the crop one always associates with irrigation. Irrigated paddies have been cropped once or twice and sometimes three times a year for generations to support a large part of the world's population and yields have not declined even though little fertilizer has been applied to the land. Considerable ingenuity has been used both in developing the machinery for lifting water into the paddies, and also in developing the level terraces on the hillsides such as one finds in Luzon (Philippines) (Plate 2.17). These terraces were constructed well before the birth of Christ, and before levels were introduced. They had no metal implements or means of raising water. Similarly in Sri Lanka, many tanks remain, some covering up to 1,800 ha, built about 500 B.C. to supply water for irrigation and domestic use.

Despite this universal picture of flooded paddies, rice is not an aquatic plant but is water tolerant, and uses no more water to produce a kilo of grain than wheat or other cereals. Given adequate oxygen in the irrigation water rice thrives if water is present for most of the growing season. Efferson (1952) suggests that one of the main functions of water in rice paddies is to control weeds which would otherwise choke the crop. Transplanted crops are generally considered to outyield broadcast crops, but Rosher has shown that, at the present level of farming, in

Plate 2.20 Irrigation a rice field in Uttar Pradesh with a water-wheel driven by draft animals. (Courtesy FAO)

Trinidad the yields from broadcast rice are equal to transplanted rice where snail and weed control is efficient during the early growth.

Dry seed will not germinate under 25 mm of soil which is covered with water, therefore, when dry seed is drilled or broadcast and harrowed in, only enough irrigation water should be applied to ensure germination. Ideally about a month after germination 150 - 200 mm of irrigation water should be applied but where weeds occur, of which *Echinochloa* spp. are about the most serious, these may have developed in the warm damp soil to such an extent that the crop is choked. Under these conditions the irrigation water level should be kept just above the *Echinochloa* but not high enough to submerge the rice. This gives excellent weed control, but now this weed can be killed with propanil which makes it unnecessary to use so much water for weed control. The water saved by this change could be used to extend the irrigation area or for double cropping. In most areas, including Sri Lanka and Thailand, where rice is sown broadcast, it is pre-germinated and sown on the puddled surface or even on standing water. Alternatively the rice plants may be raised in nurseries and transplanted into puddled fields. The application of irrigation water after the establishment of the transplanted seedlings varies between areas, largely according to the supply of water available. In general the water level is maintained at about a third of the height of the crop, kept high in the early stages to support the seedlings and lowered at tillering. Seldom does the water depth exceed 300 mm and twice or more during the growing period the land is completely drained. This aerates the soil, aids mineralization and offers the opportunity for fertilizer application. Too high a water level during growth weakens the plant, reduces tillering, retards flowering and causes uneven ripening and lodging. After flowering the paddies are gradually dried out. Rice is highly sensitive to moisture deficiency between flower initiation and heading (Plate 2.20).

Floating rice varieties are grown in many valleys and deltas in Asia and Sierra Leone subject to seasonal flooding where the rice grows up to 6 m, its growth keeping up with the rising flood. A third of the rice area of the Central Plain of Thailand is floating rice (see p. 104).

Water requirement of rice Kung (1966) summarized the extensive work carried out in Asia.

	Average	**Extremes**
Water requirement from transplanting to harvest	800-1,200 mm	520-2,549 mm
Transpiration	200-500 mm	132-1,180 mm
Evaporation	180-380 mm	107-797 mm
Percolation	200-700 mm	32-1,944 mm

The loss by percolation varies widely because of high loss in sandy soils compared with paddies on clay.

Studies in Thailand show the loss by transpiration is as would be expected, small after transplanting, increasing rapidly to tillering, reaching a peak at heading and flowering, finally decreasing gradually as the crop ripens.

Of the 1,240 mm of water, Kung quotes as the requirement for the complete growth cycle of rice grown in parts of South-East Asia, 40 mm are for the nursery, 200 mm for land preparation and 1,000 mm for irrigation of the paddy field.

Where growers have not ready access to irrigation water and rains are uncertain, growers store the rain which falls as deep water in their paddies in case of need. The water height cannot be varied to suit the different growth periods. While work in the Philippines shows this is no disadvantage, in Japan 25 mm depth of water outyields a 125 mm cover. Provided that the soil is completely submerged, trials in Malaysia show that yields are inversely related to water depth (Matsushima 1966).

Where water supply and drainage can be completely controlled, the following are given by Kung (1966) as the desirable water depths for the different growth stages.

1. *During transplanting* 2–3 cm Deeper water encourages deeper transplanting which delays the development of new root systems.
2. *After transplanting* 5–8 cm. To facilitate root growth and re-establish the seedlings. Drain gradually at maximum tiller stage. Weed and apply fertilizer and return the water as the soil surface starts to crack.
3. *Panicle primordia development stage* 5–8 cm. Shortage at this stage and just following will cause sterility and reduce yields.
4. *After full flowering.* Drain gradually.

The irrigation water must not contain more than 600 ppm of dissolved salts, particularly in the early growth stages. If some water of high salt content must be used, it should not be applied to the dry soil. Irrigation water should be between 20 and 30 °C. Cold water from mountains or wells should be held in shallow basins to warm up prior to use, but not so long as to become stagnant either in the basins or paddies.

In West Africa, particularly Sierra Leone, some mangrove swamps were empoldered and used for rice cultivation but sterile areas developed in these polders associated with high concentration of sulphate and iron and a very low pH (Dent 1947) (see p. 161).

In Malaysia where there is a big project to increase rice production and convert Malaysia from a rice importer to a rice exporter, the water requirements of paddy have been comprehensively studied (Matsushima 1966). The critical period for water deficit or excess water is 20 days before and 10 days after heading. The transpiration need is fairly constant at 5 ml per day for each gram of dry matter produced. For maximum yield the soil should be dried at least 1 month before transplanting (see. p. 20). The water level should completely cover the soil but should be no deeper than necessary. The water temperature is rarely too high in Malaysia.

By applying this knowledge to use the water more scientifically provided the water supply can be completely controlled, paddy yields could be maintained using 35–40 per cent of the water used at present, or yields could be increased with a saving of 15–20 per cent of the water used.

Sugar cane

Over 1,250 mm of well-distributed rainfall is considered to be necessary for good yields of sugar cane, though economic crops are grown in areas receiving much less, as in Barbados, where certain estates receive an average of only 1,000 mm. In many areas cane yields can be increased by irrigation (Plate 2.21), particularly where water can be applied in a

Plate 2.21 Irrigation sugar cane. Jamaica. (Courtesy West Indies Sugar Co.)

period of drought. The amount required is a local matter and depends upon the climate, soil and crop condition.

Sugar production along the sub-humid coastal belt on the West coast of Mauritius is dependent upon irrigation, as the rainfall is only around 1,000 mm per year with wide variations due to cyclones. However, with the high solar energy from the cloudless skies, and the higher temperatures, irrigation will give the highest sugar yields, particularly as the soils in this area are generally more fertile than in the wetter areas due to less leaching. The reverse conditions apply in the cold wet cloudy super-humid zone where a quarter of the sugar cane is grown, an area better suited to tea. With furrow irrigation in Mauritius one cusec a flow of 0.28 m^3/s will irrigate about 30 ha. The gravelly soils are too porous for flood irrigation but well suited to overhead spray irrigation.

Provided that plenty of water is available at the required season and the land is reasonably flat, surface irrigation is often the simplest and cheapest method of irrigation, the cane being planted on ridges or in furrows, standard methods of cane planting. As irrigation is less efficient than rainfall, roughly 50 mm of irrigation water is needed to replace 25 mm of rainfall. With surface irrigation in the dry season, up to 100 mm of water is applied every 10–21 days. Before the cane has closed over the soil, the evaporation loss may account for a quarter of the water applied. Cultivation after irrigation is necessary to prevent a deterioration of the structure of the topsoil.

Where it is desirable to make the maximum use of the available water, and the land is undulating and subject to erosion, overhead sprinkler irrigation is to be preferred. Fertilizers may be applied in the irrigation water. There is a high capital outlay for pumping machinery and transportable piping which is easily damaged if mishandled. Sena sugar estate in Mozambique installed a sprinkler irrigation system covering 6,700 ha of cane, using water from the Zambesi. A sprinkler system could be installed on the existing field layout using 25 per cent less water than with surface irrigation. The water is desilted in reservoirs before use.

Silt and debris, eroded from the hills, often occur in the irrigation water and can be troublesome if they silt up the water channels or wear away the pumping equipment by abrasion. Due to the additional weight of the silt, the pumping efficiency of the machinery is reduced. In such cases it should be screened out. If the silt can be carried to the fields without trouble, it may be a source of fertility.

As cane requires dry sunny weather with cool nights to mature, irrigation water should not be applied during the 6 weeks before harvest.

Tea

Tea likes about 100 mm of water each month. In many tea areas there is a loss of water by run-off and drainage in the wet season and a shortage in the dry season. With the new high yielding clones well fertilized, irrigation, usually by overhead sprinklers, has become profitable. A tea area in Uganda, with an annual rainfall of about 1,000 mm a year with two dry seasons, raised yields from 1,300 to 1,700 kg/ha to 2,000 kg by irrigation. The production was more evenly spread over the year giving a better utilization of the factory.

A similar response to irrigation was obtained in Tanzania where yields were doubled and stem canker (*Phomopsis theae*), a disease of dry areas, was reduced. As the roots of established seedling and clonal plants went down 4.5 m and dried the soil to 3 m there was little advantage in irrigating until 3 to 4 weeks had passed without rain (Kasasian *et al*. 1978).

Sprinkler irrigation is recommended for tea on the undulating lands of North-East India which are subject to serious droughts.

Besides increasing the yield and spreading the crop, irrigation is desirable when replanting valuable clonal material where the rainfall is unreliable. Sprinkler irrigation can also be used for regular applications of nitrogen and to correct deficiency diseases such as magnesium which occur with high production.

Ellis (1967) has pointed out that in Central Africa yield is directly proportional to solar radiation, but in Malawi half the solar radiation occurs in the sunny dry season when soil moisture is limiting. He suggests that if this solar radiation could be fully utilized by irrigating in the dry season, 80 to 100 per cent increase in yield might be possible, with a better distribution of the crop throughout the year and an improvement in the quality of the tea.

Irrigation of clones in Malawi resulted in better root growth both in depth and lateral spread after 20 months in the field. This was reflected in much better top growth, measured by weight of prunings and the cross-section of the stem (Willatt 1971).

This suggests that areas too dry but sunny and otherwise very suitable for tea, as occur in Sri Lanka, might be profitably planted to high yielding tea with irrigation. Under these conditions, blister blight, where it occurs, might be of no importance, which would reduce production costs.

Tobacco

Though irrigation normally gives only a small yield increase, it has attracted a growing interest among tobacco growers, largely on account of the high cash return from this crop. In Queensland it is desirable to harvest the crop before the wet season starts as this is often accompanied by strong winds. This means planting out in the dry season, and consequently about 90 per cent of the tobacco area is irrigated. Irrigation stabilizes production with more reliable returns and does give an increased yield. Much of the tobacco land is furrow irrigated, but some growers use sprinklers. Irrigation after planting out ensures a better stand, and subsequent waterings give a more even ripening and improve the grade. As the tobacco plant is very susceptible to waterlogging, overwatering, particularly when the plants are small, must be avoided. Irrigation must be regular and moisten the entire root zone. From about 5 weeks after planting, the plants must not be allowed to wilt and growth never checked for lack of water.

The quality of the water is very important as too high a chloride content lowers the leaf quality. Irrigation, even with water free from chlorides can increase the uptake of chlorides from the soil, and as the chloride content of the irrigation water increases, so does the leaf chloride content. In Queensland where 620 mm/ha of irrigation water may be applied to flue cured tobacco in one season, the upper limits of water suitability are 25-40 ppm chloride depending upon the effects of the environment to keep the cured leaf content below 2 per cent. The level of chloride in the irrigation water is high in South Africa. Rainfall is important in removing chlorides from the soil by leaching (Akehurst 1981).

An adjustment of the fertilizer practice is necessary to obtain maximum yield of a good grade leaf with irrigation. The irrigation water should not be infected with nematodes.

Tobacco draws mainly on the top 0.5 m for its water requirements and has not a high water requirement, recovering well from drought. Thus frequent light waterings of 25-50 mm are preferred, though up to 100 mm may be applied under certain conditions. With sprinkler application, a low rainfall rate for a fixed amount of water is desirable - 8-25 mm/hour is usually applied, but the rate should not exceed the capacity of the soil to absorb water and much slower application may be necessary on certain soils. A lower rate involves higher pressures, and a higher rate requires more frequent moving of the equipment and a more severe effect on the soil from the pounding droplets. High pressure, high-trajectory sprays need more power, suffer greater wind disturbance and have a more serious effect on the soil surface. Small rotary sprinklers are more satisfactory. A low rate of application may encourage blue mould.

The depth of water penetration into the soil must be checked the day after irrigating. The soil surface should be disturbed as soon as possible after each watering, and the crop kept weed free. Soil moisture strain is most serious when the tobacco crop is knee-high.

Irrigation equipment is probably economic for permanent nurseries fumigated after each crop.

Portable irrigation should be a useful asset where a dry period is liable to coincide with planting out. Irrigation would ensure a more complete stand, which would not only reduce the cost of filling the gaps, but bring all the crop to harvest at the same time.

Plate 2.22 Tea bushes after pruning. Sri Lanka. (Courtesy Sri Lanka Tea Centre.)

PRUNING*

Pruning is directed to replenishing bearing wood, removing old useless wood, weak shoots and any diseased branches, shaping the tree, and keeping the centre of the tree open. It is often used to shape a tree and keep the crop within easy reach for easy harvesting. Removal of very low branches is often desirable to help weeding. (Plate 2.22).

Frequently pruning is badly carried out, sometimes because the method used with one tree crop is applied to another where the needs are different. Cocoa produces the crop on the old wood, coffee on the new growth, and citrus needs hardly any pruning at all. A sharp cutlass can be used very efficiently on cocoa but is not good for coffee. Young trees are better pruned with secateurs or a sharp pruning knife to avoid damage. When training young trees of all types, the ease of spraying, where necessary, should be borne in mind and branches growing into the spraying lanes cut back.

* See Brown, G.E., 1972

Even with well established crops there is frequently controversy over the best pruning system, as in East Africa where the single stem system (topped about 2 m) and the multi-stem system are both practised with *arabica* coffee. The vertical stems are usually non-fruiting with the berries on the laterals. Comparison of single stem and multiple stem pruning on clones of *arabica* coffee at Lyamungu (Tanzania) showed 10 of the 12 clones yielded better on the multiple stem system than on the single stem system. Multiple stem pruning controls biennial bearing of vigorous clones to some extent (Fernie 1965).

The maximum yield of a stem on robusta coffee is in its third year, therefore stems should not be removed before they are 4 years old. If a constant number of shoots, about four or five, are maintained around the tree and the oldest cut back each year, there will be a constant regeneration of productive branches. Pruning should be low, about 30 or 50 cm from the ground, so that the cherry is produced at a reasonable height for picking. Pruning after picking at the start of the rains is the most practicable and the best time for sucker regrowth. Excess suckers must be pulled off as soon as possible. When robusta trees have been neglected, heavy pruning is necessary, but one long 'breather' should be left until the required number of well spaced new shoots have been selected. Complete cutting down of bushes is liable to cause the loss of some plants. The way in which a coffee tree is pruned is more like pruning a banana stool than a cocoa tree.

Bananas are pruned to remove the old stems that have borne fruit, and reduce the number of young suckers in order to produce bunches of the size required for the market at the most profitable time of the year.

The method of training 'nganda' coffee trees in Uganda by bending them outwards, which was practised before the arrival of the Europeans (Thomas 1947) is ideal for a peasant system of culture as it requires a minimum of pruning, and gives a good ground cover for weed control. A single tree may have at 6 m spread and this large surface is covered with berries.

A number of tree crops such as tea and coffee store their food reserves in their roots. These reserves vary between years and seasons depending upon the balance between photosynthesis and demand of the crop. In Sri Lanka the tea bushes over 900 m above sea level store adequate starch in their roots but at the lower elevations the starch reserve is frequently deficient. If the tea bush is deficient in reserves when pruned, it cannot form new foliage and thus is killed. Once the post-pruning loss of tea bushes was explained and dissociated from fungal infection, a system of pruning was designed by which sufficient leaf was left on the bush to enable photosynthesis to take place and the canopy to be replaced in the absence of reserves, as shown in table 2.18.

The starch content of the roots at pruning time is readily shown by staining a section with iodine.

Table 2.18 Effect of three pruning systems on the loss of tea bushes at low altitude in Sri Lanka

Type of prune	Average number of leaves per bush	Number of deaths
Clean prune	3.0	67
Cut-across (low)	41.6	27
Rim lung	200.5	8

Source: Gadd (1949).

Recently it has been suggested that while the 'lungs' minimize the risk of 'die-back' and death from pruning, contributing to the carbohydrate supplies and the utilization of reserves, they exert a depressing effect on the buds developing on the frames. The balance of these effects determines how long the lungs are retained on the recovering bush (Nagarajah & Pethiyagoda 1965).

Tea bushes recovering from pruning may die from the effects of defoliation by blister blight attack if reserves are low or depleted by a succession of attacks, hence the importance of efficient fungicide protection of pruned tea. Removal of tea prunings in North-East India reduces the yields of subsequent crops.

While mature citrus, well trained in the nursery, requires little pruning apart from the removal of dead or diseased wood, vigorous lemon trees require heavier pruning. Branches of mandarins showing die-back from overbearing should be cut out. Oversized trees should be brought back to normal size over a few years. Rejuvenation by heavy pruning is hardly likely to be successful.

When cocoa is propagated from a fan cutting the growth habit is different to that of a seedling. Eventually a cutting will produce a chupon at the base and this should be encouraged and trained to produce a seedling-type tree. An old practice in Trinidad was to allow a basal chupon of an old tree to develop into a new trunk. In this way the life of the tree could be prolonged without the cost of replanting. This should only be done if the tree is as good as the clonal replacement material available.

Mention should be made of the importance in tobacco cultivation of removing the terminal bud when the required number of leaves have developed and later the buds which then develop in the axils of the leaves. This is a weekly operation up to harvest. Topping allows a limited number of leaves, 8–12 or sometimes more, to mature and prevents seed formation. Where this is not carried out, as in the case of Turkish tobacco and tobacco grown for nicotine extraction, the leaves are thinner and narrower. When the terminal bud is removed the cut surface may be treated with a desuckering oil which runs down the stem and prevents the axillary buds developing, eliminating weekly desuckering by hand. Chemical sprays are available which when applied to the leaves prevent tobacco buds growing. Fewer heavier leaves are left for fire-curing, compared with flue-curing, and air-curing is intermediate.

HARVESTING AND PROCESSING

Harvesting of farm crops of temperate climates is now highly mechanized but tropical crops offer the same problems as horticultural crops. Some progress has been made with the mechanization of tea plucking. Cotton picking, maize, rice and sugar cane harvesting can be completely mechanized. This applies to a minor hectareage, usually flat land and the equipment is expensive and complicated. It is difficult to visualize the harvesting of any of the tree crops being mechanized, and in many crops selection of the ripe crop is important, as in tobacco, cocoa, cotton, coffee, etc. Sorghums, millet beans and rice, when grown in large flat fields, can be harvested with modified equipment originally designed for temperate crops, but the largest part by far is harvested by a hand knife or a form of sickle, frequently a single head at a time.

Hand harvesting is very laborious but frequently unavoidable. It may account for 15 per cent of the cost of production of maize or 40 per cent in the case of tea, but it has certain advantages

1. Regular harvesting could ensure that only crops at the correct stage are harvested. This will maintain the quality as with tea, coffee, cocoa and tobacco crops.
2. Regular harvesting prevents the build-up of certain diseases and the crop loss from them, e.g. cocoa - black pod control.
3. Hand harvesting enables the farmer to eliminate rubbish. This is important in the case of cotton.
4. Hand harvesting enables the farmer to select his seed for his next crop more carefully. This applies particularly to cereals.
5. Hand harvesting is important in relation to root food crops: sweet potatoes, yams, cassava, tannias, which are best stored in the ground and dug as required. More labour goes into harvesting sweet potatoes than any other of the growing operations. Hand harvesting is the only practical method with intercropping as done by small growers.

Some examples of the amount of crop harvested by one person in a day are:

Cassava	200 kg	
Cocoa	Collect 1,500 pods Open 1,500 pods	33 man days per tonne

Coconuts	Collect 1,000 fallen nuts.
*Coffee (*robusta)	35 - 45 kg cherry picked on task, but only about one-third of this is picked when day rates are paid.
Cotton	10 - 14 kg seed cotton picked per day in a rain grown crop; 20 kg in Kenya.
Grapefruit	1,000 - 1,500 picked and crated, much more if shaken and harvested.
Groundnuts	1/75 ha lifted and gathered.
Maize	270 - 360 kg of dry cobs picked and stripped.
Oil palm	Cutting 100 - 200 bunches (1.5 - 2.5 t). Cutting and carrying one-third less. Cutter/carrier pair of workers 1.25 - 2 t.
Oranges	1,500 - 2,000 picked and crated, much more if shaken and harvested.
Pigeon peas	70 - 90 kg.
Rice	1/25 ha.
Rubber	350 - 500 trees tapped and collected (6 am - 10 am). It is general to pay tappers on a poundage basis, according to the dry rubber content of the latex collected.
Sugar cane	Cut and strip 2 - 3 t of a 75 t/ha crop. Burning before harvesting should mean a higher output.
Sweet potatoes	136 - 225 kg
Tea	14 - 36 kg green leaf according to the available crop.
Yams	450 kg

These figures vary with the crop and the incentive to the labourers. In many cases labour is paid by the task rather than the day.

After harvesting the process of preparing many crops for marketing or storage is of considerable importance. Food for storage, such as maize, should be well dried either on drying racks, a reed table, a barbecue, or, if there is no alternative, the bare area in front of the house (Plate 2.14). Harvested cassava roots begin to decay within about 48 hours. If not disturbed from the plant they can be stored longer in the ground.

The amount of interest shown in treating a cash crop before selling frequently depends upon the grading system in operation. For example it is impossible to grade dried robusta coffee cherry, hence there is no financial incentive for growers to pick only ripe cherries and to dry the berry on specially constructed tables. On the other hand cotton, unless picked over to remove those bolls damaged by stainers, may not be purchased, or purchased at an inferior price. The curing of tobacco and fermenting of cocoa are two very important processes which frequently result in a wide price differential between peasant grown and processed, and the higher quality estate product.

The results that education can achieve in maintaining a high standard for native produce is demonstrated by some of the high grade coffee marketed by the African growers in Kenya, and by the quality of cocoa produced by some of the small growers in Trinidad. One cannot expect growers to take extra trouble to produce a higher quality product unless there is a worthwhile cash premium for the extra quality.

Coffee berries are either dried, usually in the sun, then hulled, or pulped and fermented to remove the outer flesh, and the resulting parchment dried and hulled. From 100 kg of fresh berries, about 23 kg of parchment is produced which gives 20 kg of clean coffee. Dry processed arabica has a coarse hard flavour and robusta often a mustiness due to incomplete drying. Pulped and fermented coffee has a milder or neutral flavour and sells for a premium price. However it takes 20,000 litres of water to process a tonne of coffee, though this can be reduced to 5,000 litres by recirculation. Water is not always available near to the coffee and wet coffee berries are heavy to transport, so it is not always possible to produce the higher valued product. By drying the berries the weight to transport is reduced to about a third as 100 kg of berries give 35-40 kg of dry cherry or 'buni'. It is much easier to grade fresh cherry and ensure that only red berries are picked. These give both a higher out-turn of clean coffee and a better quality bean.

Harvesting and threshing of rice

Like most other crops, once rice is ready for harvesting the quicker it is in the store the better, and the sooner the land can be prepared for the next crop. Much of the rice grown in the tropics is the tall *indica* which lodges more than any other cereal crop, making harvest more difficult and causing a loss of crop. The crop is cut with a sickle, scythe or knife and gathered for threshing by hand or with animals (Plate 2.23). The four main methods of hand threshing are:

1. Treading on the heads.
2. Beating the heads into a tub, against a screen, or on to a threshing ladder.
3. Beating the heads to remove the grain.
4. With a pedal operated threshing drum.

Animals are used either directly to tread out the grain or to pull an implement such as a sledge across the heads.

There is no difficulty in threshing rice mechanically with equipment

Plate 2.23 Threshing rice. Philippines. Courtesy FAO.

designed for other cereals, or even combining the crop. The difficulty has been that paddy is grown by a large number of small farmers with little capital for expensive heavy machines. Small threshing machines are available such as the one from Garvies in Aberdeen (Plate 2.24) which is of simple design and easily transportable, weighing about 150 kg including the air cooled 150 cc four stroke petrol engine which drives the threshing drum. It is ideal for the small farmer and fits in with normal harvesting methods. The machine is operated by five men; two to collect the paddy to maintain a steady flow to the thresher, one to feed the machine, one to remove the straw from the box and one to bag the grain. Following reapers round the field the output is not so great but 12 to 18 reapers are needed to keep the machine busy. Output is greater, about 0.5 t/hour, when threshing standing heaps of paddy. It has been estimated that this machine besides saving labour and making the work of threshing much easier, by reducing losses produces 5 - 10 per cent more grain than hand threshing from a similar amount of crop. Less than 1 per cent of the grain is broken. This thresher can also be used for sorghum, soya beans, beans and mung and possibly coffee.

The International Rice Research Institute (IRRI) has developed a portable thresher weighing about 100 kg which can be carried by two

Plate 2.24 Small portable rice thresher designed by NIAE, Silsoe, England. (Courtesy National Institute of Engineering)

people. It is powered by a 5 hp air cooled engine, and has a winnowing fan to combine threshing and winnowing into one operation. Two or three people can thresh and bag about 500 kg hour with less than 2 per cent grain separation loss. This is being manufactured in the Philippines. The IRRI have also developed a portable grain cleaner with a plywood body and powered by an electric motor. It can clean 1 t per hour.

Machines like these must be used more in future to reduce the wastage and take much of the hard labour out of farming (Plates 2.25 and 2.26). It is important that they are adaptable for handling the new rice varieties now being developed, as some of these are proving difficult to thresh in the traditional ways.

Mechanization of processing

A considerable amount of labour is involved in processing crops after

Plate 2.25 Rice winnowing Ilocos, Philippines. (Courtesy Philippines FAO).

harvest (see Table 2.19). Where this concerns food crops it falls largely on the women.

Small hand hullers are available for both maize and groundnuts. Between 1955 and 1959 small hand-operated groundnut shellers were introduced into Nigeria and as a result the percentage of exports in the special grade (over 70 per cent whole nuts) rose from 2 per cent to 98 per cent of the output.

The possibilities of developing some of the rudimentary native techniques are illustrated in the development of the Tropical Stored Products Centre groundnuts sheller.

A home-made groundnut sheller used in Zambia consists of three tree branches, a perforated dustbin lid and a small piece of metal attached to a wooden handle. The branches formed a tripod for the lid which was concave side upwards with the broken edges of the holes on the convex side. The small piece of metal also had holes punched in it with the jagged edges on the convex side. The unshelled groundnuts were put in the lid and the piece of metal moved backwards and forwards over them. This sheller produced 80 per cent unbroken nuts, which is better than the available manufactured hand shellers. This idea has now been developed into a commercially available machine of more sophisticated

Plate 2.26 Winnowing of paddy by hand. Hmawbi farm school, Burma. (Courtesy FAO)

design, but basically operating in the same way so that nuts fall from the machine as soon as the shell is broken by the rubbing action.

Two hand-operated groundnut decorticators, Skylux and Kano with an automatic feeder designed by the Tropical Products Institute, will give 90 per cent whole kernels when correctly adjusted.

A simple groundnut stripper (Collins & Coward 1971) was developed in Gezira costing £8 with an output of 400 kg of fresh pods in a 6 hour day, equivalent to an output of more than ten persons hand stripping. It also has the advantage of preserving the foliage intact for cattle feed (Ramly 1966).

A mechanical cocoa pod breaker has been developed in Costa Rica which breaks 2,800 pods an hour, nearly eight times as fast as manual breaking. Some of the beans are lost with the shell, but the beans are not damaged and ferment normally.

The traditional method in West Africa of extracting oil from the pericarp of the oil palm is to boil the fruit for several hours, pound it with a heavy pestle, and extract the oil from the pulp with water and skimming. Alternatively the fruit may be extracted by treading in a clay

Table 2.19 Labour requirement in man-days per hectare for the production of the main food crops in the forest zone of Zaire

	Maize (1st crops in rotation)	**Maize** (after regrowth of cassava)	**Rice**	**Cassava**	**Plantain**
Share of initial preparation	17-25	17-25	17-25	17-25	17-25
Weeding after maize harvest	—	—	4	4	4
Weeding after regrowth of cassava	—	40	—	—	10
Planting including preparation of planting material	20	20	30	35	15
Weeding	10	20	6	6	6
Harvesting	25	25	40	50	20
Total labour growing crop per hectare	72-80	122-130	97-105	112-120	72-80
Transport	7	15	5	40	(included in harvesting
Construction of supports for drying	7	15	—	—	—
Husking	3	5	—	—	—
Grinding and storage	20	40	20	—	—
Winnowing	3	4	30	—	—
Bagging	3	4	3	—	—
Peeling	—	—	—	99	12
Soaking and drying	—	—	—	—	—
Total man-days/ha	115-123	205-213	155-163	301-309	84-92
Av. yield in kg/ha	1,000	2,000	1,000	10,000	4,000
Total man-days/ 100 kg	12	10	16	3	2

Notes: The estimate is based on the fallow land cropped with maize, groundnuts, rice, plantains and cassava. The labour of clearing initially is divided equally between the crops, though it could be argued that plantains and cassava, which occupy the land longer than the other crops, should carry a higher proportion. The first half of the table is based on a hectare of cultivation irrespective of the yield, whereas the figures in the second half are based on the crop yield rather than the hectareage.

Source: Geortray (1956).

hollow or a canoe, as is common in the Niger delta. This is a short labour intensive process which recovers at best 50-60 per cent of the oil, and considerable hydrolysis and oxidation of the oil takes place in the process.

In the past 25 years improvements have been made in small-scale processing giving a higher extraction percentage of better quality oil. A hand-operated press was developed to give 60-70 per cent extraction, and at the end of the Second World War the Pioneer Mill was introduced, which was the first attempt at small-scale mechanical processing though it used a lot of hand labour. This mill used centrifugal extractors to recover about 85 per cent of the oil. The modern factories are still more efficient and produce an even better quality product.

The Japanese have been introducing into the tobacco areas of the Far East, cheap electric driers which use hot air or infra-red to replace the traditional sun of air drying. The traditional coconut driers which used non-flaming materials such as coconut shells are being replaced by diesel fuelled hot air driers. The traditional ways of processing rubber sheet have been replaced, by shipping latex, or by producing a crumb from the latex by adding castor oil, or by comminuting sheet rubber.

It is at this part of the crop growing process that mechanization can possibly be most effective. Besides hand shellers, winnowers, threshers, grinding mills, coffee hullers, rice hullers and palm kernel crackers are available. The machinery, where small, can be individually owned, or, where more advanced equipment is necessary, on a cooperative basis as in the case of the cotton ginneries and coffee mills in Uganda. The requirements for labour to cultivate certain crops in East Africa are shown in Tables 2.20 and 2.21.

Storage losses

Conservation of food in store is the easiest, cheapest and most profitable way to increase food supplies. The post-harvest loss and wastage caused by rats, moulds and insects frequently amounts to about a quarter of the crop, which could be reduced at little cost. Many of the traditional granaries are well designed to reduce such losses and in these the loss may be less than ten per cent. Before they are refilled with the new harvest they should be thoroughly cleaned out, repaired, and if possible treated with a safe insecticide such as malathion or lindane. Granaries should be weatherproof and protected against rats (Plates 2.27 and 2.28).

Crops going into the store such as cereals, beans and groundnuts should be very well dried, clean, and free from broken grains and pods, as these are more easily attacked by insects and go mouldly more quickly than whole grains. The groundnut pods should be removed from the

Table 2.20 A comparison of the labour requirements, yields, crop values and costs for cotton grown on the flat and on ridges at Kibos in Kenya[a]

	Plot I–cotton on ridges	**Plot II–cotton on flat**
Labour requirements, man-days per hectare		
Ridging and initial weeding	49.0	36.6
Marking rows and planting	17.4	29.3
Thinning	11.7	14.3
Resowing	9.5	9.3
1st weeding	50.7	57.8
2nd weeding	52.5	70.8
Spraying	6.5	6.0
Picking	112.9	132.6
Cutting stalks and clearing	9.8	12.4
Total man-days	320.0	369.1
Yield and grade of seed cotton		
Yields (kg/ha)	1,495	1,729
% Grade A	96.2	97.2
Value of crop and production costs (shillings per hectare)		
Value of seed cotton	1,759	2,046
Cost of ploughing and harrowing	160	160
Cost of labour @ 2/- per man-day	640	738
Cost of insecticide[b]	148	148
Profit (value of crop less production costs)	811	1,000

[a] Each of the two plots was of approximately 0.2 ha.
[b] Four sprays of DDT at weekly intervals followed by four sprays of Sevin at weekly intervals.

Source: K. J. Brown (1972).

haulms and thoroughly dried in the sum on mats. Thorough drying soon after harvest is essential to prevent the development of aflatoxin, a highly toxic substance produced by a common mould *Aspergillus flavus* which grows rapidly on moist groundnut kernels at tropical temperatures. Undamaged groundnuts in their shells are very resistant to insects. After shelling, mouldy and discoloured kernels should be taken out and used on the fire, not thrown down for the poultry around the compound.

If rice is too dry at harvest, the heads will shatter and grain will be lost so it is usually cut at about 30 per cent moisture. If it is then stored in heaps it will heat up and discolour. The heads must be dried to 14 per cent moisture before storing. It is particularly important to dry and store rice well as, unlike most cereal grains, it is eaten whole, not as flour, and the appearance and flavour when cooked must be correct.

The traditional method of sun drying rice in the open air either

Table 2.21 Labour requirement for the cultivation of crops in Uganda in man-days/ha

	Cassava	Sweet potatoes	Maize	Beans (Phaseolus)	Sim-sim	Groundnuts	Cotton
Cultivation	50-60	50-60	50-60	50-60	50-60	60-100	50-60
Collection of planting material and planting	20	75	20-40	40-50	15-20	40-50	20-25
Weeding	125-135	50-65	80-120	75-85	75-100	100-110	75-150
Harvesting	75-100	225-250	40	60	125-150	125-150	75-100
Total	270-315	400-450	190-260	225-255	265-330	315-370	230-375

Notes: Clearing elephant grass and rough digging need 40-60 man-days.
One man-day is 7 hours.
These figures vary widely from those of Zaire, Congo.
The work-output in the Congo was undoubtedly higher from paid labour.

Plate 2.27 Storage of maize cobs. Ivory Coast.

before or after threshing may lose 10 to 25 per cent of the grain. A low cost solar drier has been developed in Thailand, especially for drying the second rice crop which is harvested in a rainy period of the year. The air is channelled up through the rice using both the prevailing wind and convection. As the rice is protected from the rain by an opaque roof there is no deterioration from rewetting or overdrying.

A storage shed for 300 t of green coffee has been developed in Malagasy which uses solar energy to keep the crop in good condition. The air is heated in the space between the black-painted double roof and circulated by a fan through the store. Except in very wet conditions the coffee is kept below 12 per cent moisture and in dry weather coffee with 15 per cent moisture can be dried. This could be applicable to other crops (Stessels & Fridmann 1972.)

In Sri Lanka it is a traditional practice to place some fresh leaves of the lime tree (*Citrus aurantifolia*) in stored rice to protect it from *Sitophilus* weevils.

Cowpeas treated with a small amount of groundnut oil are protected against weevils for over 6 months. This treatment does not affect the taste or germination capacity of the cowpeas (IITA 1976).

Plate 2.28 Granaries made in red earth around a village on the border of the Ivory Coast and Upper Volta. Old sorghum stalks on the ridges.

Finger millet (*Eleusine coracana*) if well dried can be kept for 10 years or more in traditional mud-and-thatch granaries. Maize and sorghum, on the other hand, can lose 10 per cent by weight in 6 months even though dried before storing (Plate 2.29).

Cassava*, which deteriorates 2 or 3 days after digging, and sweet potatoes are best stored in the ground. When sliced and dried in the sun they are soon attacked by boring insects. Cassava is processed and stored as gari in West Africa (see p. 130). Yams may be stored in the ground for short periods. As the stems are cut, the tubers may sprout. All the soil should be removed from harvested yams which should be stored in a dry shaded place. A good method used in West Africa is to tie them to a framework of poles in such a way that they do not touch and the air can circulate freely around them. If the upright poles are living trees such as *Ficus* spp. the structure is permanent and a natural shade is provided. Any sprouting or rotting yams can be seen easily and removed. This method is effective in the dry season, but once the rains start the yams deteriorate rapidly.

Beans, cowpeas, grams and pigeon peas are all readily attacked by

* The storage of cassava was reviewed by Ingram and Humphries (1972).

Plate 2.29 Grain storage cribs. Malawi. (Courtesy Tropical Products Institute)

boring insects. After 6 months' storage under typical conditions in Uganda about half the *Phaseolus* beans have holes in them.

Where malathion and lindane have been used regularly to control storage pests, resistance has developed. A newer, safe insecticide, pirimiphos-methyl (Actellic®), when applied to bags in which rice is kept, has controlled resistant strains of *Tribolium castaneum* and protected rice for 6 months. The new synthetic pyrethroids should also prove useful.

Peasant farmers often preserve their seed by drying it well and then mixing it with dry ashes. Frequently, seed is stored in the smoky dry area above the cooking fire.

Gourds are often used to store seeds. They should be no larger than needed, and free from cracks. Air exchange is further reduced if the gourd is oiled. The carbon dioxide produced by the fungi makes conditions unsuitable for further mould growth, and the insects are deprived of oxygen.

The new systemic fungicides, thiabendazole and benomyl, are being used to prevent storage and ripening rots of fruits including oranges and bananas. Storage rots of root crops may be similarly reduced using low concentration dusts of these systemic fungicides.

The handling and storage of crops off the farms is dealt with by Hall (1970).

3 CROP IMPROVEMENT*

Plant breeding offers the cheapest way to increase crop production in the tropics, and it formed the basis of the 'Green Revolution'. This increase may result from: a better adaptation to the environment; a resistance to a pest or disease which previously caused a significant loss of crop; or a development or growth habit which results in a higher proportion of the crop being garnered. The new varieties may also have a greater value from an improvement in quality, such as appearance, flavour or fibre strength; a greater uniformity making for easier handling or processing; or better shipping and storing properties.

There is, however, not only a need to develop improved varieties of all crops but also to produce and distribute clean quality seed and planting material of these varieties. Often this essential is lacking in crop improvement programmes. Without good drying and storage facilities it is difficult to produce in the wetter tropics quality seed with a high germination capacity and free from pathogens.

SELECTION

Selection has been practised ever since man took up a settled form of agriculture. Cultivated crops originated in man's selection and propagation of a more useful form of wild plant. These selected plants may have arisen through a chance combination of genes in a natural cross; by mutation, a natural change in the gene structure; or by polyploidy, the replication of whole sets of chromosomes, a process which sometimes gives rise to new and useful characters. Several valuable citrus cultivars including Washington Navel orange, Marsh grapefruit and Shamouti orange arose by mutation. Selection depends for its success upon the variation that occurs in all field crops, other than those planted to a single clone which can be regarded as a single plant, though mutations do occur in clones and may respond to selection. Natural selection is occurring continually: those progeny unable to withstand competition,

* See also Simmonds (1979)

susceptible to a prevalent disease, or unadapted to the ecological conditions where they are growing, are eliminated. During the course of natural selection some characters, like response to fertilizer, or resistance to a new disease may have been lost in the development of the main varieties. Man, choosing his seed or planting material for the next crop, is mass selecting.

Distribution of Crops

As tribes have migrated they have carried with them their favoured crops and returning travellers, including such unlikely people as slave traders, have brought back plants from other lands. After Columbus's first voyage in 1492 there was a speedy exchange of crops between the Old and New World. On his second voyage in 1493 Columbus took sugar cane and citrus to the Caribbean, and brought maize and sweet potatoes to Spain.

This interchange of planting material was at its zenith during the late eighteenth and the nineteenth centuries, when botanical gardens flourished throughout the tropics and plants were introduced, studied, and if they thrived, propagated. For dicotyledonous crops Purseglove (1968) gives the examples of how the main areas of production of some major economic crops are far removed from the region in which they originated, and lists the many factors that have influenced this (Table 3.1). An example from recent times is the post-war development of cocoa in Malaysia, where the best planting material was introduced into an area free from many of the pests and diseases but with a history of cultivating tree crops for high production. Black pod is as yet unimportant and 'die-back', which was serious in early plantings, has been

Table 3.1 Centres of production of dicotyledonous crops

Crop	Country of origin	Country of maximum production
Black wattle (*Acacia mearnsii*)	Australia	S. and E. Africa
Citrus (*Citrus* spp.)	S.E. Asia	United States
Clove (*Syzygium aromaticum* (syn. *Eugenia caryophyllus*)	Moluccas	Zanzibar
Cocoa (*Theobroma cacao*)	S. America	Ghana
Coffee (*Coffea arabica*)	Ethiopia	C. and S. America
Groundnut (*Arachis hypogaea*)	S. America	India, China
Nutmeg (*Myristica fragrans*)	Moluccas	Grenada, W. Indies
Pyrethrum (*Chrysanthemum cinerariaefolium*)	Dalmatia	Kenya
Rubber (*Hevea brasiliensis*)	S. America	Malaysia
Soya bean (*Glycine max*)	N.E. Asia	United States

reduced in importance by the importation of Upper Amazon planting material which is hardier than the original Amelonado introduction. Consequently yields over 2,000 kg/ha of dry beans are possible on inland soils in West Malaysia (Mainstone, B.J. 1976). The oil palm industry of Indonesia is reputed to be based on four palms originally introduced to the Botanic Garden of Bogor, Java (Buitenzorg), and the rubber trees of Sri Lanka and the Far East have been largely derived from the 2,397 rubber seedlings raised at Kew from the 70,000 or so seeds which Wickham collected in Brazil in 1876 (some may have come from Cross' collection), the Malaysian rubber industry being founded on only 22 of these seedlings sent from Kew to the Singapore Botanic Gardens in 1877 possibly with 5 that may have survived from the 50 sent out in 1876. These were supplemented later (1888) by seed from Sri Lanka produced from the original Wickham introduction. It is interesting that more recent introductions of both oil palms and rubber have been disappointing in comparison with the selections from the original material. Plans are in hand to introduce more breeding material into Malaysia from Brazil, but there is a risk of bringing in South American leaf blight, or a seed-borne virus disease which could destroy the rubber industry of the East. The bulk of the coffee in South and Central America is thought to have come from a single seedling sent from Java to the Amsterdam Botanic Gardens in 1706 (Krug 1959). It is remarkable how many of the commercial crops, particularly tree crops, have been developed from a very restricted introduction of planting material. There is tremendous scope for widening the genetic base to introduce desired characteristics such as disease and pest resistance, earliness, quality, yield, etc.

Modern plant breeders requiring additional characters or greater variability, search for them in the original home of the crop or in those regions where it has been longer estabilished. Pound's expedition up the Amazon provided cocoa breeders with valuable new material including a source of resistance to witches' broom, black pod and *Ceratocystis*, and Simmond's expeditions to the Far East and with Baker to East Africa promise to provide sources of resistance to both Panama disease and Sigatoka for incorporation in a marketable banana.

In any crop where selection has not been practised, there is considerable variability. This may cover growth habit, growth period, disease resistance, fruit size, seed colour, fibre quality and yield. Because yield is a significant factor in all crops it is worth while examining some of the variations that have been recorded in unselected populations of certain crops.

One of the earliest studies was on rubber (Whitby 1919). Among 245 trees recorded, the rubber content of the latex varied from 23 to 55 g/100 c.c. with a mean of 36 g; 1,011 trees had an average yield of 7.1 g of latex per day, but individual yields varied from 2 to over 27 g per day. Nearly a third of the yield was contributed by 10 per cent of the trees and 14 per cent of the trees did not yield sufficient to cover the cost of tapping. The four best trees yielded over 40 g of rubber per day. Though some of these differences were caused by differences in soil fertility and competition, the relative yields of the trees were unaffected by season. Similar studies in the Dutch East Indies, Sri Lanka and Malaysia, confirmed that the same variability occurred in the rubber trees in these countries.

A survey of coffee in Sri Lanka (Anstead 1921) showed that practically half the crop came from 10 per cent of the trees and the 70 per cent of the trees that were poor bearers contributed only 31 per cent of the crop.

In the case of tea in Sri Lanka, Kehl (1950) found that the vast majority of bushes growing on tea estates are low yielders largely due to the planting of unselected material, but also diseases, bad pruning and poor pockets of soil. Figure 3.1 shows the variability of 0.8 ha of good uniform tea (1,000 bushes) at St Coombs. In Java, Wellensiek selected clones giving twice the average yield.

A cocoa survey carried out in Trinidad in 1935 (Shephard 1937)

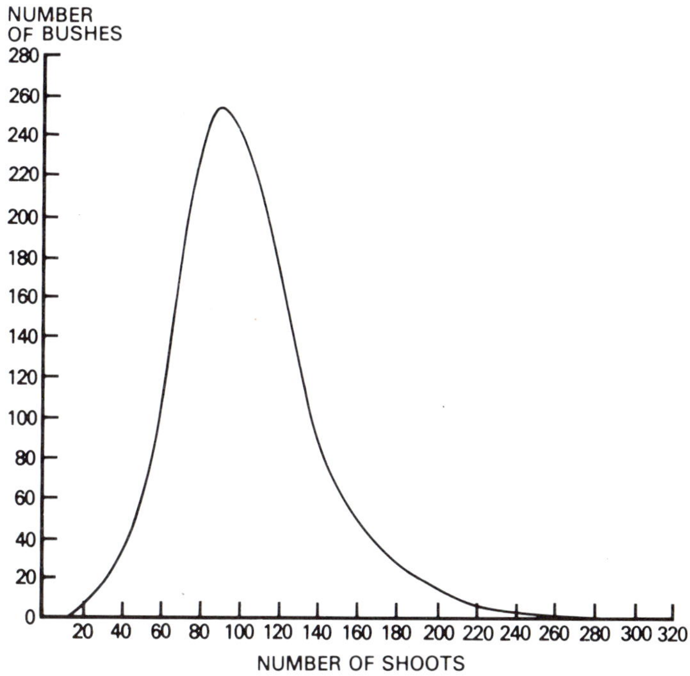

Fig. 3.1 Variation in tea bushes. Sri Lanka. (Kehl 1950)

showed that 42 per cent of the cocoa trees on good soils and 75 per cent of the trees on the poor soils were uneconomic, but 20 per cent of the trees on fertile soil and 6 per cent of those on poor soil gave good yields. The best 25 per cent of the trees yielded from 58 per cent (on the good soils) to 84 per cent (on the bad soils) of the total yield.

Similar relationships apply to all crops and therein lies the scope of the breeder whose object is to replace this variable population with a more uniform high yielding population, while retaining variability for certain other characters. In many crops big improvements were easily made, but further advances were difficult and depended upon the introduction of new material, skilful crossing and selection.

Mason (1938) made the observation that primary selection, that is selection of superior plants from a large field of plants, gives a greater step in improvement than selection from small plots at an experimental farm. Many of the important cottons have originated from single plant selections from a field planted with the normal seed distributed from a ginnery. These include: BP52 selected by Stephens and developed by Nye into the main cotton of Buganda; U4 upland cotton resistant to jassid, selected by Parnell in South Africa which spread throughout all South and Central Africa as far north as Zaire; and V135, a single plant selection made by Harland in St Vincent in 1916 which became the important Sea Island cotton.

An example of a major food crop improvement is the discovery in the USA in 1964 of the high lysine mutants opaque-2 and floury-2. These characters have been introduced into African maize varieties. This could make a significant improvement to the quality of diet in Africa, Central America and much of South America. Millet cultivars in India have a total protein varying from 10 to 23 per cent and lysine 0.9 to 3.1 per cent (Indicative World Plan FAO 1970, p. 107) which could provide a similar improvement to the diet of the people in the African savanna regions.

In all selection work there should be a sound basis for selection of certain characters, not just because the breeder believes it to be desirable. This probably applies more to animal than plant breeding.

Phenotypic variation has genetic and environmental components; selection acts upon phenotypes and is effective only as far as the genetic component allows. Selection methods vary greatly in the efficiency with which they isolate superior genotypes, ranging from comparatively crude eye selection to such refined techniques as the progeny-row method of Hutchinson and Panse (1937). This method, developed primarily for cotton, is applicable to many crops and depends upon a statistical test of the performance of progeny for its efficiency.

The approach to the improvement of a crop depends upon its mode of propagation as shown in Fig. 3.2. The following possibilities occur.

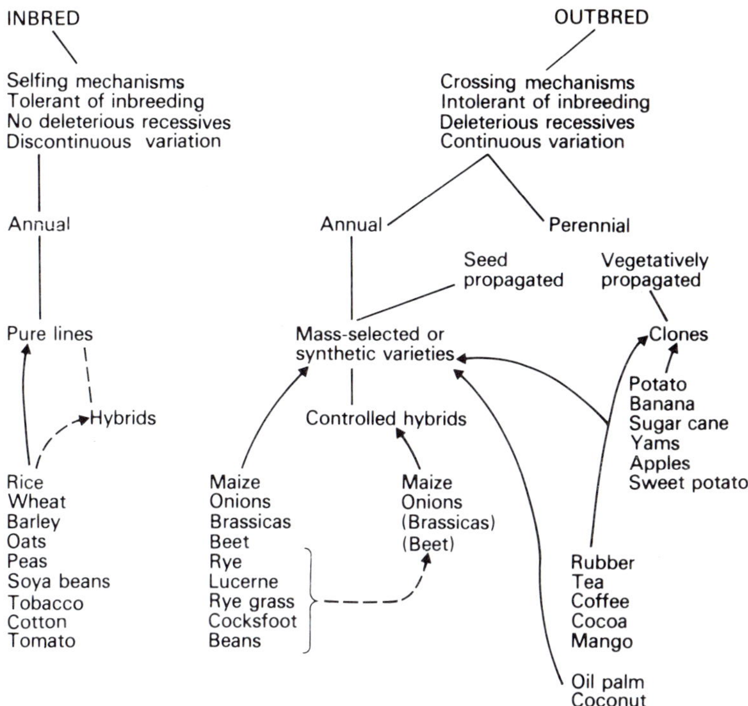

Fig. 3.2 Breeding and propagation systems in crop plants. (Simmonds 1969)

Some plants may be propagated both by seed and vegetatively. Improvement in such cases is generally simplified. (See Table 3.2).

Plants propagated by seed

Seed normally selfed, e.g. groundnuts, rice, *arabica* coffee

Such a crop is largely a mixture of pure lines. New lines or varieties can be created by artificially crossing and selecting new lines among the progeny in advanced generations. Naturally occurring mutations may contribute a desirable characteristic. Over 30 mutations have been recorded in arabica coffee, mainly in Brazil, and some crossing occurs. There is much to be said for planting such crops to mixtures of lines all of high yield and quality but variable for resistance to diseases. As cross-pollination of rice does occur the panicle should be bagged when breeding.

Seed normally crossed

1. Largely self-sterile, e.g. *robusta* coffee, tea, cacao,
2. Can be selfed artificially - maize.

Such populations are very heterogeneous. Where vegetative propagation is possible, selection can be carried out, and the superior material increased vegetatively. If the crop is a vegetative part, such as the tea leaf, self-sterility is unimportant, but in a crop grown for seed, such as coffee and cacao, mixtures of clones must be planted. The productivity of such mixtures should be studied as there is no guarantee that a crossing of two high yielding clones will be high yielding or maintain the other desirable qualities of the parents. If the production of seed is essential for propagation this group provides a major problem in crop improvement as in the case of the oil palm and coconut where no method of vegetative propagation has yet been developed, apart from tissue culture in the case of the oil palm.

Seed either selfed or crossed, e.g. cotton, Tjir I rubber clone

In most cases steps can be taken to ensure that selfing occurs. This is either done by covering the flower with a polythene bag or as in the case of cotton, tying the flower bud before it opens. New varieties are developed by artificial crossing, selfing to obtain segregation and homozygosity, and finally multiplying in segregated plots. Cultivars of such crops tend to become mixed fairly rapidly in cultivation and it is desirable to have a continuous wave of replacement seed coming forward from segregated areas, particularly where uniformity of product is important as is lint length in cotton.

Plants propagated vegetatively*

Continued vegetative propagation appears to lead to increased sterility as in the sweet potato and cassava.

Fertile seed formed, e.g. tea, rubber, sugar cane

New varieties can be formed by controlled crossing, selection and subsequent vegetative propagation. The phenotypes selected will probably be highly heterozygous, which will be maintained by clonal propagation.

No fertile seed formed, e.g. banana, seedless citrus

Once the limit of selection among the existing clones is reached, in the

* For a full survey of the vegetative propagation of the majority of tropical crops, see Fielden (1940).

absence of natural mutations, the only approach is a cytogenetic and physiological study to discover the cause of infertility and a method of overcoming it. The absence of seed may be a desirable feature, as in the case of the banana and seedless grapefruit. In this case it is necessary to produce viable seed which will give a plant not only having a wide range of desirable characters, but a plant that will not set any viable seed itself. Sweet potato clones which to not flower normally can be induced to do so by grafting on to morning glory under short-day treatment.

Table 3.2 gives a summary of the aims and basis of breeding for the main tropical crops.

PROBLEMS AND ACHIEVEMENTS

When considering the possibilities of producing better crop varieties success is most likely where there is a greater genetic variability in the plant material available to the breeder.

Gene banks

As improved varieties of a crop are developed, the genetic base of that crop is narrowed as material is discarded. In order to maintain the source of material to supply the answers to new problems facing plant breeders, gene banks have been established for many crops. With rapacious man destroying tropical vegetation on an ever-increasing scale, it is becoming more urgent to collect the wild and related species and a wide range of cultivars, in central gene banks before they are lost for ever. The IRRI in the Philippines has more than 14,000 rice varieties in its world collection but it is estimated that there are a further 20,000 cultivars which could be added. In Central America there is a collection of bananas at Lancetilla in Honduras. Turrialba in Costa Rica has the largest *Coffea* collection in the world; and at Summit in Panama is a remarkable collection of palms (Leon 1971). Material is freely available from these centres to breeders throughout the world. Similar collections now exist for many other crops, including sorghum, cowpeas and pigeon peas. Unfortunately much time and care is needed to maintain these collections which are only rarely required but the need when it arises may be critical. With material from such collections it becomes possible to redesign a plant as in the case of the new rice varieties, introduce more genes for pest and disease resistance, or increase the protein content of a cereal. In the future some of the traditional tall tree crops, such as coconuts and rubber, are likely to be

Table 3.2 Summary of basic facts for improvement of tropical crops

Crop	**Latin name**	**Chromosome number** (basic, diploid)	**Normal method of propagation**	**Is there a practical method of vegetative propagation?**	**Valuables part of plant**
Abaca	*Musa textilis* Née	$2n = 2x = 20$	Vegetative. Suckers	—	Fibres from sheaths.
Banana	*Musa* cultivars L.	Triploids ($2n = 3x = 33$) most important. Diploids and Tetraploids (rare) occur.	Vegetative	—	Fruit
Beans	*Phaseolus* spp. L.	Mainly $2n = 2x = 22$ Some grams $2n = 24$	Seed	No	Seed
Cocoa	*Theobroma cacao* L.	$2n = 2x = 20$	Seed	Yes. Stem or single leaf cuttings.	Seed
Cassava	*Manihot esculenta* Cranz	$2n = 4x = 36$?diploid	Vegetative. Stem cuttings.	—	Tuber
Coconut	*Cocos nucifera* L.	$2n = 2x = 32$	Seed (Fruit)	No	Fruit
Coffee arabian	*Coffea arabica* L.	$2n = 4x = 44$ (Tetraploid)	Seed	Yes. Cleft grafting. Stem cuttings.	Seed (fruit a drupe).
Coffee-robusta	*Coffea canephora* Pierre ex Froehner	$2n = 2x = 22$	Seed	Yes. Stem cuttings.	Seed
Cotton-American upland	*Gossypium hirsutum* L.	$2n = 4x = 52$ Allotetraploids	Seed	No	Seed coat fibres.
Cotton - Sea Island and Egyptian	*Gossypium* - old world cottons, *barbadense* L. *G. herbaceum* L. *G. arboreum* L.	$2n = 2x = 26$.			

Is production of seed necessary?	Characters required in addition to yield	Diseases to which resistance is required	Remarks
No	Fibre quality	Bunchy top. Mosaic	Some seed produced
Undesirable	Consumer acceptability. Transportability	Panama disease. Leaf spot	Hard seed in a soft fruit, very rare
Yes	Flavour Appearance	Anthracnose, Rust, Root rots, Bacterial blights	Wide range of testa colours.
Yes	'Quality'	Swollen shoot. Black Pod. Witches' broom	Widespread incompatibility in Trinitario group. Incompatibility in Amazonian clones near centre of origin
No	Low alkaloid content	Mosaic. Bacterial wilt	Fertile seed produced but sterility common. Flowers unisexual. Crossing general. Many varieties rarely flowers
Yes	Copra yield Oil content	'Red Ring' Lethal yellowing	Crossing general in tall palms
Yes	Flavour	Leaf rust. Berry disease	Normally selfed. 7-9% natural cross-pollination
Yes	'Quality'	Very resistant	Self-incompitible
Essential	Lint quality Ginning out-turn. Insect resistance Glandless seeds - gossypol free.	Blackarm - bacterial blight. Leaf curl. *Verticillium* wilt	Up to 15% out-crossing which is easily prevented. Adequate seed produced. Seed varies from naked with a tuft of hairs to fuzzy

Table 3.2 Summary of basic facts for improvement of tropical crops (continued)

Crop	**Latin name**	**Chromosome number** (basic, diploid)	**Normal method of propagation**	**Is there a practical method of vegetative propagation?**	**Valuables part of plant**
Citrus	*Citrus* cultivars L.	$2n = 2x = 18$	Budding on to rough lemon or sweet orange - resistant to Tristeza. Often grown from seed.	—	Fruit
Ground-nuts	*Arachis hypogaea* L.	$2n = 4x = 40$	Seed	No	Seed
Jute	*Corchorus capsularis* L. and *C. olitorius* L.	$2n = 2x = 14$	Seed	No	Phloem fibres from stem
Maize	*Zea mays* L.	$2n = 2x = 20$	Seed	No	Seed
Oil palm	*Elaeis guineensis* Jacq.	$2n = 2x = 32$	Seed. Tissue culture of meristem now achieved.	No	Fruit
Rice	*Oryza sativa* L. 2 types — *Indica* and *Japonica*	$2n = 2x = 24$	Seed	No	Seed
Rubber	*Hevea brasiliensis* (Willd ex Adr. de Juss.) Muell. -Arg.	$2n = 2x = 36$	Seed or bud grafting.	Yes	Latex from the cortex of the trunk.

Is production of seed necessary?	Characters required in addition to yield	Diseases to which resistance is required	Remarks
Preferably absent.	Flavour Colouring properties Absence of seeds	Tristeza. *Phytophthora* root rot. Canker	In certain citrus varieties seed from nucellar not sexual embryo thus 'clonal' seed
Yes	Uniformity of maturity Seed dormancy	'Rosette'. Rust. Leaf spot	Strongly selfed
No	Fibre strength	*Rhizoctonia*	Cross pollinated
Yes	Better storing properties High lysine	Rust *Helminth-osporium* and other leaf diseases. Many insect pests	Crossed. Mainly wind pollination 'Synthetics' open pollinated
Yes	Fertility Drought resistance.	*Fusarium* wilt.	Monoecious. Cross-pollination by wind general. Now always grown as *tenera* palms.
Yes	Uniformity for milling. Absence of red colour. Appearance after cooking. Deep water tolerance. Response to fertilizer	Blast and virus diseases.	Mainly self-fertilized. Crosses between *indica* and *japonica* races may be sterile.
No, only to provide root stock.	Bark renewal. Absence of low branches. Wind resistance	South American leaf disease. *Oidium* *Phytophthora*.	Most clones self-sterile. Some produce 'clonal seed' by self-fertilization and transmit high production

Table 3.2 Summary of basic facts for improvement of tropical crops (continued)

Crop	**Latin name**	**Chromosome number** (basic, diploid)	**Normal method of propagation**	**Is there a practical method of vegetative propagation?**	**Valuables part of plant**
Sorghum	*Sorghum bicolor* (L.) Moench	$2n = 2x = 20$	Seed	No	Seed
Sim-sim sesame	*Sesamum orientale* L. syn. *S. indicum* L.	$2n = 2x = 26$	Seed	No	Seed
Sisal	*Agave sisalana* Perrine	$2n = 5x =$ 138-149 Pentaploid $x = 30$	Vegetative bulbils or suckers.	Normal	Leaf fibres
Sugar cane	*Saccharum officinarum* L.	$2n = 80$ to *136* $x = 10$	Vegetative stem cuttings.	Normal	Stem
Sweet potato	*Ipomoea batatas* (L.) Lam.	$2n = 6x = 90$ Hexaploid $x = 15$	Vegetative stem cuttings.	Normal	Tuber
Tea	*Camellia sinensis* (L.) O. Kuntze	$2n = 2x = 30$	Seed or rooted cuttings.	Yes	Leaf
Tobacco	*Nicotiana tabacum* L.	$2n = 4x = 48$ (allotetraploid)	Seed	No	Leaf
Yam	*Dioscorea* spp. L.	$2n = 20$ to 144 $x = 10$ (Asian, African) $x = 9$ (American)	Stem cuttings or tubers	Normal	Tuber

Is production of seed necessary?	Characters required in addition to yield	Diseases to which resistance is required	Remarks
Yes	Yellow colour. High lysine. Striga resistance. Resistance to bird damage.	Head smut	Compact heads largely selfed. Up to 5% crossing recorded. Bagging ensures selfing. Hybrid sorghums developed.
Yes	Non-shattering. Oil content.	Phyllody virus.	White seeded varieties for food. Predominantly self-fertilized, some cross fertilizing by insects. Black seeded varieties for oil.
No	Fibre strength	*Phytophthora* rot.	Flowers after 7 years or more then plant dies. Produces some viable seed.
Undesirable	Sugar content. Non-inversion. Self-stripping	Virus, smut, insects, particularly borers	Large number of canes raised from seed in breeding but small proportion superior
No	Flavour	Mosaic Black rot	Many varieties self-sterile but set seed when crossed. Seedlings show wide variation of leaf and root form.
No	Flavour. Vegetative habit	Blister blight	Largely self-sterile
Yes	Leaf quality. Less nicotine	Very many including *Cercospora* Blue mould	Normally self-fertilized. Some crossing, prevented by bagging
Yes	Low alkaloid content. Flavour	Tuber rots. Virus disease	All species dioecious and highly sterile. Fertile seed of *D. rotundata* obtainable

grown as dwarf varieties, maturing earlier and easier to maintain; particularly if they have to be sprayed.

Disease resistance (See also Ch. 4)

Breeding of crop varieties resistant to economically important pathogens gives an immediate yield response, and saves the cost of fungicide spraying if normally carried out. Resistance to airborne pathogens depends either on one or a few genes (oligogenes) which may confer immunity to the existing races of the disease, or it is controlled by several genes of small individual effect (polygenic resistance) which confers no immunity but gives the crop variety resistance to many races. Varieties with oligogenic resistance initially show a marked increase in yield and the area planted to these varieties is extended. In a few years, however, a new race of the disease develops and resistance breaks down. On the other hand, varieties with polygenic resistance show field tolerance and a low but acceptable incidence of disease with the old and new races of the pathogen. Simmonds (1969) predicts that in time oligogenetic specific resistances to airborne fungi will be a thing of the past: polygenic 'field resistance' is harder to handle but it is much more stable and effective. Simmonds also points out that the same principles apply to diseases caused by bacteria, viruses, soil-borne fungi and nematodes, but have not worked so dramatically in practice, oligogenic resistance to these diseases being more effective and stable though breakdowns have occurred.

The following deals briefly with some of the problems and achievements of the main tropical crops.

Banana*

The requirement is a banana resistant to the two major diseases, Panama and leaf spot, with the shipping and 'consumer acceptability' qualities of 'Gros Michel'. Recently estate packaging of banana hands has become the normal shipping method. This may alter the requirement of bunch form and ripening characteristics from previous requirements.

Early breeding work in Trinidad and Jamaica produced the varieties IC 2 and S 19 in Jamaica, tetraploids ($2n = 44$), from pollinating 'Gros Michel', a triploid, with the wild *Musa acuminata*, a diploid wild species resistant to diseases. The 'near success' of these varieties controlled breeding policy for a number of years using different parents and

* For a full account see Simmonds, 1966; Menendez & Shepherd, 1975.

crossing with and among the resultant tetraploids. Repeated disappointments resulted in a decision that the ultimate solution lay in crossing 'Gros Michel' or 'Highgate', a semi-dwarf mutant, with a male parent resistant to the diseases, with a reasonable bunch formation, and parthenocarpic, so a series of new male parents was created by crossing the wild *Musa acuminata* with an edible diploid 'Pisang lilin' from Malaysia.

The labour and cost involved in producing the seedlings from the cross with 'Gros Michel' is shown by Simmonds. Despite some knowledge of the factors influencing female fertility only one or two seeds are produced per bunch. The germination of the seed is poor and over half the seedlings are heptaploids and are discarded. Only 9 per cent of the seeds obtained in the breeding programme gave tetraploids, giving an approximate yield of 60 seedlings per hectare of 'Gros Michel', all the fruit of which is destroyed in the search for the seeds. The effort of extracting 600 seeds from a hectare of bananas is tedious and tremendous. If the figures for success in the early days of sugar cane breeding in Java are any guide, where only 1 seedling in about 10,000 was above average, this could mean searching through 100 ha or about 2,000 tonnes of pollinated bananas to obtain the right seedling, but the theoretical genetic constitution of the seed leads one to expect a much more productive return of effort. The current trend in banana growing is the high density planting of 'Robusta', a short Cavendish clone which is resistant to Panama disease but very susceptible to leaf spot and nematodes.

Cocoa* (Theobroma cacao L.)

Cocoa is a single variable species with a large number of localized populations. When all cacao varieties are considered, this species exhibits one of the largest ranges of observable variation known in plants, which offers great scope for producing better varieties. Three main types are grown, all of which can inter-cross. Type 2 dominates the market.

1. *Criollo.* Long pointed pods mainly red or yellow with broad beans, white or pale violet cotyledons. High quality. Mainly grown in Venezuela, Colombia and Central America, but introduced to other areas. Not very robust and a low producer. Mainly found in mountainous or drier areas.
2. *Amazonian forastero* (Amelonado). The main cocoa of the world. Grown in West Africa and Brazil. More rounded, flat surfaced,

* See Wood (1975).

smoother pod with a thick shell and flat beans with purple cotyledons. Fast growing, high yielding. Ordinary quality.

3. *Trinitario.* A heterogeneous population probably arisen from a mixture of the two preceding groups. More robust and a better yielder than Criollo. New Guinea is now the largest producer of Trinitario cocoa.

Cocoa yield is very subject to environmental differences, but among the hybrid and variable low yielding populations of cacao found in the Caribbean area (Trinidad, Venezuela, Colombia and the Dominican Republic), are a number of high yielding trees, superior to the average on a genetic basis, not due to fortuitous soil or position in the field. By selecting such trees and propagating them vegetatively the yield of new plantings made with such selections will be much higher than average. The cocoa grown in the Bahia zone of Brazil and in West Africa does not have such a large variation.

In 1930 a search for good clones was started in Trinidad, reported as follows:

Altogether about 250,000 trees have been surveyed and some 250 are on the provisional lists. It is hoped by the end of 1934–35 crop the number found which will yield at the rate of at least one ton to the acre will exceed 100. (Pound 1934, p. 11).

One ton of dry cocoa per acre (2.5 t/ha) was the selection standard based on the production of 50 pods per tree per year with a pod value* of 7.5 at a spacing of 12 ft by 12 ft (3.6 m × 3.6 m). From many thousands of trees surveyed, 100 trees were selected which formed the Imperial College Selections (ICS 1 to 100). These were propagated by rooted cuttings.

Unfortunately many of these selections which are described in the Annual Reports of the Regional Cacao Research Scheme around 1934 were disappointing either as yielders, susceptibility to witches' broom or cocoa beetle (*Steirastoma breve*), or because of their poor flavour. ICS 1 for example which has yielded up to 1,800 kg of dry cocoa per hectare on sandy loam soil compared with 350–600 kg average for mixed trees, suffers from wind damage, poor drainage, thrips, and needs a good shade cover. ICS 89, a slow grower and late yielder, and ICS 95, which is probably the best clone selected in Trinidad bearing well in a variety of environments, are suitable for most cocoa soils in Trinidad. ICS 95 has proved to be the best clone planted in Surinam. Four of the clones ICS 39, 40, 43 and 60 are large seeded Criollos which with ICS 95 were the five highest yielding selections. Few of the ICS clones yielded half the target of 2.5 t/ha. The breeding of cocoa is shown schematically in Fig. 3.3.

* Pod index or pod value is the number of pods required to make one pound of dry cocoa, and can vary from 6 to 22 pods.

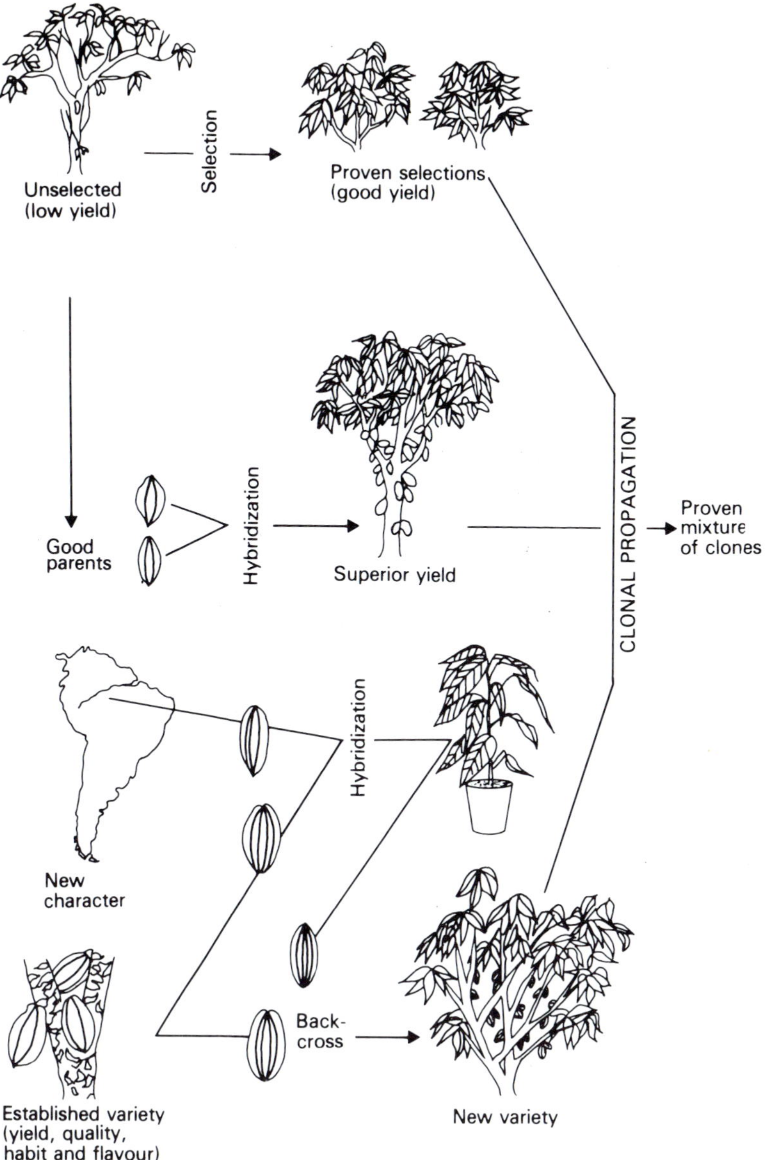

Fig. 3.3 Plant breeding – cocoa.

Vegetative propagation of clones needs an efficient organization and close supervision both during propagation and planting out. Propagation from seed is easier and cheaper. Certain trees are self-incompatible and this has been used to produce hybrid seed similar to the method used with maize. Unfortunately the yield or appearance of the parents has little relation to progeny. As the selected clones are strongly heterozygous for many characters, crossing these selections produces considerable variability in yield and plant type. It is therefore necessary to create good parents, less heterozygous by inbreeding, and combine the ideal parents in a way which produces the desirable characters in the progeny together with hybrid vigour. This is a very slow process with a tree crop but some reasonably good yielding hybrids have been produced from first generation inbreds.

The comparison of clones and their seedling progeny is shown in a plot planted at Santa Cruz (Trinidad) in 1963.

Results for the 7 month crop up to early 1967 were (Freeman 1967):

	Clone TSA 644	**Clone TSA 565**	**Seedling progeny TSA 644 × TSA 565**
Number of trees	194	197	171
Pods per tree	12.4	9.8	7.6
Pod index (Pods/lb dry beans)	8.2	6.8	8.2
Dry cocoa (kg per tree)	0.73	0.64	0.41
Dry cocoa (kg/ha)	650	607	390
Bean wt. (grains dry)	1.2	1.5	1.2
Beans per kg (approx.)	831	688	831

The development of seedlings is as yet not sufficiently advanced to justify the planting of seedling material as opposed to clones, except where the after-planting management is not adequate to care for vegetatively propagated plants. The main breeding programme at the Cocoa Research Institute at Tafo, Ghana, is aiming at producing varieties to be grown from seed.

Yield, bean size and flavour are not the only characters of concern to the cacao breeder. Diseases and pests are serious in all areas and cause a heavy loss of crop and trees. To find a source of resistance to witches' broom disease which was becoming serious in Trinidad, Pound went to Ecuador in 1936 and up the Amazon the following year, and in 1942. The material he collected is now the basis of all cocoa breeding programmes.

In 1937 Pound observed differences in resistance to witches' broom (*Marasmius perniciosus*; syn. *Crinipellis perniciosa*) in trees on a plantation in Ecuador. These clones, SCA 6 and SCA 12, proved to be practically immune to this disease. Though high yielding with many pods, the pods and beans are commercially undesirably small. Some resistance also occurs in some of the ICS selections including ICS 1 and ICS 95. SCA 6 in particular has been shown to transmit this resistance to its progeny (Bartley 1957). SCA 6 also shows resistance to black pod which is again often transmitted to its progeny. IMC 67, which Pound collected from Iquitos in the Upper Amazon, has no resistance to black pod but has resistance to *Ceratocystis*. Due to the existence of physiological races of black pod, the resistance of a clone varies from country to country. For example, Catongo is resistant in Brazil but susceptible in Trinidad.

Ceratocystis wilt caused by the *Ceratocystis fimbriata/Xyleborus* complex has become a major disease in the Caribbean area. Resistance fortunately occurs in SCA 12, IMC 67, and two Venezuelan selections, Panaquirito 87 and Santa Cruz 10. Criollo varieties on the other hand appear to be entirely susceptible. Current progress in Venezuela was reviewed by Dominguez (1971).

As an example of the current breeding programme in Trinidad, Freeman (1967) gave the development of TSH 991.

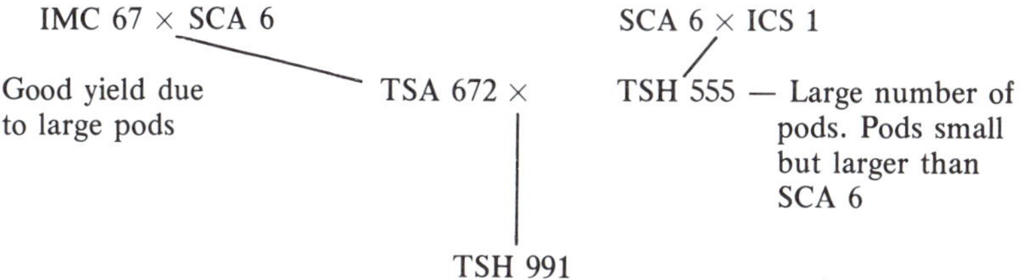

The seven trees of TSH 991 clone at 3 years old yielded at the rate of 2,240 kg dry cocoa per hectare. The pods are large with only about 44 beans per pod, and 19 pods for 1 kg of dry cocoa. This clone is resistant to witches' broom, *Ceratocystis* wilt and *Phytophthora*. About 200 other clones developed by hybridization are resistant to *Ceratocystis*.

Progress is slow. It takes 10 years, as shown below, to get the initial data on a clone after the cross is made.

A cross made in 1967
Seed sown in 1968
Seedling selection 1972
Clonal cuttings 1973
Initial data on new clones 1977

Testing for resistance to *Ceratocystis* and black pod is carried out in the laboratory with bark strips from 6-month-old seedlings inoculated and incubated for 72 hours. This gives good correlation with field performance and allows susceptible clones to be discarded early. Clones were distributed as rooted cuttings up to 1957, after which hybrid seedlings were produced.

In West Africa selections have been made among the Amelonado cocoa of West Africa and clones for crossing have been introduced from other countries, particularly Pound's selections introduced from Trinidad, mainly to counter 'swollen shoot' disease. The Upper Amazon types come into bearing in the fourth year compared with the seventh year for the local Amelonado cocoa. The early bearing character of the Amazon tends to appear in their hybrids. The Amazon selections introduced into West Africa appear to be more susceptible to black pod and canker than the local Amelonado types.

Vegetative propagation.

Efficient methods of vegetative propagation have been developed largely in Trinidad. Archibald (1955) summarized the experience in Trinidad and West Africa.

The work of Evans and his colleagues in the early 1950s at the Imperial College of Tropical Agriculture (ICTA) showed that the main points for successful vegetative propagation were as follows:

1. Sufficient leaf area to maintain carbohydrate production for the developing root system - 6-8 leaves are usually left on a cutting, with half of each removed to reduce mutual shading.
2. Adequate light. Usually about 5 per cent of full sunlight.
3. A steady air temperature of 80-84 °F. (27-29 °C)
4. A completely saturated atmosphere.
5. A rooting medium which allows free drainage and adequate aeration, but holds sufficient water to keep the cells of the leaves fully turgid. Sawdust is good. Waterlogging will cause the cuttings to rot.
6. Absence of parasitic fungi, bacteria and nematodes.
7. Presence of adequate growth hormone. The ends of the cutting should be treated with a mixture of β-indole-butyric acid and α-naphthalene-acetic acid.

The various designs of propagators are described and illustrated by Hardy (1960). Under optimum conditions a prolific root system develops in 2 to 3 weeks. The rooted cuttings are taken from the rooting bin after 4 weeks and hardened off for 2 weeks. About half the original cuttings should produce usable plants.

Bud grafting, one of the earliest techniques used for the vegetative propagation of cocoa, has not been used to any extent. Van Der Burg (1969) describes a new technique used in Tanzania which is much cheaper than growing from cuttings. Tissue culture has yet to produce a cacao tree but the signs are hopeful.

As with most other tree crops, the use of polybags reduces the cost by eliminating the need for repotting and causes less root disturbance.

Coconuts (Cocos nucifera L.)

The coconut is monoecious and declinous (i.e. the stamens and pistils are on separate flowers but on the same plant) with a greater number of male than female flowers on the spadix. In the tall palms, which are most frequently grown for copra production, the flowering periods of the two types of flowers do not overlap and outcrossing is general. In dry weather the pollen may remain viable for a longer period and fertilize the female flowers on the same spadix but this is not thought to be common.

On the other hand, with the dwarf palm, the interval between the male and female flowers opening is shorter and the female flower remains receptive for a longer period. Consequently self-fertilization is general (Jack & Sands 1929). While the dwarf palms are quicker to come into bearing (3 to 4 years) and easier to harvest, they yield inferior copra and are shorter lived. The tall and dwarf forms can be crossed easily. Early results from Malaysia show that the Malayan Dwarf × West African Tall hybrids yield earlier and better than the Malayan Tall coconut. This hybrid has the best frond production rate - hence a higher bunch number, more female flowers, and better nuts/palm and copra/palm ratios. Hybrids with the Rennel Tall, a large fruited coconut, gave the highest copra/nut return (Chan 1979).

There is no known method of vegetative propagation of the coconut palm; tissue culture is being tried.

Where selection has been applied, the standard practice, and one recommended for many years, has been to select seed nuts from high yielding palms with a desirable habit and other characters. The weaknesses of this system are obvious. The male parent is unknown and the high productivity of the mother palm may not be transmissible but a result of a fertile patch of soil, reduced competition due to an edge effect, or loss of nearby palms, or a chance combination of genes which is not transmitted on crossing.

In Sri Lanka the suggestion was made (Cheyne 1952) that equally good results can be achieved by a careful selection of the best nuts from the heap harvested from the best blocks on the best estates. This is a much simpler method which avoids recording, but suffers from the weakness of the mother palms selection method together with the absence of any information about either parent.

A further proposal in Sri Lanka was to establish an isolated seed garden, planted with controlled crosses made with high yielding palms. The success of this method depends upon the best crossed with the best producing the best, but the results of such crossings in other crops have sometimes been disappointing, and the method needs experimental validation.

The only effective method of selecting mother trees would appear to be by progeny testing. The process of discovering male and female parents with specific combining ability will be very slow but some mother palms will transmit the high yielding character irrespective of the male parent and certain controlled crosses will give rise to particularly high yielding progeny. A minimum of five plants should be in the progeny row which should be replicated at a different site. As it takes a tall palm 7 to 10 years to come into bearing, progeny testing is a long process.

Harland has suggested that the process of identifying male transmitters can be speeded up by using the dwarf palms, which come into bearing very early, as female top-cross testers. Each bunch on the chosen dwarf female parents would be pollinated by a different high yielding tall male. Since the dwarfs are largely self-pollinated their progeny are reasonably homogeneous. The progenies of single bunches on a palm, each from a different male, can be studied to identify the most prepotent males. Progenies from 5 to 10 seedlings will be adequate to make a preliminary and tentative separation.

A programme of selfing of progeny tested palms which transmit high yield to develop pure lines with this character followed by the planting of isolated seed gardens of pure lines which cross particularly well might give outstanding yielders in the same manner as 'hybrid maize', but this would take a very long time to develop.

Selection in the nursery may be of value. The percentage discards on nursery selection should be determined in progeny testing of controlled crosses and those giving a high percentage of rejection discarded. It is obvious that much time could be saved in establishing a source of reliable seed if means could be found of propagating the coconut palm vegetatively or by tissue culture, as is now possible with oil palms. The date palm (*Phoenix dactylifera*) and certain other palms produce basal suckers which can be used for vegetative propagation. Narayana and John (1949) record rare cases of this happening in coconut. They also recorded the formation of bulbils by the female flower (vivipary) in a manner similar to sisal. These should be followed up and research effort devoted to attempting to propagate the palm vegetatively. Harland also suggests that use could be made of the tendency of the coconut to form adventitious roots from the trunk under certain circumstances. If this causes rejuvenation of the palm, the newly rooted proven palms could be transplanted to an isolated seed garden.

As most of these methods are long-term projects it is important that nursery selection, which is currently playing such an important part, should be soundly based on a true correlation of seedling and adult characters.

Breeding work in Jamaica

The important disease of coconuts in Jamaica is 'Lethal Yellowing' which has been slowly creeping across the island over the last century. In the 1970s this disease reduced copra production from over 20,000 t to under 5,000 t. The cause is uncertain, but the result is rapid death. It has been shown that all the three colour forms (yellow, green, red) of 'Malayan Dwarf' coconuts are resistant to 'Lethal Yellowing' (Whitehead 1966). This resistance, which is the only defence against this disease, appears to be transmitted to the F_1 generation. The yields of Malayan Dwarf palms in Jamaica are at least comparable with the yields from the local tall palms. This is sometimes deceptive in the field as the nuts ripen quicker on the dwarfs which are thus always carrying a much smaller proportion of their annual yield than tall palms. Dwarf palms, which grow to 8-10 m after 20 years and begin bearing about the third year when 1 m high, can be planted closer together and are much better than tall palms, which grow up to 20-30 m, for replacing hurricane losses, due to their earlier maturity. As the available plant material for breeding is limited the Research Department of the Coconut Industry Board of Jamaica have developed methods of freeze drying of coconut pollen and have exchanged pollen with all the major coconut growing countries. This saves at least 4 years in the introduction of new characters (Whitehead & Thompson 1966). By selection it should be possible to raise the nut and copra qualities of the dwarf palms to that of the tall palm. Dwarf coconuts become profitable within 6 years of planting. As the dwarf palms are mainly self-fertilized such improvement will be maintained, whereas the progeny of tall × dwarf hybrids are very variable. The tall × dwarf crosses may benefit from hybrid vigour. Linked with the Jamaica breeding programme is a UK-based research project on the vegetative reproduction of the coconut palm by tissue culture (Schwabe 1973). Progress will be slow, and the aim may not be achieved, but if it is the reward could be very significant to those to whom the coconut is so important.

Coffee - arabica

Coffee arabica the only tetraploid (2n = 44) coffee species widely cultivated, is the only self-compatible *Coffea* species. Interspecific hybrids with *C. canephora* (robusta coffee) have been produced ($2n$ = 33) but these are of little value due to their low fertility associated with their triploid nature. Improvement has therefore been confined to this one species, though efforts are continually being made to introduce the marked disease resistance of many of the diploids.

As *C. arabica* is predominantly self-fertilized (over 90 per cent at Campinas, Brazil; Krug 1959), selected trees should be reasonably homozygous and their progeny raised from seed resemble the mother trees. Indeed arabicas are so clone-like in character that they offer little potential for selection.

Techniques have been developed for crossing varieties. Selection should continue for a few generations after such a cross to stabilize more homozygous varieties for propagation. In Brazil work has been directed to the improvement of yield under a variety of growing conditions. The arrival of *Hemileia* leaf rust in 1970 has changed priorities.

Considering the importance of arabica coffee to the economies of certain countries, surprisingly little breeding work is carried out. The largest programme has been carried out at Campinas in Brazil. The most important strains developed at Campinas belong to the 'Mundo Novo' cultivar, the best strains (e.g. CP 387-17 and MP 388-6-20) yield about two and a half times as much as the *typica* variety that it has largely replaced, and 80 per cent more than the high yielding *bourbon* variety. The beans are also larger and there are more flat beans than with other varieties.

The first distribution was in 1939 but it was widely planted on a large scale (Krug 1959). No significant differences between the yields of single and double hybrids and the best Mundo Novo progenies have been noticed. There is no evidence for heterosis in these crosses (Carvalho *et al*. 1978).

In the Far East, including India and Sri Lanka, resistance to coffee rust has been a major objective. At Balehonuur (Mysore, India) selections of S.26 after selfing gave S.288, released in 1938 which was crossed with the famous Kents variety to give S.795, released in 1946. These varieties, S.288 and S.795, carry some resistance to *Hemileia* and are popular in South India where at least nine races of rust are thought to exist. Back-crossing of robusta × Kents to Kents and interspecific crosses are being tried to build up high yielding resistant trees of desirable quality. Breeding for resistance is made more difficult by the existence of physiological races, more than 25 are known world-wide, and also by basing resistance on unreliable single genes. Some of these coffee varieties are being used to replace robusta in parts of India, but in Indonesia in the early 1900s the industry was saved by the reverse process, the arabica trees dying out and being replaced by robusta trees. Resistance to the two races of rust in East Africa was obtained by crossing the Geisha type from Ethiopia with the existing varieties. Blue Mountain K.7 is resistant to race II and S.L.28 to coffee berry disease (*Colletotrichum*) which has become serious in East Africa since 1950. San Ramon a dwarf arabica was introduced into India in 1953 where its resistance to rust has been improved. Three selections have resulted from this breeding programme at Chickmagolur (Srinivasan &

Vishveshwara 1979). When hybrid material field resistant to coffee leaf rust and coffee berry disease has been developed it will be necessary to top-graft existing trees. The wild *C. arabica* trees collected since the war in Ethiopia, are remarkably diverse in their characters and offer a wealth of new material to breeders.

Coffee - robusta

Coffea canephora (Pierre ex Froehner) is a diploid ($2n = 22$) species which came into prominence when the rapid spread of coffee rust eliminated the arabica coffee from Indonesia in the early 1900s. Robusta coffee is generally considered to be self-sterile, but Thomas (1947) records that when planning to move the agricultural research station from the Kampala plantation to Kawanda in 1935, the best trees on the plantation were selfed and their progenies planted at Kawanda though one wonders why when Thomas also records that the traditional planting method, developed long before the arrival of the European in Uganda, was not to grow coffee from seed but from large woody cuttings. Vegetative propagation has played little part in the development of coffee growing.

One reason was the belief that cuttings will not form tap roots and are therefore vulnerable to drought and wind. Given suitable rooting conditions, cuttings will develop tap roots as easily as seedlings. Another erroneous reason was that robusta is largely self-incompatible and a field planted to a single clone would set little crop. As a defence against pests and diseases and other risks, it would in any case be advisable to plant a mixture of clones not a single clone. In Uganda vegetative propagation was started on a large scale, using soft-wood orthotropic (upright growing) cuttings, 6-8 cm long with one pair of leaves reduced to half their area. These root in 10-12 weeks. Some clones root much more easily than others. Similarly in the Ivory Coast, robusta clones are vegetatively propagated on a large scale (Capot 1966). In Indonesia many of the robusta coffee trees were grafted largely to have a rootstock resistant to nematodes.

Ferwerda (1959) states that open pollinated progeny of good clones yield about twice as much as the original unselected seedlings, and the yield from single or multiple crosses from outstanding clones shows an even larger improvement. On the other hand experience in Java from a wide selection of material shows that less than 5 per cent of the mother trees selected resulted in a really good clone, and due to pronounced self-incompatability monoclonal blocks set little seed. Selection of clones that together give a high yield is a difficult and long process.

In many countries the seed used for raising new plants is collected from good mother trees or from mixed clonal plantings. This cannot

Plate 3.1 A heavy crop of Robusta coffee in Buganda. (Courtesy Department of Information, Uganda ; photograph Ron Ward.)

guarantee good progeny. The problems of improvement are similar to those involved in coconut breeding, though coffee is a little simpler in that vegetative propagation is possible.

The selection work of Thomas (1940) illustrates what can be achieved by selection and careful observation. Some of the selections were upright in growth, others spreading. The latter *nganda* type are ideal for peasant cultivation (Plate 3.1). Vigour after 10 years was one basis for selection. One selected mother tree died at 120 - 150 years of age.

For most of the countries growing robusta, the present variable planting material is satisfactory as pests and diseases are of minor importance.

Cotton

The world's cotton production is based on the seed fibres of 4 of the 39 species of *Gossypium* as shown in Table 3.3. The wild species of *Gossypium* do not have spinnable fibres on their seeds. Cotton species are basically tropical perennials and the early cultivation systems ratooned the crop, but today they are grown as annuals.

Table 3.3 The four commercial species of cotton

Species		**Contribution to world production**	**Lint length**	**Cytotaxonomy**	**Main areas of growth**
G. hirsutum L. American Upland	New World or American species	≏95%	Medium-long staple $1\frac{3}{8}''$	Tetraploids $2n = 4x = 52$	USA USSR
G. barbadense L. (syn. *G. peruvianum*) Cav. Sea Island Cotton		≏5%	Extra long $1\frac{3}{8}''$ and over		Egypt Sudan USSR Peru
G. herbaceum L.	Old World or Asiatic species	Less than 1%	Short staple less less than 7/8"	Diploids $2n = 26$	India and Pakistan
G. arboreum L.					

After: Phillips (1976).

Breeding technique

The cotton flowers open when the stigma is receptive, and the pollen, which remains viable for 12 hours, is shed at the same time. The pollen is too heavy and sticky for wind dispersal and cotton is generally considered to be self-pollinating. Crossing, caused by visiting insects, mainly bees, ranges from zero to 50 per cent in unusual circumstances, but where insecticides are sprayed cross-fertilization by insects is minimal. Negligible crossing occurs with a 100 m isolation zone.

Pollination of cotton is easily controlled by sealing the flower bud before it opens, using wire, paper clips, string or a bag to exclude insects. The seed from flowers which were selfed should be kept separate from the seed from the same plant where the flowers were not tied and may have been cross-fertilized. A single plant selection will yield about 100 g of seed but often much more. For crossing, the flower bud can be emasculated the day before the pollen is released. Flower buds at this stage are easily identified. The pistil of the emasculated flower is then protected by a short length of drinking straw. A cross should yield 20 - 30 hybrid seed in the boll.

As cotton is predominantly self-fertilized the commercial crop is a mixture of relatively homozygous lines. Mass selection is used to maintain the variety and to establish pure lines. There is sufficient natural cross-pollination to maintain heterozygosity and some heterosis.

Methods of improvement

1. *By introduction.* Most of the areas of existing cotton production are based on introduced plant material. Further introductions are largely confined to special material for hybridization.
2. *Selection.* Many major advances have been derived from single plant selections from a large commercial crop (e.g. BP 52 in Uganda). Selections are normally grown together in progeny rows, each row from a single plant selection which has been selfed. The flowers of this row are protected from crossing to maintain the selection. Recurrent selection has been used successfully to improve fibre length.
3. *Hybridization.* Both natural and controlled hybridization have been used. Interspecific hybrids, *Gossypium hirsutum* with a diploid, produces a triploid F_1 which when treated with colchicine produces a hexaploid. By selection, these rapidly produce some tetraploids. Alternatively, the chromosome number of a diploid cross can be doubled to produce a tetraploid to cross with *G. hirsutum*.

Yield potential

The cotton breeder is aiming at an increased yield of a specific quality of lint. As lint yield is determined by the number of bolls per unit area, seeds per boll, and lint per seed, increased yield can be achieved by improving any of these three.

Maintenance of stock

To maintain the type once decided it is necessary to have replacement waves of selected seed. One progeny row of 20 plants from a single plant selection if carefully used can plant 0.1 ha the next year, but it will be 4 years before the seed supply is adequate to plant an area of 16,000 ha, at a multiplication rate of 20 times per annum.

The genetic base for the cotton grown in many countries is generally very narrow. For example, Egyptian cotton which is a form of *G. barbadense* is said to have originated from a single plant in a Cairo garden in 1820 - Gumel's cotton. In future the genetic base of all cultivated cottons will be broadened.

Disease resistance

Breeding of tolerance to certain diseases is included in all programmes. *Verticillium* wilt (*V. dahliae*) is probably the most serious disease worldwide. Unfortunately, tolerance to this disease is associated with excessive vegetation growth at the expense of yield. There are also pathogenic races of *Verticillium*.

High resistance to bacterial blight or blackarm (*Xanthomonas malvacearum*) can be achieved by a combination of numerous major and minor genes.

Insect resistance

As cotton is attacked by such a large number and variety of insects, more insecticide is sprayed on cotton in the world than on any other crop. Not only is this very expensive but also a serious pollutant of the environment, so there is double need to introduce cotton varieties which are resistant to pests. The best successful example is the development of resistance to jassids (*Empoasca* spp.) based on the selection and breeding of the hairy variety, U4, by Parnell in South Africa in the early 1920s. The jassid is now no longer a serious cotton pest in tropical Africa. Resistance to a number of other pests does occur in different species of *Gossypium* but as yet has not been introduced into commercial varieties sufficiently successfully to reduce the need for spraying.

Adaption to mechanization

With the increasing cost and the difficulty of obtaining hand labour for picking there is a need to breed varieties suitable for mechanical harvesting. Yield must be maintained at the wider spacing, the bolls must be dispersed on the crop in a way suited to the machine, and the bolls must mostly ripen together which may reduce pest damage.

Maize [Zea mays L.]

As maize is naturally cross-pollinated the crop is highly heterozygous. In most tropical countries mass selection is practised. Mass selection has been responsible for the improvement of primitive maize and has given the wide diversity of forms which occur. Having reached a certain stage, mass selection can only maintain the standard without having any permanent effect on the improvement of the crop. Selections on the basis of a single good cob may produce a lower yield than plants having two medium sized cobs.

Maize has however attracted as much attention as any crop due to the development of 'hybrid maize'. Though the first commercial hybrids were released in Connecticut in 1921 they were not generally accepted until more productive material was produced. In Iowa in 1933 less than 1 per cent of the maize acreage was planted to hybrid seed, but in the next 10 years it replaced all other planting material. Initially hybrid maize seed raised yields by 30 per cent over the older varieties, and in succeeding years a further 20 or 30 per cent increase was achieved (Sprague 1960). In 1969 over 100,000 ha of hybrid maize were grown in Kenya which yielded about 20 per cent more than the varieties replaced.

From 1969–73 the area under hybrid maize in Kenya increased from 100,000 to 300,000 ha.

To produce hybrid maize seed, pure lines are first developed by continued selfing. Continued selfing concentrates the desirable characters and the undesirable recessives can be discarded. At each selfing approximately 50 per cent of the remaining heterozygous loci are converted to the homozygous state. Two such pure lines are then crossed, and two of the F_1 generations similarly crossed and the resultant double cross seed is distributed as commercial hybrid seed. Since the late 1950s single cross hybrid seed has become more popular in the USA. This is more expensive but with the new inbred lines can yield more. Success lies in combining the right parents for the F_1 generation and the double cross. Thousands of pure lines have been developed but less than 50 are important in the production of hybrid maize. Once a successful cross is discovered it can be repeated to order. In the absence of vegetative propagation the breeder has no alternative method of creating a heterozygous population of predictable performance. The seed harvested from the hybrid corn crop isolated from other maize, if planted, produces less heterotic progeny that may show no superiority over fields planted to open pollinated seed.

Hybrid seed has the great disadvantage for more widespread use, that fresh seed must be purchased annually and this is expensive due to the care and knowledge needed to produce it. A local seed industry is necessary to supply the seed. High yielding maize demands a high standard of cultivation and a high level of fertilizer usage.

Most of the maize grown in the tropics is grown under peasant conditions and hybrid maize is not suited for this system. Of the large number of pure lines developed, undoubtedly some are superior to the natural population from which they were derived. If planted in large blocks, pure seed, the product of continued selfing, would maintain itself in a fairly pure condition for a number of generations. The Rockefeller Foundation agricultural programme in both Mexico and Colombia is based on hybrids of lines selfed for a single generation. The performance of S_1 lines in hybrid combinations is essentially the same as the hybrids from the same lines selfed over a number of generations (Sprague & Miller 1952).

This Mexican programme is essentially designed to develop synthetic varieties. A synthetic variety is one which is maintained from open pollinated seed following its synthesis by hybridization in all combinations among a number of genotypes, which can be clones, inbred lines or mass selected populations. The difference between synthetic varieties and varieties from mass selection or line breeding is in the choice of genotypes. These have been tested for combining ability, and only those which combine well in all combinations are put in a synthetic variety (Allard 1960). Put simply, the chosen genotypes have been progeny tested. Synthetic varieties are of greater value in the tropics than hybrid maize as the seed can be saved from one crop to the next for 3 years; whereas with hybrid maize it is generally necessary to replace the seed for each crop, and the maintenance of seed supplies needs close supervision.

'Synthetics' have a lower yield ceiling of 3 to 5 t/ha compared with 5 to 6 t from the best 'hybrids', but they have a wider range of adaptability. Current breeding work on these varieties is designed to reduce plant height to avoid lodging and increase plant population; to introduce multiple ear characters; to improve resistance to rust, downy mildew, stalk rot, leaf blight and stem borers; and to raise protein content and quality (FAO 1970, p. 110).

Currently, maize produces less protein for the same amount of calories than wheat, millet or sorghum. Though rice has a lower protein content than other cereals it is of better quality due to its amino acid pattern. Lysine is the limiting factor in all cereal proteins.

The discovery (Mertz *et al.* 1964) of a mutant gene Opaque-2 that changes the protein composition and increases the lysine and tryptophan content of maize endosperm focused the efforts of maize breeders on increasing the quality as well as the quantity of the maize yield. The original Opaque-2 maize varieties had a low yield, a greater susceptibility to ear rots, and a soft floury texture which was not popular, and greater weevil infestation. Breeding programmes have now produced Opaque-2 varieties without these defects. It is anticipated that by the mid-1980s varieties with 0.55 per cent lysine and 0.13 per cent tryptophan and an acceptable yield will be available. Such a development would have a profound impact in feeding the population of Africa, South and Central America, where maize is a main part of the diet. The same technique can be used to improve the protein quality of sorghum.

There is, however, a very good case for planting a very mixed population of varied ancestry as a form of insurance against new races of disease or other calamities. *Puccinia polysora* (maize rust) which swept through West Africa around 1950, would have been held more in check if the seed had been from a wider origin as it might then have contained a source of resistance which would have increased in importance by natural selection. Resistance to this new race occurs in the Caribbean maize population but was apparently not introduced with the original seed.

Under conditions of low fertility hybrid maize and other improved varieties may not yield as much as the local well adapted varieties.

Since the end of the nineteenth century when the stress in the USA was on exhibition ears, tillering has been bred out of maize, each plant producing a single ear. Harland in 1967, lecturing in Trinidad, questioned the wisdom of this, particularly for the peasant farm. Two or more ears per plant, even if they were smaller, could give a higher yield as well as providing an insurance should ears be lost from rots or insect damage. The production of a maize-sorghum hybrid at the Centro Internațional de Mejoramiento de Maiz y Trigo (CIMMYT) opens up the possibility of combining the favourable characteristics of both species in a new food crop (CIMMYT Review, 1977).

Oil palms* (Elaeis guineensis Jacq.)

The oil palm, like the coconut, is monoecious and diclinous. The male and female inflorescences open at different times and rarely is the plant self-fertilized. The proportion of female to total inflorescences (sex ratio) which is important in determining yield, is partly genetic and partly controlled by the environmental conditions when the flowers differentiate. Dry, sunny conditions give a high proportion of female flowers, and in the rainy season, when solar radiation is low, a higher proportion of male flowers. Shade and the loss of leaf area through pest attack also give a low sex ratio. In young plantations in Malaysia with a high sex ratio, assisted pollination is carried out to increase yield. This is not necessary in West Africa where the sex ratio is lower. Normal methods of vegetative propagation cannot be applied to palms, so until recently all multiplication had to be by seed. At the Unilever Research Laboratory in England, a multiplication technique has been developed using tissue culture (Jones 1974). A wide variety of young tissues can be developed including root and leaf, which after passing through a callus stage develop roots and differentiate into young palms. The first leaves of the clonal plant are narrow like a grass but later normal leaves are produced. There is a wide variation in the ability of different genotypes to develop roots, some are easy to root, others very difficult but a sufficient number of selected palms can be propagated in this way for the technique to be extended to production scale, and Unilever are establishing a laboratory in Malaysia to provide clonal planting material for their estates.

There are already field trial plots with 20 genotypes established in Malaysia from plantlets propagated in this way in the UK laboratory. It is anticipated that these selected clones will yield 20 per cent more than the best material currently used, and bunch yields of 40 t/ha and palm oil yields of 10 t/ha are being considered potentially possible.

Not only is this technique ideal for producing the maximum planting material from a selected palm but it can be used to propagate F_1 crosses which would segregate if multiplied by seed. For example crosses of

* See Hartley (1977).

Elaeis guineensis with *E. oleifera*, the South American oil palm which has a more liquid oil due to the higher content of unsaturated fatty acids can now be multiplied in this way. The possibilities of using tissue culture for breeding oil palms and other tropical crops is reviewed by Jones (1977). The same breakthrough has not yet been achieved with the coconut palm, the small plantlets dying off at a young stage.

A palm growing well will produce an inflorescence in the axil of each leaf, a series of one sex, followed by a series of the other sex, in an unpredictable pattern.

The female inflorescences develop into bunches of stone fruit (drupes), consisting of a thin outer epicarp or exocarp, an oily mesocarp from which much of the oil is extracted, a hard stony endocarp which is the shell of the palm oil kernel, and a large endosperm from which oil is also extracted (see Fig. 3.4).

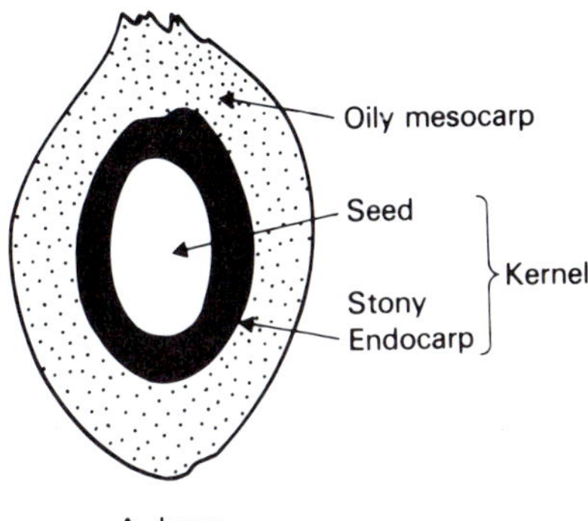

Fig. 3.4 Section of oil palm fruit.

The oil palms of West Africa can be grouped into three main forms by their fruit forms (Fig. 3.5):

1. *Dura:* This form has a fairly thin mesocarp and a thick shell (endocarp) with a large kernel. Though the mesocarp thickness varies, this form is not a high oil producer but gives a good kernel yield.
2. *Tenera:* This form has a thin kernel in an easily cracked shell surrounded by a thick mesocarp. This form is a good oil producer. With the greater return from the palm oil than kernels, *tenera* is the most desirable type for production.
3. *Pisifera:* This form has a small kernel without a shell and a thick mesocarp. It is often female sterile which causes a lot of bunch failure. The fruits ripen badly and frequently rot before reaching maturity. Though useless as producers these palms are valuable for breeding to increase the oil yield. Beirnaert and Vanderweyen (1941) showed that *tenera* is a hybrid between *dura* and *pisifera.*

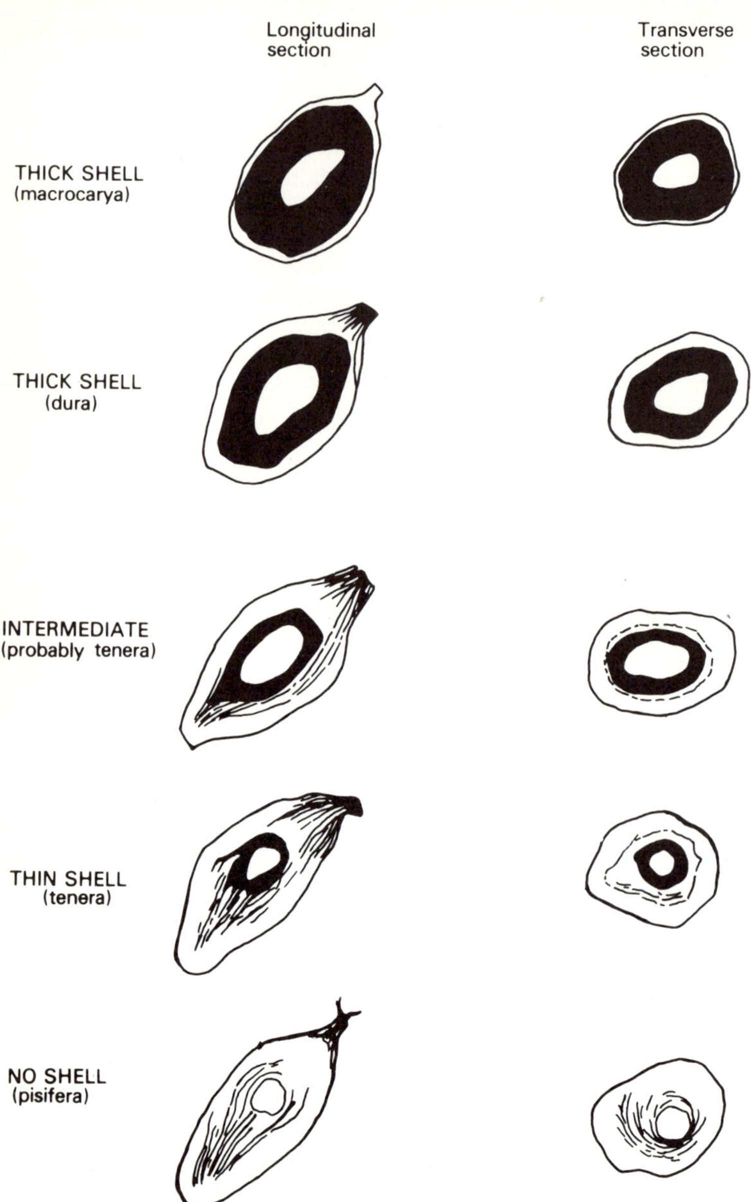

Fig. 3.5 The principal 'forms' of oil palm fruit, distinguished by their internal appearance. (West. African Institute for Oil Palm Research)

Plate 3.2 Oil palm. A Deli fruit. (Courtesy West African Institute for oil palm research.)

The Far East grows the Deli *dura* palm (Plate 3.2) which is well adapted to the climate of that region (Toovey & Broekmans 1955) and is mainly descended from the four palms originally planted in the Botanic Gardens, Bogor, in 1848 and later planted in ornamental avenues on the tobacco estates of Deli on Sumatra. About 1870 it was taken to the Singapore Botanic Gardens and spread as an ornamental in Malaysia. It has darker green leaves than the West African palm and bears a few very heavy bunches. The flowering habit is also different and the fruits are larger and more yellow in colour. The mesocarp which has a lower oil content makes up 60 per cent of the fruit weight. From 1958 to 1961 *dura* × *tenera* material was planted in Malaysia rather than Deli dura previously planted. From 1961 *tenera* material has been planted on a large scale.

The yield of the oil palm is determined by:

1. The successful production of female spadices which is dependent upon growth (Mason & Lewin 1925). The sex of the flowers is determined by a variety of factors inherited and environmental as already discussed.
2. The number of fruits set per bunch. In the Far East the short 'dumpy' palm sets fruit poorly without hand pollination. An insect (*Elaedobius kamerunicus*) is an active pollinator in Cameroon.
3. Fruit size which is determined again by genetical and environmental factors.
4. The ratio of mesocarp to kernel.
5. The oil content.

Dura × *pisifera* crosses in West Africa have given 100 per cent *tenera* form and the Deli palm crossed with *pisifera* pollen has given the Deli *tenera* form (Hartley 1958). This indicates the possibility of producing Deli *tenera* palms with characteristics superior to the existing Deli material. The Deli palm may also be more resistant to wilt and deficiency diseases.

As the *pisifera* type is a poor fruit producer, no selection of the male parent is practicable; in effect, therefore, the selection of breeding material must be based on progeny testing and the methods discussed under coconut seem applicable.

Ollagnier and Gascon (1965) reported that the first inter-origin crosses between Deli from the Far East and *pisifera* or *tenera* from Africa planted in the Ivory Coast gave an average yield of 5.5 t/ha of bunches more than the Deli and 2 t more than the Africa × Africa strains. Gray (1965) discussing this superiority of the inter-origin crosses, considered it warranted further investigation. It could be due to a combination of high bunch weight with high bunch number, in which case selection should be for additive genetic variance rather than combining ability of Deli × *tenera* crosses. On the other hand this superiority might be due to environmental adaptation of the Deli. The Deli *dura* has a much higher genotypic yield potential than is apparent in the adverse conditions of West Africa.

Breeding work is concentrated on palm oil production (Plate 3.3) as in the *tenera* fruit now planted there is four to eight times as much palm oil as palm kernel oil. The main approach has been to produce improved *tenera* by crossing *dura* and *pisifera* parents selected on the *tenera* progeny they produce. One *dura* female parent will produce about 1,000 seeds per bunch which will provide planting material on fertilizing with *pisifera* pollen for 4 ha. Six bunches or more should be produced each year. Selections of *tenera* can be selfed and crossed to produce

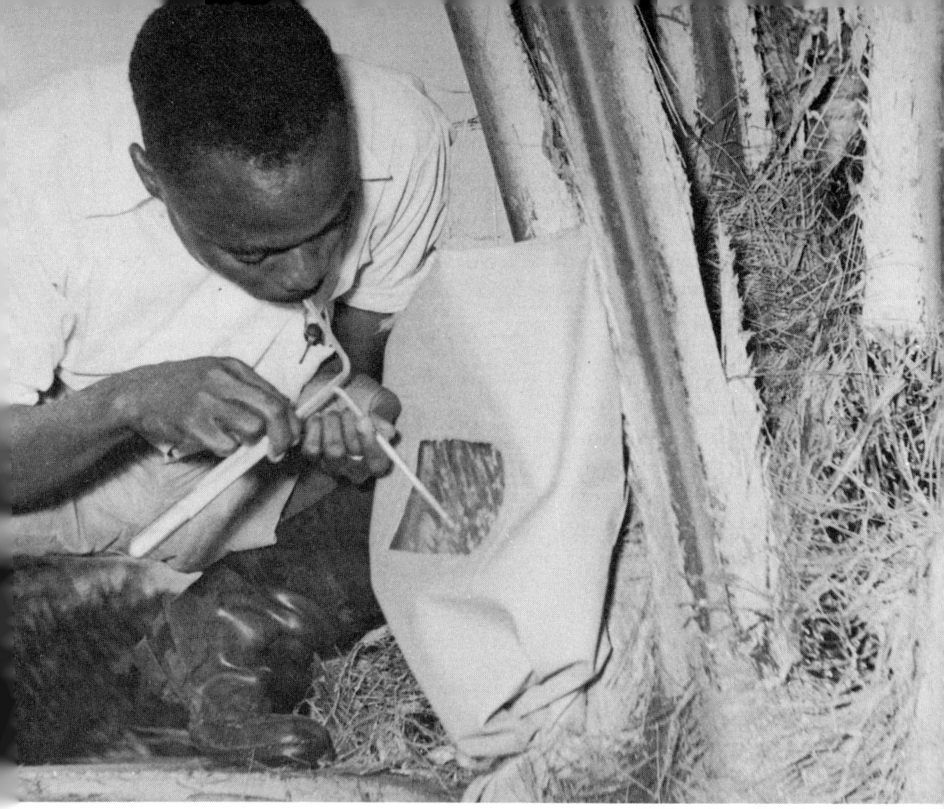

Plate 3.3 Hand pollination of oil palm. Benin, Nigeria. (Courtesy West African Institute for Oil Palm Research).

dura and *pisifera* parents which can be recombined to give improved *tenera*. Inbred *dura* and *pisifera* types are also being produced to eliminate undesirable characteristics. Interspecific hybrids, *E. guineensis* × *E. oleifera* have been produced and back-crossed to obtain shorter mature palms.

The selection of the oil palm has been studied by Devuyst (1953). He considered that concentration on increasing the mesocrap content of the nut had produced little effect on yield due to a lack of improvement in bunch weight and suggested the elimination of mediocre producers in the nursery. He found a positive correlation between bunch production and the number of leaves and spear leaves present in the crown of the palms during the first two productive years, after which the correlation is lost. The frond length is the main factor in determining leaf surface which is correlated with bunch production. Provided that all the conditions in the nursery are equal, Devuyst suggested that it was possible to select in a nursery at one year on a basis of height and thus select for future yield. Hartley (1977, p. 394) however does not consider his case has been made out.

All poorly developed, deformed, abnormal and stunted plants should be rejected in the pre-nursery and nursery stages. Also nursery seedlings which grow erect, have a drawn or bunchy habit, narrow or short leaves, or leaves at an unusual angle, and diseased seedlings should be discarded from the nursery. This may mean allowing for about 20 per cent extra plants for discarding by selection. The dry heat (38–40 °C) treatment of seed gives a high percentage of seedlings (85–90 per cent). Of the germinated seed 75–80 per cent should be available for planting out.

The low yield of the palm groves of West Africa is partly due to poor planting material, mainly *dura* which gives a low yield of oil per bunch, also slow soil exhaustion, oil palms, overcrowding, tapping for wine, leaf pruning and incomplete harvesting due to the tree height.

Rice [Oryza sativa L.]

The varieties of *Oryza sativa* are divided into two geographic races or sub-species. The chromosome numbers of the two races are identical. A sterility barrier appears to separate the two forms, though many crosses exist.

1. *Japonica* group which includes the varieties native to Japan, Korea and North China, more adapted to long days.
2. *Indica* group which includes varieties from India, Sri Lanka, Taiwan, South China and the Philippines.

Some intermediate types are found in Burma and Java.
The races differ in the following characteristics:

Character	**Indica**	**Japonica**
Climatic zone	Tropical monsoon	Temperate
Photoperiod	Usually very sensitive	Not sensitive
Day length	Usually short	Not sensitive
Tolerance to unfavourable conditions	High	Moderate
Vegetative period	Long	Short
Response to fertilizer	Poor	Good
Lodging	Susceptible	Resistant
Seed dormancy	Present	Absent
Grain shape	Usually long, narrow, flat in section. Broad forms occur	Short, thick, round in section
Diversity of forms	Much	Moderate

Character	Indica	Japonica
Awns	Generally absent	Sometimes present and well developed
Plant colour	Light green	Dark green
Texture of plant tissues	Soft	Hard
Plant height	Tall	Short
Tillering	Many, usually spreading. Erect forms occur	Moderate, erect
Leaf hairs	Dense	None
Flag leaf	Long, narrow and drooping after flowering	Short, fairly broad and erect after flowering
Last stem node	Extruded	Not extruded
Ear number	Large	Large
Ear length	Medium	Short
Ear branching	Moderate	Sparse
Ear density	Moderate	High
Ear weight	Light	Heavy
Shedding of grain	Susceptible	Resistant
Husk hairs	Sparse, short, thick	Dense, long
Endosperm	Translucent	Chalky
Breakage during milling	High	Low
Yield potential	Medium	High

(After Purseglove 1972)

Traditional rice varieties, and there are more than 14,000 of them, are mainly of the *indica* type, well adapted to a traditional agricultural system. Their characteristics have been developed over centuries of mass selection by small farmers with paddies where the water and weed control is poor and fertilizers rarely used. These farmers must grow their crop in the monsoon season when the light intensity is low and the day length and temperature are both decreasing after germination. Seedling growth is rapid and this gives the plant a good root system to take up nutrients. The vigorous initial growth enables the rice plant to compete with the weeds, but it has no relation to eventual grain production. On the contrary, it causes mutual shading of the leaves, with reduced nutrient uptake and photosynthesis, and premature death of the leaves. Respiration losses increase with plant weight and this, together with the reduced photosynthesis from mutual shading, causes a low net assimilation in the period after flowering which is critical for grain production (Fig. 3.6). These effects may be accentuated by the ap-

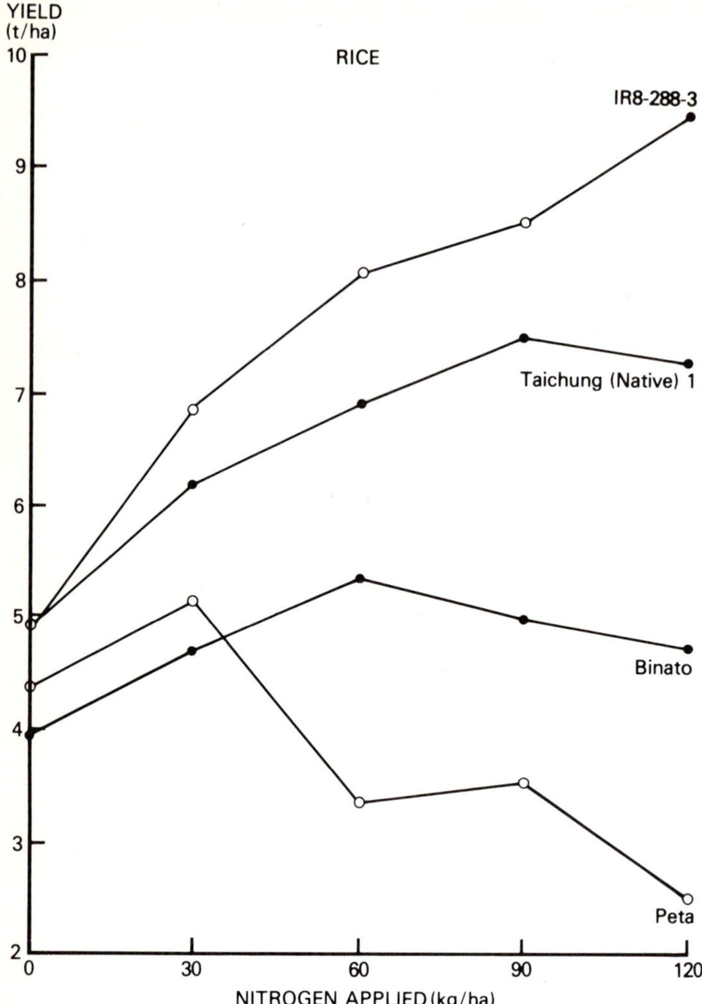

Fig. 3.6 Response of four rice varieties to nitrogen. [Source IRRI]

plication of nitrogen fertilizer and close spacing, which increase plant size and mutual shading. Lodging is common and if this occurs before flowering it induces sterility which can cause a loss of half the potential crop or more. As the grains fill out and the panicles become heavy, the high winds and monsoon rains flatten the crop if it is still standing. If nitrogen is applied shading and lodging are increased. A laid crop cannot photosynthesize properly so the grains do not fill out. However, as the grains remain dormant for some time after ripening, the panicles do not sprout. The grain-to-straw ratio at harvest is low. Despite all these drawbacks they yield and these traditional varieties have been selected for cooking and eating qualities.

The temperate zone cultivars are adapted to high light intensities and fertile soils and give much higher yields. These differences are independent of whether the temperate zone cultivar is a *japonica* or *indica* variety (Jennings 1966).

It was obviously necessary to develop a new plant type for the tropics with the following desirable characteristics.

1. Early maturity and insensitivity to day length. Varieties maturing in 100-120 or 130 days from seedling to harvest, non-sensitive to day length allowing planting at any time of the year and double or treble cropping. Insensitivity to photoperiod is largely controlled by a single recessive gene. Earliness is dominant. Both characteristics are fixed early in segregation. Taichung Native 1, Belle Patna, and some American hybrids are a good source for these genes.
2. Short narrow upright leaves to make maximum use of the light with minimum respiration losses. A short broad flag leaf is desirable.
3. Short strong straw with moderate tillering to reduce lodging even with heavy fertilizer application. Short stature in Taiwan *indicas* including Taichung Native 1, has made a big contribution to the IRRI programme. This characteristic is basically conditioned by a single recessive gene which has no unfavourable linkages. This gene is being transferred to many popular tropical varieties. Short internodes and erect small leaves also help to prevent lodging.
4. Nitrogen response. This is partly associated with resistance to lodging.
5. Seed dormancy after harvest to prevent sprouting if it is wet at harvest time.
6. Resistance to shattering of the heads to prevent losses after ripening.
7. Resistance to diseases, particularly blast, leaf blight and viruses. Blast resistance is complicated by the existence of a number of races.
8. Tolerance or resistance to insect attack, especially borers.
9. Milling and cooking quality.

The IRRI breeders have successfully developed this desirable plant type and exceptional results have been obtained very quickly in the breeding programme as shown in the Table 3.4.

In 1962, the first year of research at the International Rice Research Institute (IRRI), a cross was made between the semi-dwarf *indica* variety, Dee-geo-woo-gen, a parent of Taichung Native 1 the popular semi-dwarf bred in Taiwan in the mid-1950s and Peta, a tall traditional Indonesian variety grown in the Philippines. From this cross came the

Table 3.4 Grain yield in kg/ha. IRRI dry season, 1966

	Nitrogen applied (kg/ha)				
	0	30	60	90	120
Tjera	4,264	5,961	4,969	3,741	3,017
Peta	4,425	5,223	3,389	3,555	2,546
Taichung Native 1	4,945	6,298	6,977	7,517	7,340
IR 8	4,924	6,996	8,099	8,591	9,477

Notes: Tjere Mas and Peta are two locally recommended typical tropical varieties. Taichung Native 1 is a dwarf variety from Taiwan.

famous IR8 (Plate 3.4), the first named variety to be released by the IRRI and which established yield records in many parts of Asia. IR8 matures in 125 - 130 days, is widely adaptable, grows about 100 cm tall, tillers well, is resistant to lodging, responds well to nitrogen, and has a high ratio of grain to straw. Unfortunately it has little resistance to pests and diseases. The grains are medium-long, bold and chalky, and prone to breakage in milling. Having a high amylose content it cooks dry and fluffy as preferred in India and Pakistan but not in the Philippines and Indonesia. Rice is often taken into the field and eaten cold during a break from cultivating, but IR8 is not appetizing cold as it forms a sticky mass.

Other IR varieties followed which corrected some of the weaknesses of IR8, and these semi-dwarfs were distributed throughout the rice growing areas of the world, either for growing as varieties or for use in local breeding programmes. The major characteristics of these varieties are given in Table 3.5 and the resistance ratings in Table 3.6 (p. 272).

By 1972 about 10 per cent of the world's rice area was planted with semi-dwarfs.

However, nature has a way of striking back. The brown plant hopper (*Nilaparvata lugens*), which not only causes hopperburn but is a vector of 'grassy stunt' virus and ragged stunt, a virus disease first reported in the Philippines and Indonesia in 1977, was only a sporadic pest in the tropics, but is now the most serious rice pest in tropical Asia. The early rise in importance of this pest was countered by planting more IR26 which was resistant to the brown plant hopper, and IR26 quickly became an important rice in the Philippines and certain other parts of Asia. However, it was soon noticed in Kerala in India that the hoppers were killing IR26, and the same was true elsewhere in Asia. From tests it became clear that there are at least four biotypes of this insect against all of which none of the IRRI varieties was resistant. All the modern varieties depend for resistance upon a single gene which is very liable to break down as the hopper population moves to biotypes that can break the plant's resistance. Breeders are now trying to develop varieties with

Plate 3.4 One of the first of new varieties, IR8, bred at the International Rice Research Institute in the Philippines. In contrast to the common tropical rices it is only 100 cm (39 in) tall. (Courtesy IRRI)

more than a single gene for resistance, and also develop field-tolerant varieties.

The seriousness in 1979 of this pest upsurge is shown by the appeal for food relief for nearly 2 million people in Vietnam, as half the rice in the Mekong delta was seriously affected by floods or hoppers.

It is suggested that this is a 'man made' pest outbreak. The traditional varieties had been selected by the farmers for resistance. The new semi-dwarf varieties are shorter and leafier providing a micro-climate more favourable to the pest. With multi-cropping the food for the insect is available all the year round and the irrigation ditches which are never drained provide a means of movement from field to field. It is also suggested that the increased use of non-selective insecticides has killed off many of the natural predators. The brown plant hopper should be an important warning to all involved in large-scale development in the tropics.

The IRRI have a breeding programme incorporating semi-dwarf

Table 3.5 Major characteristics of varieties named by IRRI

Character	IR8	IR5	IR20	IR22	IR24
Growth duration					
Dry season (Dec. seeding)	125 days	135 days	120 days	115 days	125 days
Wet season (June seeding)	130 days	145 days	135 days	130 days	125 days
Sensitive to photoperiod	No	Weakly	Weakly	Weakly	No
Grain					
Length	Medium	Medium	Medium	Long	Long
Width	Bold	Bold	Slender	Slender	Slender
Appearance	Some white belly	Some white belly	Translucent	Translucent	Translucent
Head rice recovery	Low	Moderate	High	High	High
Amylose content	High	High	Moderately High	High	Low
Gel consistency	High	Low	Medium	High	Low
Gelatinization temperature	Low	Inter-mediate	Inter-mediate	Low	Low
Seed dormancy	Moderate	Moderate	Moderate	Moderate	Moderate
Seedling vigor	Very good	Very good	Very good	Good	Good
Height (cm)	90-105	130-140	110-115	95-105	100-110
Tillering ability	High	High	High	High	Moderate
Lodging	Resistant resistant	Moderately resistant	Moderately	Resistant	Resistant

characteristics into varieties that can adjust to varying water depths, and also to develop varieties tolerant to flooding for the extensive areas where the water supply is not controlled.

Other programmes include the improvement of protein content in the rice, and tolerance of soil deficiencies and toxicities where the current semi-dwarf varieties cannot be grown.

Rubber (Hevea brasiliensis (Lq. v.))

Natural rubber provides about a third of the total world supply of rubber, and the whole of this commercial natural rubber production is

IR26	IR28	IR29	IR30	IR32	IR34
130 days	105 days	115 days	106 days	140 days	120 days
130 days	105 days	115 days	109 days	145 days	125 days
No	No	No	No	No	No
Medium	Long	Medium	Medium	Long	Long
Slender	Slender	Slender	Slender	Slender	
Translucent	Translucent	Opaque	Translucent	Translucent	Some white belly
High	High	High	High	High	High
High	High	Waxy	High	High	High
Medium-low	High	Low	Low	Low	High
Low	Low	Low	Inter-mediate	Low	Low
Moderate	Moderate	Moderate	Moderate	Moderate	High
Good	Good	Very good	Very good	Very good	Very good
100–110	100–100	90–100	95–105	100–110	120–130
High	Moderate	High	High	High	High
Moderately Resistant	Moderately resistant	Moderately resistant	Moderately resistant	Resistant	Moderately resistant

Source IRRI (1975a)

founded upon an unselected sample of wild genotypes collected by Wickham in 1876 in Brazil.

The variation which occurs in the production of individual trees established from seed was recorded in the early days of the Far East rubber plantations. Furthermore the yield of trees grown from seed produced by high yielding trees though greater than the yield of unselected seedlings was not markedly so. Even when the poor yielders had been removed to ensure that both parents were above average yielders the result was disappointing and the progeny very variable. Between 1910 and 1916, a method of budding was developed in West Java by van Helten in cooperation with two estate managers, Bodde and Tas. In 1918, the first budded plantations were planted in Sumatra and Java.

Table 3.6 Resistance ratings of IRRI varieties*

Variety	Diseases				Insects				Soil problems			
	Blast	**Bacterial blight**	**Grassy stunt**	**Tungro**	**Green leaf-hopper**	**Brown plant-hopper**	**Stem borer**	**Gall midge****	**Alkali injury**	**Salt injury**	**Zinc defi-ciency**	**Phos-phorus defi-ciency**
IR8	MR	S	S	S	R	S	MS	S	S	MR	S	MR
IR5	S	S	S	S	R	S	S	S	S	MR	R	MR
IR20	MR	R	S	R	R	S	MR	S	S	MR	R	R
IR22	S	R	S	S	S	S	S	S	S	S	S	MR
IR24	S	S	S	MR	R	S	S	S	MR	MR	S	MR
IR26	MR	R	MS	R	R	R	MR	S	MR	MR	S	R
IR28	R	R	R	R	R	R	MR	S	MR	MR	R	R
IR29	R	R	R	R	R	R	MR	S	S	MS	R	R
IR30	MS	R	R	R	R	R	MR	S	MR	MR	R	MR
IR32	MR	R	R	R	R	R	MR	R	S	—	—	—
IR34	R	R	R	R	R	R	MR	S	S	S	R	R

R = resistant MR = moderately resistant MS = moderately susceptible S = susceptible ** Rated in India

Source: IRRI (1975a) * Rated in Philippines

Prior to bud grafting open pollinated outbred seed was used in planting. The first tapping results from budded plantings were published in 1922, but the development was treated with reserve up to about 1928. The need for caution in recommending new clones and the advisability of not planting up too large an area with a single clone is illustrated by RRIM 501 which was tapped from 1935, recommended for many years and removed from the recommended list in January 1960, due to widespread trunk snap, particularly in the windy areas. This clone had been yielding over 1,120 kg/ha in the second year of tapping.

Clonal seed

Local research stations give lists of recommended clones for bud grafting and producers of clonal seed which can be identified by their colour, shape and markings. Seed from an area planted to budded clones is too mixed in its heredity, to be clonal seed, and too unpredictable in its performance. Such seed should be proven by the performance of its progeny before being planted on a large scale. Certain clones, of which Tjir 1 is probably the most outstanding, are self-compatible. If this clone is planted in an isolated monoclonal block, the seed will produce trees which equal Tjir 1 budded material in performance (Edgar 1960). Due to the necessity of proving the progeny, the development of clonal seed is much slower than budded material. Tjir 1 has proved such a successful parent that seed from a mixed stand of Tjir 1 with modern clones should provide reasonable seedlings. A comparison of clonal seedlings in Malaysia (Anon 1966b) showed that Tjir 1 is out-yielded by trees from seed of modern polyclone gardens, by RRIM 501 and PB 5/51 monoclone seed and by illegitimate seed collected in RRIM 600 series clones trials. RRIM 600 illegitimate seedlings averaged 1,063 kg/ha for the first 2 years, compared with 747 and 680 kg/ha for two sources of Tjir 1 which is no longer recommended.

Seedlings compared with clonal buddings

Advantages	**Disadvantages**
Easier to bring into bearing	Yield more affected by height of cut
More robust	Potential productivity lower
Bear a year earlier	Probably more susceptible to diseases (clones selected for resistance)
Recover better from damage	Planted in a denser stand and thinned out, which increases establishment costs
Often bark thicker and more easily tapped by unskilled labour	Tapped less severely particularly in early years Lack of uniformity makes tapping far more difficult

Plate 3.5 Stripping rubber bud patches in the field. Malaysia. (Courtesy Dunlop Rubber Co. Ltd.)

Trees grown from clonal seed currently available cannot compete in yield with trees budded with the best clones, and about 90 per cent of the rubber planted in peninsular Malaysia in recent years has been budded material.

Clonal budding

Single component clonal trees can be produced by rooting cuttings (Levandowsky 1959) or marcotting. Unfortunately such trees do not produce a tap root, and as the rubber tree is 20 - 25 m tall with a heavy crown the leverage on the roots in a strong wind is considerable, and without a good anchorage they are easily uprooted. (plate 3.5 and 3.6). Buddings are normally tapped when they reach an average girth of 50 cm at 100 cm above the ground compared with 50 cm above the ground in the case of seedlings. Crop improvement has already made a remarkable difference in yield (Figs. 3.7 and 3.8) and the limit has not yet been reached (see Edgar 1960).

A three component tree can be produced by grafting a high yielding trunk onto a seedling rootstock, and top-grafting this with a crown which is resistant to wind damage or leaf diseases as required locally.

Plate 3.6 Young rubber shoot of RRIM 519 growing from cutback seedling stock. Malaysia. (Courtesy The Rubber Research Institute of Malaysia)

Such trees have not been widely planted. AVROS 2037 from Sumatra, is a good yielding clone, useful for crown budding to reduce wind damage and to increase the yield of other clones.

Disease resistance

South American Leaf Blight (SALB) (*Microcyclus ulei*) (syn. *Dothidella ulei*) has been the main reason why rubber plantations have not been successfully established in South America. Fortunately this disease has not yet been reported in the Far East, but with modern air travel the introduction of SALB is a constant threat to the industry. There are at least four pathological races of *M. ulei* and no selection of *Hevea* is resistant to all four (Wycherley 1976). Besides selections of disease-

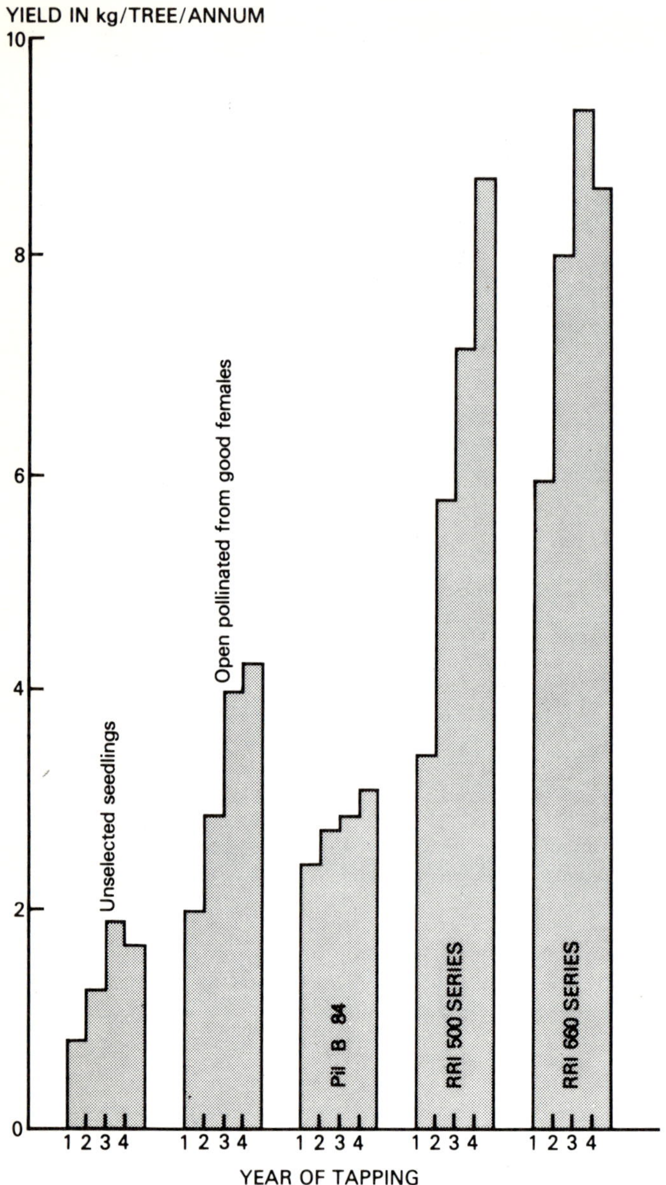

Fig. 3.7 The improvements in yield of *Hevea brasiliensis* in Malaysia.

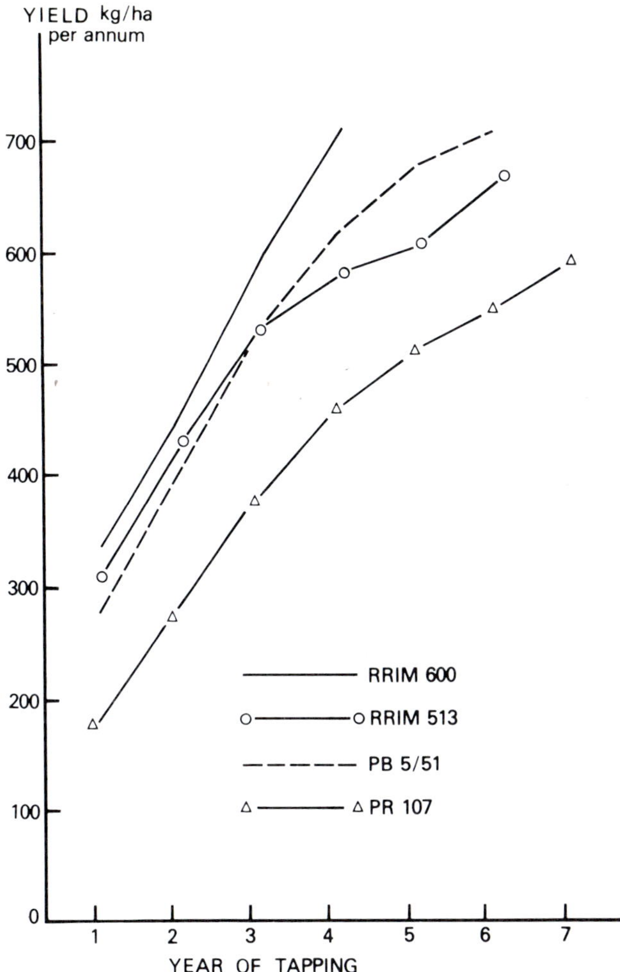

Fig. 3.8 Yield of four widely planted rubber clones. (The Rubber Research Institute of Malaysia)

tolerant strains within *Hevea brasiliensis*, resistance has also been introduced from *H. benthamiana*. New genetic material is regularly being sought in Brazil, to give not only higher yield potential but also wider tolerance to SALB and if possible resistance to *Oidium*. Resistance to *Phytophthora* is already present in certain clones.

Strict plant quarantine precautions are taken with the plant material introduced into Malaysia but there is a serious risk of SALB being carried as spores on the clothing of the people who have collected the new plant material, which could be disastrous to both Malaysia and Indonesia.

Selection in nursery

Before budding was developed, Cramer developed the 'Testatex' method of grading seedlings in the nursery. A series of cuts are made in the seedling bark and the amount of latex exuded after half an hour is assessed. This method is, however, only useful in eliminating the poor producers (Dijkman 1951).

Seedlings are now raised in polythene sleeves filled with soil. All slow growing, poorly developed and yellow seedlings should be discarded whether the seedlings are for planting out or budding.

Green budding

Green budding which was developed by Hurov (1961) in Sabah is the budding of 3-month-old rubber seedlings about 60 cm high rather than year old seedlings, using buds taken from vigorous 6- to 8- week-old leafy terminal shoots. The bud patch is applied near the base of the young seedling and bound with clear plastic budding tape. The stumped buddings can be transplanted about 1 year after sowing the initial seed, saving a year on the conventional technique.

Experience in Malaysia has shown that the saving in time between planting and tapping is dependent upon the season and is generally less than the time of advancement of budding. Green budding in Nigeria has shown an average of 83 per cent success. Falling off of the polythene strip due to the weather has been the main cause of failure. Leaf pruning of budwood 7 days before budding was more successful than pruning just before budding. Green budding is expected to reduce the pre-tapping period of rubber in Nigeria by 6–8 months (Togun and Ajibike 1964).

Green budding was also used by Hurov on a number of fruit trees in Sabah. It was described by Topper (1957) for use in cacao but does not appear to have been accepted as nursery practice, which is surprising in view of the bias certain cacao planters have against the rooting system of vegetatively propagated material. It should also be useful for citrus and tea clones. Crowns can be green budded on rubber 2 m tall.

A wide variety of clonal planting material is still used in rubber plantations including bare-root budded stumps, polybag buddings, soil core whorled buddings, mini-stumps and stumped buddings (Ooi Cheng Bin 1977).

Latex production

Rubber latex is an aqueous suspension of around 40 per cent dry matter which is 90 per cent hydrocarbon. It flows from the latex vessels which

Plate 3.7 Budded rubber tree just tapped, latex dripping into cup. Note bark consumption marks on the panel. (Courtesy The Rubber Research Institute of Malaysia)

are concentric cylinders of modified sieve tubes in the phloem formed from the cambium. The number of vessels per ring and the number of rings depends on the clone, and the age and thickness of its bark. As the girth of the tree expands, the outermost rings of virgin bark are ruptured and cease to function. There are, therefore, more functional latex vessels nearer to the cambium, and the tapping cut must go as near as possible to the cambium without damaging it. The number of functional vessels is greater in renewed bark. There is no movement of latex until the bark is injured, as in tapping, when the latex flows to the cut from the heart-shaped area which surrounds it. Most of the latex comes from the latex vessels below the tapping cut. Tapping is done at the bottom of the panel by cutting the bark away with a curved blade which makes a sloping cut which opens the tops of the vessels coming from the base of the trunk. (Plate 3.7). The bark regrows. Girth and vessel ring numbers and bark thickness are therefore important in determining yield.

Indeed, as was established in the late 1920s, no character shows a more consistent correlation with yield than the number of latex vessel rings. The diameter of the latex vessels is less important. The number of latex vessels differentiated each year appears to be a clone characteristic largely independent of the increase in girth or age. Wycherley (1969) suggested more conscious attention should be paid to this characteristic in selecting parents. Selection in the nursery on the basis of latex vessel ring count has been considered impractical, due to the small number of vessels too easily influenced by other factors at that age. As significant correlations have been obtained between the number of latex vessels in the mature tree and the number in the 3 year-old nursery buddings, by using green budding Wycherley considered that early selection on the basis of latex vessels could be developed.

Latex flow is controlled by the internal water balance which is why tapping is done as early as possible in the day when the turgor pressure in the latex vessels is greatest.

Latex flow continues until the ends of the vessels collapse and the latex coagulates to block the flow.

Improvement of rubber

The unique way in which rubber is produced makes many of the criteria for selection different from other crops. These may change in future with the greater use of ethylene to maintain latex flow (see p. 365) which may produce economic yields from ordinary planting material; and the possibility of developing dwarf trees.

Selection of new clones has been based on the following:

1. *Vigour of growth:* This is assessed by girth and yearly girth increments which are correlated with yield. If less vigorous seedlings had been discarded in the Malaysian programme, RRIM 600, one of the most successful clones would not have been retained.
2. *Yield of dry rubber per hectare:* Records are made on one or two tapping days a month by coagulating in the cup.
3. *Latex quality:* It should be white in colour, free flowing and not coagulate prematurely. A maximum of 6 per cent resin is desirable otherwise the sheets are slow to dry, and soft.
4. *Bark characteristics:* The bark should be reasonably thick with an abundance of latex vessels and without too many hard stone cells in the layers. The latex should flow quickly. The renewed bark should form evenly and quickly after tapping.
5. *Habit:* Vigorously growing trees tend to form a large dense crown which is desirable for shading out weeds, provided that it is not too heavy for the trunk and cause it to snap. Trunk snapping is too prevalent in some modern clones, e.g. RRIM 513 and 603.

   Certain forms of wind damage are associated with particular

branching habits. A narrow angle fork rivalling the leader should be avoided. Where the water table is high, the root penetration will not be sufficient to anchor clones with a heavy head.

6. *Absence of fluting on the trunk.*
7. *Association of clone with different soil fertility levels.*
8. *Disease resistance:* This is particularly important in South America against *Microcyclus ulei* (formerly *Dothidella ulei*). Resistance to pink disease *Corticium salmonicolor* has become more important.

Development of rubber clones in Malaysia

The national production of natural rubber in Malaysia rose from 0.59 million t in 1954 to over 1.50 million t in 1977 as a result of the massive replanting programme that followed the Mudie Mission in 1954. During this time the average yield per hectare of rubber in tapping on Malaysian estates doubled to over 1 t/ha; 95 per cent of the estates and 80 per cent of smallholdings are replanted with high yielding clones. A replanting cycle of 20 years rather than 30 years is aimed for, but currently 3 per cent of the trees are being replanted annually (Anon 1977).

Unselected seedlings rarely yield more than 550 kg/ha per year whereas modern clones are yielding up to 1,500 - 2,000 kg/ha on a field scale, and the maximum potential yield has been estimated as 5,600 kg/ha per year.

Progress in breeding new clones is very slow, it takes 25 years from the original hand pollination until a clone is approved for large-scale planting, and the recommended clone is probably the only survivor from 2,000 originally prepared, selected in a series of field plots of increasing size.

For example RRIM 600, the most popular clone in peninsular Malaysia since 1964, was a hand-pollinated cross (Tjir 1 × PB 86) made in 1937. In 1955 it was recommended for small block planting, in 1963 for up to a third of the total replanting area, and from 1967 for large-scale commercial planting. More than half the area planted with budded material between 1967 and 1973 was planted to this variety. In 1975 it declined to 28 per cent of the planting with GT 1 increasing from 12 per cent in 1973 to 25 per cent in 1975, with a growing interest in PR 261 from Java (Leong and Mayakrishnan 1975). These three clones have yielded well over two tonnes per hectare per year.

RRIM 600	-	2,940 t/ha 14th year.
GT 1	-	2,340 t/ha 8th year.
PR 261	-	2,420 t/ha 8th year.

Early results from RRIM 730 suggest this may yield better than RRIM 600, and PB 280 which yielded 1,006 and 1,725 kg/ha in the first and second years of tapping may yield more than both (Anon 1977).

The current recommendations in Malaysia, Enviromax, are to maximize the yield potential of a particular environment. Some of the yields obtained in the first years of tapping on commercial estates are given in Table 3.7. All these clones are resistant to wind damage, which may contribute to the high yields

Table 3.7 Commercial Tapping Yields of six clones.

Clone	**Yield** (kg/ha per year)	
	First 5 years tapping	**6–10th year tapping**
PB 28/59	1,532	
RRIM 527	1,398	1,816
RRIM 628	1,386	2,029
GT 1	1,379	1,860
PB 5/51		1,787

Source: Anon 1976.

Dwarf rubber trees

Late in 1969 at Sungei Buloh in Malaysia a seedling from a RRIM 605 × Ford 351 was discovered with a compact habit. Buddings of this clone were only half the height of normal clones growing alongside. Unlike the normal trifoliate leaf, this clone had from two to five leaflets, mainly five. A dwarf rubber tree half the normal height, less susceptible to wind damage, possibly earlier maturing, and easier to spray could be a major advance in natural rubber production.

Sugar cane*

Sugar canes are wind pollinated outbreeders; highly polyploid and many aneuploid. Regular Mendelian segregations are neither expected nor observed. Being outbreeders and clonally propagated, they are highly heterozygous and intolerant of inbreeding. Crosses between clones therefore display great variability and it is among such F_1 progenies that cane breeders seek new varieties (Walker and Simmonds 1981). The object of cane breeding is an economic yield of sugar sustained over a series of ratoons. This implies a high yield of quality juice with a low fibre content to facilitate milling, produced under the conditions of soil, light, fertilizer and water which the crop will be grown. They must also be resistant to diseases such as smut, rust, downy mildew, red rot, gummosis, leaf scald and the viruses including mosaic, Fiji disease, and grassy shoot, according to their local importance. If

* For a full account see Stevenson, 1965

canes are to be harvested by hand, sharp leaf edges, irritant leaf hairs, and steel-like stems hard to cut are undesirable. The canes should not lodge. Millions of sugar cane seedlings are raised annually, but the difficulty lies in choosing the superior crosses which combine high sugar content with disease resistance and ease of harvesting and milling. The new clones are nobilized seedlings ($2n = 100-125$). Though it was only 8 years after the POJ 2878 seedling was produced that it became the dominant variety in Java, today it takes 10 to 14 years for a cross to be introduced commercially and the life of a modern cane variety may be no more than 15 years. Modern canes are founded on about 20 noble canes and about 10 *spontaneum* or *spontaneum* derivatives.

The genus *Saccharum* has five species of interest to cane breeders, with somatic numbers varying from $2n = 40$ to $2n = 164$ or more for odd forms.

Saccharum officinarum. [Noble canes.] Predominantly $2n = 80$, some atypical canes over 80 included often for convenience, e.g. Loethers $2n = 99$. Large bright coloured soft thick canes with good sugar quality and low fibre. Poor resistance to most of the main diseases, except gumming disease (*Xanthomonas vasculorum*) and smut (*Ustilago scitaminea*) to which many noble canes are resistant. The Otaheite, synonymous with Bourbon, Lahaina (Hawaii), Vellai (India), was the first noble cane cultivated world-wide, followed by the Cheribon series from which most of the commercial varieties have been bred.

Saccharum spontaneum $2n = 40-128$. Wild canes of Pacific and Asia, e.g. 'Glagah' of Java. Source of resistance to many major diseases including 'Sereh', mosaic, gumming, red rot (*Physalospora tucumanensis*), downy mildew (*Sclerospora sacchari*). Many forms are susceptible to smut and 'Fiji' disease. All perennial grasses ranging from dwarf to tall forms. Internodes, long thin with pithy or hollow centre, colour green-yellow to white when ripe with waxy bloom. Rhizomes may become aggressive colonizers.

Saccharum barberi $2n = 82-124$. Sometimes included in *S. sinense*. Indian canes. Hard thin canes with a high fibre content. Well developed root systems. The 'Saretha' group have been the most important breeding canes. Immune to gumming and mosaic, resistant to downy mildew but many forms are susceptible to smut and red rot.

Saccharum sinense $2n = 82-124$. Chinese or Uba cane. Thin hard canes generally with poor juice quality and undesirable milling qualities. Some forms have a high sugar content. A difficult breeding cane which has given some useful breeding lines, particularly in Barbados. Immune to mosaic and gumming disease, resistant to downy mildew. Some forms resistant to leaf scald (*Xanthomonas albilineans*). Susceptible to smut and some forms susceptible to red rot.

Saccharum robustum $2n = 60-194$, mainly $2n = 60$ or 80. Wild cane from New Guinea and Melanesia. Grows up to 9 m in compact tufts. Hard woody stems with a pithy or hollow centre and little juice. Susceptible to mosaic, and often to leaf scald, and downy mildew. Some resistance to red rot. Has been used to some extent in breeding lines. *Saccharum edule* is probably a mutation of *S. robustum*.

Stevenson (1965) points out that evidence suggests that the wild canes *S. spontaneum* and *S. robustum* are the only true species in the systematic sense, the others being complicated hybrids which for practical purposes can be regarded as species. All these species intercross readily, and the commercial canes grown today are all interspecific hybrids.

There are reports of intergeneric crosses of *Saccharum* spp. with a wide range of genera including *Erianthus, Zea, Imperata, Sorghum* (Dillewijn, van 1946), *Narenga, Sclerostachya, Miscanthus* (Stevenson 1965). There is no commercial variety of sugar cane which is an intergeneric hybrid of *Saccharum*, but Grassl (1962) considers such crosses offer great promise of breakthrough in many breeding programmes. They are hardly likely to increase sugar yield, but could be important sources of resistance to pests, disease or adverse growing conditions.

The big advance in sugar cane breeding came when it was realized in 1888 by Harrison and Bovell in Barbados and Soltwedel in Java that sugar cane could set fertile seed which could be utilized to obtain better varieties. This came at a time when diseases were threatening the cane growing regions.

The original breeding stations were established where cane has a natural tendency to flower, but with modern techniques, including manipulation of the day length, irrigation, temperature and humidity, flowering can be induced, or delayed.

To facilitate crossing, both male and female arrows can be cut and kept alive in dilute sulphurous and phosphoric acid, diluted with rainwater to 0.01 to 0.03 per cent, and still continue to flower and produce viable seed. Pollen can be freeze dried for movement between countries or for storage. The arrows are kept in lanterns made from polythene or closely woven fabric, to prevent fertilization by outside pollen, and bagged after fertilization. Crosses are frequently done in a greenhouse rather than outside.

The earliest production of sugar was in the north of the Indian continent where many varieties of the thin Indian canes (*Saccharum barberi*) were grown. From earliest times the Melanesians grew noble canes as a food crop and these were transported throughout the Pacific, later to form the basis of the sugar industries of Java and Mauritius. Sugar production started in the Caribbean area in the early sixteenth century from the Creole cane, originally the 'Puri' cane of India, a thin erect greenish yellow cane with short internodes, with high juice and fibre content, but a poor yielder. This variety was the only one grown in this region for about 250 years when the Bourbon cane from Otaheite was introduced in the second half of the eighteenth century. Otaheite was the foundation cane of the Mauritius sugar industry and until it was eliminated by a disease in the 1840s. This cane was also known as 'Lahaina' in Hawaii and Java, and 'Vellai' in India. In the late nineteenth century, root diseases and gumming disease (*Xanthomonas*) eliminated the Otaheite cane from all the commercial areas and it was replaced by the Cheribon series of canes from Java, which were not only very important canes in their own right, and prior to POJ 2878 probably produced more sugar than all the other varieties combined, but have also formed the basis of many of the cane breeding programmes and contributed to most of the currently favoured varieties.

Selected noble canes were good from the point of view of yield and milling, but lacked resistance to 'Sereh', mosaic and certain root diseases. Also they were not very well suited to the continental climate of North India or to poor soil conditions.

The wild species, particularly *Saccharum spontaneum*, are resistant to 'Sereh' and other diseases and these were crossed and back-crossed with the noble varieties (*Saccharum officinarum*), a process described as 'nobilization'. Noble cane varieties still remain the major source of genes controlling sugar production, and are mainly used as the male parent. After the fourth back-cross the defects of the noble canes tend to reappear, and it is preferable at this stage to cross different nobilization lines. From 1925 hybrid varieties were grown on a large scale in Java, and these were available in time to save the Louisiana sugar industry in the 1920s from the disastrous effects of mosaic which reduced yields to 25 per cent of normal. By the end of the nineteenth century six cane breeding stations were established in Java, Barbados, British Guiana, Reunion, Queensland, Mauritius, and others, including India, Hawaii, Puerto Rico and the USA, followed early in the twentieth century.

From the thousands of crossings made each year, only about 1 in 10,000 seedlings raised is worth selection for trials, and less than 1 in 10 of these better seedlings is likely to become a commercial variety. These seedlings are all chance combinations raised from crosses based on the breeders' experience. Such a chosen cross by Jeswiet at the Proefstation Oost Java produced the most famous cane ever bred - POJ 2878 - which was selected from 2,500 seedlings produced in 1921 from crossing two varieties which had already given three promising varieties (Fig. 3.9). This cane has been grown in more countries and on a bigger hectareage than any other single variety, and has been a parent of very many commercial varieties. Jeswiet had selected POJ 2364 in 1917 as a parent which gave promising canes with a variety of male parents.

At Coimbatore in South India breeding was directed to the improve-

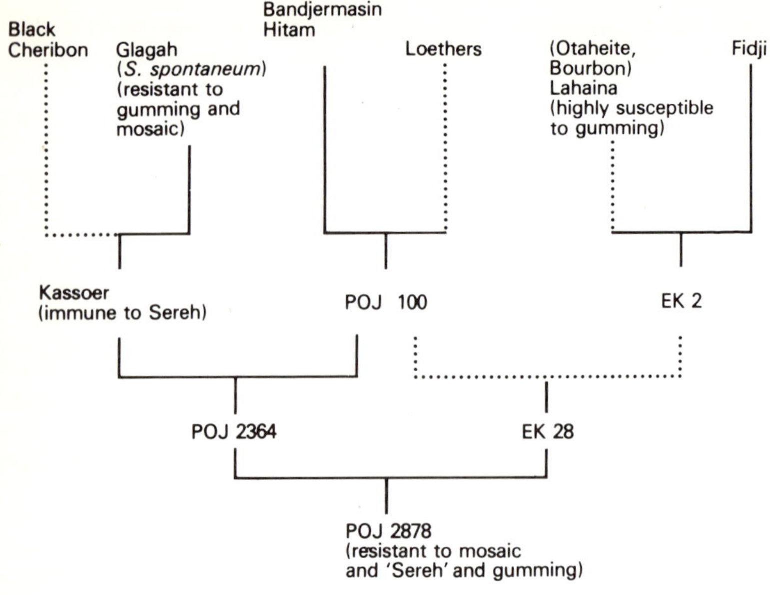

Fig. 3.9 The origin of POJ 2878 as given by Stevenson (1965)

ment of low yielding *S. barberi* and *S. sinense* to increase yield and sugar content but retain the adaptability to hot summers and cold winters. The first cross of Vellai (Bourbon) with *S. spontaneum* produced Co 205, which was a high yielder though unfortunately susceptible to mosaic. Co 205 is the only commercial example of an F_1 *spontaneum* × *officinarum* hybrid. Nobilization lines of *S. barberi* and *S. sinense* were crossed to produce some commercial tri-species hybrids.

For Queensland, cane varieties should be resistant to mosaic, gumming, leaf scald, Fiji disease and downy mildew, and be able to utilize the high light intensity to produce high yields of cane with a high sugar content. An erect growing habit suited to mechanical cutting and harvesting is also required.

Disease resistance in cultivated varieties is mainly dependent on a few major genes (oligogenic) and breaks down because the pathogen adapts to the resistant host either by mutation or recombination to form a new race. Following the cyclones in Mauritius in 1964, more virulent forms of gummosis and leaf scald appeared, to which the resistant varieties grown on 60 per cent of the cane area were susceptible. M 147/44, the outstanding cane for the drier parts of the island, which accounted for a third of the cane crushed in 1964, is particularly susceptible (Annual

Report, Mauritius Sugar Industry Research Institute (MSIRI) 1965) but has not been grown since 1973.

Current objectives

New breeding material became available following the 1957 expedition to New Guinea and nearby territories (Warner & Grassl 1958) and the collection of *S. spontaneum* material in India, and gene banks are now available in Queensland, Florida and South India to widen the very narrow genetic base.

Current objectives of breeders are towards more economical production of sugar. The yield of cane harvested is not the correct criterion for judging a new variety, but the yield of sugar per hectare. Harvesting costs are everywhere high, hence the less cane and non-sugars that must be harvested to produce a tonne of sugar the better. Varieties must be easy to harvest; hard canes are unpopular with cutters and lead to low production. Mechanization of harvesting is spreading and this will need suitable varieties. Resistance to diseases, including new races or strains, is essential, and resistance to pests such as stem borers, froghopper and beetle in Queensland desirable if not in sight. Areas where harvesting cannot be completed before the next rainy season need varieties which retain their sugar content. In Queensland, new varieties are rejected for lack of 'cover' to smother out weeds (Buzacott 1962) but with the development of persistent weedkillers, this should no longer be an essential virtue. Vlitos has suggested using X-ray micrographs of leaf sections to select those seedlings with the type of chloroplasts which are the more efficient producers of sugar, a technique which if successful could lead to a major breakthrough.

Breeding systems

Successful varieties have always been the result of chance heterozygosity. Breeders in Hawaii have introduced new noble and wild canes and crossed these with the best local varieties. Closely related varieties are often crossed to concentrate the desirable characters, but they do not favour taking this to the extreme of selfing. From a very large number of seedlings it is hoped that a superior combination can be selected. George (1962) suggests that likely commercial varieties usually appear in the first 2 years of crossing two specific parents, and hence more parental crossings should be preferable to repeating 'proven crosses' which have already given valuable seedlings.

The possibility of creating superior parents by selfing has been put forward by Stevenson (1960), who found that male sterility was no more pronounced in selfed lines than in unrelated interspecific crosses, and there is not necessarily a reduction in vigour. Indeed in some exceptional cases very vigorous progeny resulted. The canes were often thin-

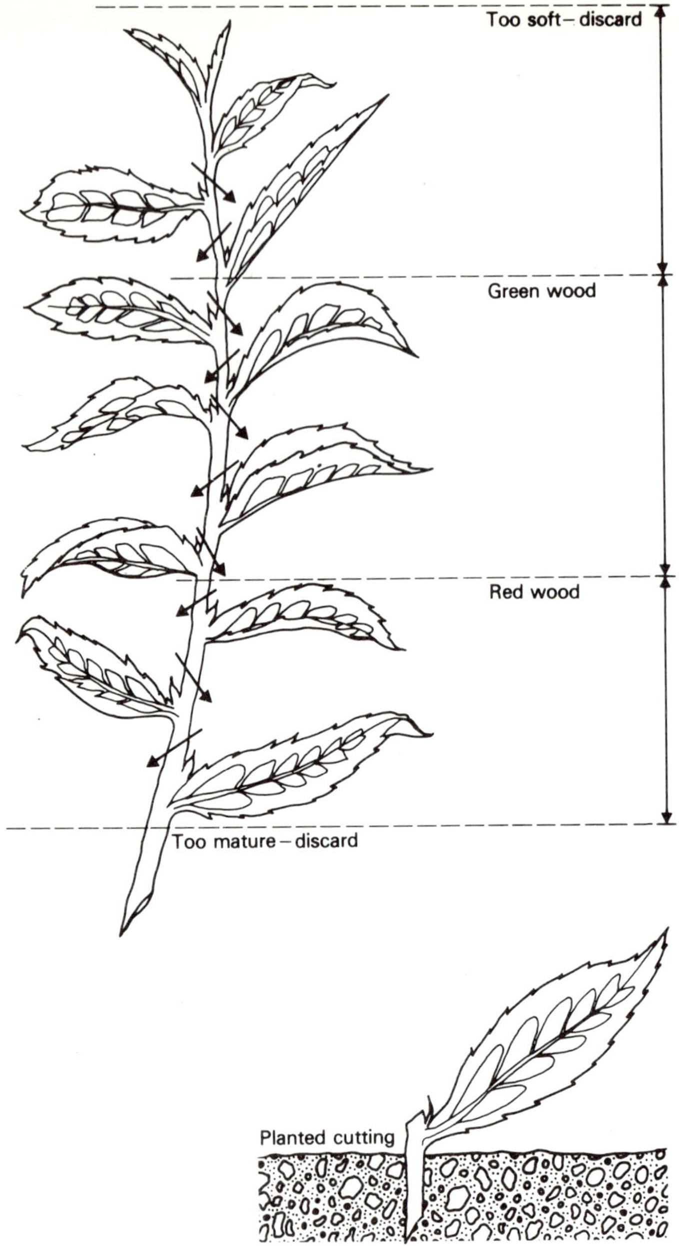

Fig. 3.10 Vegetative propagation of tea

ner than their parents. In short, the effects of inbreeding appear less deleterious than expected.

Daniels (1959) in Fiji found that selfing concentrated the genes for sugar content. High sugar content of these canes was associated with early maturity.

Tea (Camellia sinensis (q.v.))

Looking back it seems very remarkable that greater progress was not made prior to the Second World War in the development and planting of high yielding clonal material, particularly as the crop offers so much scope for rapid improvement by this method.

All the tea areas were originally planted up from seed. In many areas the seed bearing gardens were not planted to selected clones and rarely had their progeny been proven in the field. As the yield of tea depends upon vegetative growth it is quite possible that plants giving a high yield of seed are poor leaf producing strains. To obtain a good set of seed a minimum of nine clones is suggested, but it is doubtful whether this has been fully investigated. The progeny from such a mixture will be very heterozygous and unpredictable.

Budding and grafting of tea is possible but is not really practicable for the production of planting material on a large scale. Propagation by single-leaf cuttings has now been developed to a degree where the average planter can use it in replanting schemes. This is being encouraged in Sri Lanka by government subsidies for replanting schemes carried out with clonal material. Good clonal material for new plantings is also available in East Africa, Malawi, India and other tea growing countries.

Selection

As the tea flower is almost completely self-sterile, considerable intercrossing has occurred and all forms between the main Chinese and Assam types occur. The earliest selection work was probably that of Cohen Stuart at Buitenzorg in Java in 1920, carried on by Wellensiek, who chose 44,000 bushes from the million he examined and from those developed 2,100 clone mother trees. In India 1 bush in 40,000 is estimated to be superior to the average in both yield and quality (M.J.G. 1958). In East Africa 1 bush in 40,000–100,000 gives a first-class clone, including both vigour and quality. A good 6 ha field is recommended as a selection unit to provide one first-class clone and several 'good' clones (Anon. 1966d).

Early selection work in Java laid emphasis on late flowering and light green leaves, but later the emphasis was changed to yield in relation to the area occupied.

Local recommendations for the selection of clones are made by the research stations serving the tea region and cover such characteristics as:

1. A close plucking table with many plucking points per unit area.
2. A good spreading habit.
3. A good appearance with many cut branches of good thickness after pruning.
4. A good number of maintenance leaves.
5. Good recovery from pruning, not confined to the centre of the frame.
6. Low production of dormant shoots (Banjhi).
7. Internodes not too short.
8. Absence of a marked tendency to flowering.
9. Good rooting ability. A minimum of 50 per cent take. This will often be done after the field yield assessment.
10. Tasting quality of made tea. In Assam there is a direct correlation between quality and pubescence (Wight & Barua 1954) but this does not appear to apply in Sri Lanka. The pubescene is most easily seen on the undersurface of the first fully opened leaf.

Assessment is best carried out for one character at a time and different emphasis should be given to each character in the light of local experience and the final assessment made on a total score basis. Planters are continuously on the look-out for bushes with an apparent resistance to pests and diseases. Such bushes may be chosen irrespective of other characteristics.

An assessment of the yield is then made, based on about 8 or 10 consecutive plucking rounds in comparison with the average for its own section. Half the bushes are then eliminated and a further series of plucking rounds are then carried out, discarding after each series until the best 10 per cent of the trees have been chosen.

Wellensiek (Coster 1942) in Indonesia selected clones on the basis of five or six pluckings in one year. Visser (1958) found that selection based on alternate pluckings over at least a year gave as good a basis for selecting clones as all plucks, and the yield of the second series of 10 plucks in the first year after the bushes came into bearing would select the highest yielding tenth. Fresh weight yield was highly correlated with dry weight yield.

Selections by eye in Assam favoured bushes with saucer-like horizontal leaves. It is now thought that thinner leaves held more vertically are better as they absorb less of the sun's heat and remain cooler.

While considerable care is needed in selecting clones, equal care is needed in testing them out. Sufficient material should be available for planting out replicated rows of 50 to 100 bushes at the normal spacing. One bush will yield about 800 cuttings. Rows should all be planted at the same time, along with standard clones or jats chosen for improvement. Care must be taken with the less vigorous clones that may suffer unduly from competition in the early years. Small samples of tea for tasting should be made from these rows.

Selected clones should be planted in larger blocks and studied over a number of years before being planted over large areas. There are many practical advantages in having blocks planted to one clone but the planted area should be divided among a number of clones to insure against an unforeseen weakness of one clone. A blend of clones is also considered better for the production of quality tea.

The four most popular clones in Sri Lanka are T.R.I. 2023 and 2026 for the low country and T.R.I. 2024 and 2025 in the high country. These cover over half of the new clearings under the tea replanting subsidy scheme. All these record breaking clones originated from a handful of seed from a single open pollinated seed bearer at Tocklai Experimental Station brought from Assam in 1937 by F. R. Tubbs. Out of 15 germinated seeds, five record breaking clones have been produced, all with a yield potential of 5,500 - 7,800 kg/ha of made tea of reasonably good quality (Chenery 1966). These clones have now been crossed and selections will be made from the progeny. Subsequent seed introductions from the same tree have given plants which promise to provide other good clones. The average yield of tea in India rose from 596 kg/ha in 1926 to 1,500 kg/ha in 1977 by the use of high yielding clones, fertilizer, drainage, pest control and better agricultural techniques. This has raised production to 600 million kg (Quinn 1978).

The prospect of obtaining better clones from crosses between selected parents is about 1/1000 seedlings. This is much higher than selecting from a random population of seedlings as found in old established tea plantations where one good selection can be expected from 20,000 to 40,000 bushes. In Malawi, three clones, PC 79, PC 80 PB 81, are selections from crosses made in 1967. Besides high yield and quality, they are vigorous, with good rooting ability and drought resistance (Deveria 1978).

Yields of over 5,000 kg of made tea from 4 ha plots of 2023 and 2026 have been recorded in the low country of Sri Lanka and T.R.I. 2024, 2025 and some estate clones have produced over 2,200 kg/ha on up country estates (Richards 1964). Trials at the T.R.I. in Sri Lanka have shown that shade depresses the yield of certain T.R.I. clones, and also encourages blister blight.

Selection for resistance to blister blight has been in the mind of all tea growers in the areas where this disease is a problem. To date little progress has been made; this may be partly due to the fungus existing as a number of biological races, and resistance to all races not occurring in a single plant. Search for resistance in the original homes of these plants in North India and China is probably necessary when politics will permit.

Vegetative propagation of tea

The most suitable method is to use single leaf cuttings with 2 cm of stem and an axillary bud (Plate 3.8). The mother bush should be left after pruning and cut back to within 1 cm of the old growth when the cuttings are required. The time of cutting will depend upon the area; in parts of Sri Lanka it can be done during most of the year. The cuttings should be protected from the hot sun and transported in buckets of water. The soft wood at the shoot tip and the woody part at the base are discarded and the remaining red and green wood cut with a pruning knife into 3 cm lengths. The top cut is made just above the axillary bud parallel to the leaf blade (Fig. 3.10). If the axillary bud has started to grow the cutting is discarded. Lateral shoots are similarly discarded unless the maximum increase is required when everything is planted. The cuttings are prepared in the shade and planted with stem vertical and the leaf orientated north-south in shaded nursery beds. The spacing is closer if the cuttings are to be moved to baskets than if they stay in the nursery until planted in the field.

Alternatively the 2.5 cm long cuttings can be rooted in polythene sleeves, 7.5 cm in diameter and 20-25 cm long filled with a friable soil with a pH 4.0-5.0. These are kept under low shade with 25 per cent light intensity (Richards 1966).

The Tea Research Institute of East Africa have published a clear, well illustrated guide to vegetative propagation (Green 1964). The commonest fault in East Africa has been to use rich black forest topsoil for rooting cuttings. A little superphosphate is recommended for the rooting soil.

The air-moisture balance in the beds is important for good rooting. Rooting is easier in certain areas such as Darjeeling and Sri Lanka than in Assam but many estates expect about 90 per cent successful rooting.

In Sri Lanka the cutting may be planted with the leaf on the surface of the nursery bed, but in Assam this is avoided (as in Fig. 3.10) to prevent disease destroying the leaf.

Plate 3.8 Taking tea cuttings for vegetative propagation. St Coombs Tea Research Institute, Sri Lanka. (Courtesy Sri Lanka Tea Centre.)

4 CROP PROTECTION

The tropical environment which favours the fast luxuriant growth of crops and vegetation, also favours the weeds which compete for moisture and nutrients; and the parasitic fungi, insects, spider mites, bacteria, eelworms and virus diseases which make serious reductions in the crops. Without a 'close season' for plant growth, these pests may thrive all the year. After harvest, serious losses result from storage pests and rats.

These maladies may give rise to such serious epidemics that the economy of the country may be threatened. Examples of this are: coffee rust which eliminated the arabica coffee industry in Sri Lanka and is now spreading in Brazil; blister blight of tea before control measures were found, in Sri Lanka, South India and Indonesia; lethal yellowing which is destroying 100,000 - 200,000 tall coconut palms in Jamaica annually; 'swollen shoot' in West African cocoa which is estimated to have caused an annual loss of over 20,000 t of cocoa in Ghana from 1946 - 61, and over 140 million trees have been cut out in an effort to control it (Kasasian *et al*. 1978); 'wither-tip' of Dominican limes; and Panama disease of 'Gros Michel' bananas. Crop losses may be serious, but not lethal to the plant, as for example the loss of rice in hungry Asia from stem borers and 'blast'; the loss of cocoa pods from *Phytophthora palmivora* which, if uncontrolled, may account for 75 per cent of the potential crop where conditions for the disease are ideal as in Cameroon, and the loss of cowpeas in Northern Nigeria by insect pests estimated at 70 per cent. Downy mildew (*Sclerospora graminicola*) in India has become a limiting factor in the cultivation of improved) hybrids of pearl millet (*Pennisetum americanum syn. P. typhoides*).

The loss of potential crop production in Asia and Africa from insect pests, diseases and weeds is estimated at over 40 per cent (FAO 1970, p. 207). For all the developing countries this loss has been estimated to be equal in value to the proposed increase of crop output between 1962 and 1985 to meet the FAO Indicative World Plan objectives. In addition the cost of production of many crops is increased through the need to apply protection chemicals to obtain a worthwhile yield as in the case of the following crops in certain countries.

Rubber	*Oidium* and *Phytophthora*	**Arabica coffee**	Coffee berry and rust diseases
Tea	Blister blight		
Cocoa	Blackpod and capsid		
		Bananas	'Sigatoka' or leaf spot
Cotton	Jassid, aphis, whitefly, bollworms, etc.	**Tobacco**	Blue mould, aphis

Some indigenous methods of cultivation which reduce the loss of crop from pests and diseases have developed from experience rather than scientific investigation. These methods can often be improved upon by legislation, crop breeding and the use of chemicals.

A knowledge of the biology of the parasite is the basis of all control measures. Information on the life cycle of the parasite may indicate the point at which control can be best applied. Previous experience in the control of the pest or a similar pest elsewhere can be used as soon as the pest is diagnosed. The vector of a virus can often be controlled by insecticides. Seed treatment with heat or chemicals will eradicate many seed-borne pests and diseases, such as the pink bollworm of cotton and chlorotic streak virus of sugar cane.

The object of all control measures is to attack the parasite and defend and strengthen the host along the following lines.

Attack the parasite by:

1. Keeping the pest out by strict plant quarantine.
2. Use of clean seed.
3. Destroying the outbreak (Localized eradication).
4. Destroying the alternate host plants.
5. Removing its food supply by a 'close season'.
6. Rotating annual crops to prevent pest levels building up.
7. Biological control by introducing insects, fungi, viruses or bacteria which will attack the parasite.
8. Destroying the sources of infection. 'Cut and burn'.
9. Chemical attack on the parasite.
10. Use of insecticides.

Defend the host by:

1. Growing the crop at a time unfavourable to the pest.
2. Breeding tolerant varieties, or the use of resistant rootstocks.
3. Legislation restricting the varieties grown.
4. Modifying the crop environment, such as the amount of shade, the water relationship, or the soil nutrient status.
5. Applying a protective chemical.

ATTACK THE PARASITE

Plant Quarantine

Most countries restrict the import of plant material in an effort to prevent the introduction of pests and diseases. In these days of air travel and vegetative propagation, the risk of introducing a new disease is particularly serious. Apart from diseased plants being moved from country to country, viable fungal spores can be transported on the clothing of unsuspecting air travellers. Fiji disease, a virus disease of sugar cane, is in Reunion, but has been kept out of Mauritius which is visible over the sea from Reunion. It is difficult to know whether a major pest of one area is absent from another area because it has never been introduced or whether it has been introduced but has not established itself due to unfavourable conditions. South American leaf disease does not occur on rubber in Sri Lanka or Malaysia, while *Phytophthora palmivora* and *Oidium heveae* occur on rubber in Malaysia but are not as important as they are in Sri Lanka. Blister blight of tea has not been introduced into East Africa. Coffee rust (*Hemileia vastatrix*) was unknown in South and Central America until it was identified in Brazil early in 1970. In the next 3 years it spread to all the major coffee growing areas of Brazil. This rapid spread may have been favoured by the close planting giving a dense foliar canopy which makes spraying difficult. One thing is certain: the appearance of coffee rust in South America means a shortage and high coffee prices for many years to come. Sigatoka was serious in Fiji in 1913 but did not appear in the Caribbean region until 1933. In the following 15 years it spread to all the West Indian islands and Central America.

Use of clean seed

While most countries now appreciate the importance of plant quarantine for keeping pests and diseases out of a country, the equal importance of taking similar precautions when opening up a new area to a new crop, or introducing or planting a new crop on a farm is not always appreciated. Cassava mosaic is mainly spread by the indiscriminate use of infected planting material, whiteflies being inefficient vectors (Bock & Guthrie 1977). Cottonseed, heat treated to control the pink bollworm, where it occurs, and treated with a mercurial or copper fungicide to prevent the carry-over of bacterial blight (*Xanthomonas malvacearum*) is now commonly supplied from ginneries where central control makes this practicable. Similarly the sugar cane setts planted in Mauritius are normally heat treated to destroy the chlorotic streak virus, and nurseries are planted with heat-treated 'seed' (setts) to eliminate ratoon stunting. Banana planting material which is usually infected with nematodes (minute cylindrical worms parasitic in plant tissue) and borers should be pared with a cutlass or sharp knife to cut away all the discoloured flesh except the buds, until it is white, and then treated with a dilute emulsion of Nemagon® or Fumazone®, and with a soil insecticide. Alternatively nematodes can be destroyed by a hot water treatment of the banana corms and yam tubers but this is slow and must be done very carefully. Avocado pear picks up *Phytophthora* root disease so easily it should be propagated from seed taken from avocados picked from the tree not the ground, germinated in sterilized soil, and grafted with a chosen variety. Seed treatment is easy to apply and gives a large benefit at a low cost. Particular care should be taken to ensure that planting material for perennial crops is free from virus infection.

Destroy the outbreak

When a new pest or disease is discovered, rapid action can sometimes be taken to destroy the infected crop and save future crops, in the manner that the Mediterranean fruit fly was eliminated from Florida. This is normally a matter for government action, or international cooperation as in the case of locust control. Growers however can pull or dig out and burn individual infected plants before the disease spreads, and after harvest.

Destroy the alternate host plant

This is promising in theory but difficult in practice. Often the host range of a parasite is not fully known. The mealy bugs which carry the swollen shoot virus colonize a number of forest trees but it would be difficult to eliminate them. The increased planting of maize in the cotton area around Samaru in Northern Nigeria will encourage the American bollworm (*Heliothis armigera*). Experience in East Africa has shown that this pest breeds on maize, particularly at tasselling time, and then moves on to cotton to attack the bolls. *Heliothis* can be easily controlled by DDT if the spraying of a large number of small scattered plots can be organized. However, maize is also likely to encourage the false codling moth (*Cryptophlebia leucotreta*) which is much more difficult to control.

The Egyptian or spiny bollworm (*Earias* spp.) is dependent on alternate host plants to carry it over from one cotton crop to the next. Unfortunately it breeds on a wide range of plants. Species of *Abutilon* which are common in Africa and the Middle East, are a major site for the carry-over and build up of this pest. *Echinochloa colonum*, a grass weed of rice, acts as a reservoir for *Hoja blanca*, an important virus in rice in the Caribbean.

Close season

If a crop such as cotton is left in the ground after the final picking it serves as food and shelter for pests and carries them over to invade the new crop. For this reason most governments have laws under which all cotton must be uprooted and burned by a certain date, and a date is given about 2 months later when cotton planting may start again to ensure a close season. In Zimbabwe the same system was applied to flue cured tobacco, but the close season first introduced in 1933 has been shortened in recent years by growing more Turkish tobacco with slightly different growing seasons. Unfortunately this tobacco is host to the same pests, which are vectors of serious tobacco virus diseases.

When replanting perennial crops, a short period should be allowed between uprooting the old crop and replanting the new, to enable all the old roots of rubber, coconuts and tea to be pulled out or rot down, and for the nematode population to decrease. Under the subsidized tea replanting scheme in Sri Lanka, the land had to be planted with Guatemala grass for a year before replanting with tea, largely to build up the soil fertility but it has the other advantages. Some planters believe Guatemala grass reduces the nematode population. Before replanting bananas, in order to reduce the nematode population, growers in the Windward Islands are advised to keep the land free of bananas for at least 6 months during which time the land should be forked over. Food crops and vegetables can be grown on this land during the 6 months.

Rotations

Too frequent cropping of a soil with the same crop can lead to a serious build up of a plant pest in the soil. In planning the rotation, the alternative crops should not be host to the pest. Tobacco is usually restricted to one crop in 3 or 4 years to keep down the level of nematodes, and during these years other solanaceous crops such as potatoes, eggplants, chillies or tomatoes should not be grown on the tobacco land. Young shoots of Pangola grass (*Digitaria decumbens*) produce root exudates that cause the eggs of the root knot nematode to hatch. In about 5 months this grass can reduce the root knot nematode population below detectable limits (Winchester 1966). Nursery sites for tree crops and tobacco should be used in rotation if practicable, otherwise sterilized between crops to prevent the build up of root nematodes and diseases. This applies to pests which are relatively immobile like root diseases.

Biological control

1. Control of insects by insects

Many potentially serious pests are kept down to unimportant population levels by their natural parasites. Where an insect is present in a number of territories, but only a serious pest in a few, it is natural to search for the parasites which are keeping the pest in check, and introduce them to the troubled areas. The introduced predator or parasite must not itself be a potential pest. Care must be taken when introducing a parasite not to introduce any secondary (hyper) parasites with it or the value of the introduction may be lost. It is also essential that the first release should include equal numbers of both sexes.

Successful biological control appears to be limited to:

(a) *Insular areas*, particularly where:
- (i) The climate is warm and equable, allowing the parasite to multiply unchecked.
- (ii) The indigenous fauna is restricted having evolved in isolation so that the introduced parasites meet with little competition from indigenous forms. This would not apply to Trinidad, for example, due to its close association with South America.
- (iii) The area to be covered by parasite colonization is circumscribed with only a few main crops (Imms 1937).

(b) *Localized continental area*, where the crops form islands and conform as far as possible to (a) above. A good example of this is the Kenya *arabica* coffee area. Isolating geographical features such as mountains or deserts may make the area suitable for the successful introduction of parasites.

There have been a number of very successful controls by introduced parasites. The first important case of an insect being controlled by an insect was in the late nineteenth century. In 1868 the cottony cushion or fluted scale *Icerya purchasi* was introduced from Australia to California on *Acacia latifolia* plants (Marlatt 1900), and during the next few years developed into a major pest of citrus. As this pest was not serious in Australia, a search for parasites was carried out and in 1889 the vedalia beetle (*Rodolia cardinalis*) was brought from Australia and liberated in California. Very quickly the scale was reduced to an unimportant level. In 1910 the same beetle was introduced into Portugal to control the same scale on *Acacia* spp.

Other successful introductions to control serious pests include:
Cryptognatha nodiceps introduced from Trinidad to Fiji to control *Aspidiotus destructor*, the coconut scale.
Eretmocerus serius introduced from Jamaica to Seychelles to control *Aleurocanthus woglumi*, the citrus blackfly first recorded in the Seychelles in 1954.
Pleurotropis parvulus introduced from Java to Fiji to control *Prometheca reichei*, a hispid beetle attacking coconuts.
Anagyrus kivuensis and other parasites introduced from Uganda to Kenya to control the coffee mealy bug (*Planococcus kenyae*).

The egg parasite *Opius oophilus* introduced into Hawaii gives a considerable degree of control of the oriental fruit fly (*Dacus dorsalis*) and partial control of the Mediterranean fruit fly (*Ceratitis capitata*). Controlling the rhinoceros beetle with the reduviid bug *Platymerus rhadamanthus* (Douglas 1965) could be of great economic significance in the Far East if generally successful.

There are over 200 successful cases of biological control of pests and weeds in 60 countries (De Bach 1964).

Since 1929, the Department of Agriculture in Barbados has reared and released annually large numbers of *Trichogramma minutum*, a chalcid which parasitizes a large number of insects. This was directed against the sugar cane moth borer *Diatraea saccharalis*. Rearing was abandoned, as further releases were not increasing the numbers of the parasites. The natural *Trichogramma* population has increased considerably since the liberations started but *Diatraea* still remains a problem and no significant differences could be detected between the fields with and without the parasite (Metcalfe 1959). This borer is believed to be controlled by the Cuban fly, *Lixophaga diatraea*, in Jamaica.

If insecticides are not used with care, the reverse process may occur. The natural predators may be destroyed and an unimportant pest may increase to epidemic numbers. Examples of this have occurred on coffee estates in Tanzania and Kenya, in oil palm plantations in Malaysia and in cocoa plots in West Africa and Fernando Po where a persistent insecticide has been used. Sri Lanka tea planters were advised to use dieldrin to control the shot hole borer (*Xyleborus fornicatus*), but this insecticide reduced numbers of the parasite *Macracentrus homonae* which had been successfully introduced from Java in the 1930s to control the tortrix (*Homona coffearia*) and as a result the tortrix became a pest on the sprayed areas. In addition the previously unimportant twig caterpillar (*Ectropis bhurmitra*) and the looper caterpillar (*Buzura strigaria*) became prominent (Danthanarayana 1966). In Ghana it had been hoped to reduce the spread of swollen shoot by reducing the number of mealybug vectors by attacking with dieldrin the ants which attend the mealybugs. The consequent upsurge of minor pests, *Marmara* pod husk miner and the trunk borer *Eulophonatus myrmeleon* to serious levels led to the abandonment of this approach. A wide spectrum contact insecticide such as parathion could give rise to similar troubles. This has occurred in Tanzania where the normally harmless caterpillar *Ascotis selenaria reciprocaria* has attacked coffee sprayed about four times with parathion (Bigger 1966). The risk is reduced when highly selective insecticides, or systemic insecticides with little contact action are used, as the systemic effect is only toxic to those insects which attack the plants, thus the pests are attacked both chemically and by their natural enemies. There is some concern that the new pyrethroid insecticides will not be sufficiently selective and will destroy the natural parasites as well as the pests.

At Zebediela citrus estate in South Africa an intensive insecticide spraying programme was replaced by a system of integrated control. Large numbers of predators were raised and released. Most of the pests including soft wax scale and red spider were well controlled by this system, the main exception being the false codling moth (*Cryptophlebia leucotreta*). Not only did this make a large saving on insecticides, greater than the cost of rearing predators, but many birds returned to live in the plantation they deserted when insecticides were being sprayed.

2. Control of weeds by insects

Biological control of certain obnoxious weeds by introduced insects has been effective in certain cases. A large area of valuable agricultural land in Australia has been recovered from prickly pear cactus (*Opuntia* spp.) by the introduction of the cochineal insect, *Dactylopius* spp., a moth *Cactoblastis cactorum*, and a red spider. The weed *Cordia macrostachya*, introduced into Mauritius, was controlled by a leaf feeding beetle *Schematiza cordiae* which destroyed the bushes, and a seed destroying chalcid *Eurytoma attiva* which reduced the speed of replacement. *Cordia* is now unimportant as a weed but a balance between the weed and the insects remains.

3. Control of fungi by fungi

There is no case of biological control of a fungal disease by an introduced fungus or other parasite, but it is probable that in the soil, competition between fungi may keep a serious parasite in check. The non-occurrence of Panama disease on certain Jamaican soils has been attributed to the antagonism of other soil organisms (Meredith 1944). A number of actinomycetes and soil bacteria have been shown to have that effect and also to be widespread in their occurrence (Rombouts 1953). Unfortunately such a balance is easily upset by a change in soil conditions brought about by drought, irrigation or cultivation and is thus too unreliable and subject to too many varying factors which have not been analysed, to be applied to other areas. *Colletotrichum camelliae*, which cannot itself invade a healthy tea leaf, often follows an invasion of blister blight and prevents the fruiting of *Exobasidium*. This probably reduces the incidence of blister blight but is not an effective control (Gadd 1949, p. 59).

4. Control of insects with bacteria

The isolation of bacteria which cause diseases of insect pests but not of men or animals, and which can be multiplied in artificial culture for distribution, has a great attraction for the control of insects. Spore-forming bacteria are potentially the most important as these can be stored as a dry powder and survive longer after application.

Two such species, *Bacillus popilliae* for the control of the Japanese beetle and *B. thuringiensis* for the control of a wide range of caterpillars, have been developed into commercial products.

Bacillus thuringiensis produces a crystal of toxic protein at sporulation, and when this is eaten with the spores it reduces feeding and allows the bacteria to develop and enter the body cavity of the insect where conditions are favourable for the rapid multiplication of the bacteria. Of the 120 or more insect pests reported susceptible to *B. thuringiensis*, most of them are Lepidoptera but there are also a few susceptible species from the Diptera, Hymenoptera and Coleoptera. Caterpillars vary in their reaction to this bacteria. Susceptible species do not feed much after ingesting the bacteria, and the control is fairly quick.

Bacillus thuringiensis does not appear to be passed from the larval to the adult stage and to the next generation, hence there is little carry-over from season to season. Predatory and parasitic insects appear to be unaffected, and the effect of the bacillus appears to be confined to the sprayed area (Burges 1964). Similar bacterial diseases and more virulent strains will no doubt be isolated. In the Chad Republic, a spore-forming bacillus closely related to *B. cereus* was isolated from diseased caterpillars of the red bollworm (*Diparopsis watersi*), the American bollworm (*Heliothis armigera*) and the cotton worm (*Prodenia litura*) and found to be strongly pathogenic when ingested by healthy larvae (Atger & Jacquemard 1965).

5. Control of insects with viruses

It has long been known that very many insect pests are attacked by virulent viruses, indeed these may be the cause of the rapid decline of serious pest epidemics which occur without insecticide spraying. Such viruses appear to be more lethal when their insect host is under stress, as occurs when the insect population builds up in an epidemic. The application of viruses to control pest outbreaks has been developed (Thompson 1958), and the possibilities of viruses as a means of overcoming both insect resistance to insecticides, and pollution of the environment is currently being examined (Tinsley 1977).

Of the seven groups of viruses associated with insects, the Baculoviruses which are very complex appear to have no chemical, physical or biological properties in common with any known virus found in either vertebrates or plants. For this reason WHO and FAO jointly recommend that at present only the Baculovirus group should be considered as possible pesticides. Viruses are replicating systems, and once released it is impossible to restrict their spread or destroy them, and there is always the risk of a released virus mutating to a form which could be pathogenic to vertebrates or plants. On the other hand it should be realized that the viruses which can be used are existing in the natural state and are not man-created.

Tinsley has stressed the unpredictable risk of collecting virus infected and dying caterpillars, crushing them up, and spraying them on healthy colonies of the same caterpillar, an approach which has been used effectively. Such a technique can spread associated viruses pathogenic to vertebrates. Tinsley considers it is vital to isolate the virus, purify it, identify it and test it under laboratory conditions as a pesticidal agent, establish its host range, and test it for toxicology. Once the effectiveness and safety of the isolated virus have been proved, it should be increased in healthy insects raised on an artificial diet. The virus recovered from these dead hosts is then used in the field.

Only very small quantities of virus are required to treat large areas. The viruses are highly specific, and should have no effect on the natural predators of the pest. On the other hand it is necessary to maintain supplies of a large number of specific viruses, and multiply them quickly when a pest outbreak occurs unless the pest is continually present. There are problems in application, the Baculoviruses are particularly sensitive to ultra-violet light so should be applied at night. Also the purified virus, without the help of the insect body, will not adhere easily to the sprayed foliage so careful formulation of the virus spray is necessary.

In the USA a nuclear polyhedrosis virus (NPV) of cotton earworms (*Heliothis zea*) has been safely tested and cleared for field use in the USA. A virus from the army worm (*Spodoptera exempta*) has been isolated and multiplied, and is under test in East Africa. Similarly a baculovirus shows promise for the control of the coconut palm rhinoceros beetle in the South Pacific (FAO 1978a).

Insect pathogenic viruses could have an important potential in the integrated control of insect pests, provided they are used intelligently.

6. Control of insects by sterilization

The first successful applications of this technique were the releases of male screw worm (*Cochliomyia hominivorax*) pupae sterilized by radiation on the island of Curacao and in the S.E. and S.W. of the USA. (However, see C.B.C. and D.M.B. 1973.)

The technique is to raise and sterilize by cobalt radiation or chemicals, sufficient males to overflood the natural male population by about 10 : 1. The release is timed for the moment when the population is at its lowest level either on account of natural population changes, or as a result of insecticide application. It is necessary to repeat the overflooding for about three seasons. Another technique is to introduce sterility into the natural population by the use of chemical sterilants added to baits and sex attractants (Knipling 1964).

The melon fly (*Dacus cucurbitae*) was eliminated on the island of Rota in the South Pacific using 306 million irradiated pupae raised at the Hawaii laboratory. About 67 million flies were released from the air and 190 million from cages in trees, between September 1962 and July 1963. Prior to the release the natural population was reduced by bait sprays. Despite the loss of irradiated flies to predators, drifting off the island, and their shorter life, fruit infestations had disappeared by December 1962 and none was observed in 1963 (Steiner *et al*. 1965a). Prior to 1962 unsuccessful attempts had been made to eradicate the oriental fruit fly (*Dacus dorsalis*) by means of sterile flies. This pest was however eradicated in a campaign started in November 1962. Cane fibre squares soaked in a mixture of 97 per cent methyl eugenol (a male attractant) plus 3 per cent naled (an organophosphorus insecticide) were dropped on 15 occasions from aircraft at a rate of 50 squares per square kilometre at 2 to 3 week intervals. Larger squares were suspended in trees and renewed monthly. After the tenth application no male flies were trapped (Steiner *et al*. 1965b). This technique of using a sex attractant with an insecticide makes a great economy in the amount of insecticide required in comparison with the normal insecticide spraying, to reduce the insect population.

Both these techniques rely on the area to be treated being isolated from a source of reinvasion by the pest, and the whole of the isolated area being treated. Steiner's laboratory in Hawaii where the larvae for sterilization are raised is modestly equipped and the cost of this technique is small in comparison to the annual cost of insecticide. The biology of the pest and population trends must be studied before an eradication campaign is started. Where the insects only have a single mating rather than multiple matings the chances of successful eradication are increased.

7. Control of insects with sex attractants, toxicants and sterilants

These alternative methods of pest control, reviewed by Knipling (1963), have not yet been used on a practical scale in the tropics.

8. Control of insects by fungi

Fungi also parasitize insects but no practical use has yet been made of this in tropical agriculture, though they no doubt contribute to the natural control of insect populations. The increased use of fungicides to control diseases may have destroyed the entomogenous fungi and resulted in increased populations of certain pests.

9. Integrated control

With the growing realization of the importance of using insecticides in a manner which will control the pest, without destroying the predators or parasites of either the pest or a potential pest, the idea of integrated control has gained wide support. Highly selective insecticides such as Dipterex® or Gusathion® are used which have little effect on the pests' natural enemies and assist these to control the pest. Mineral oil is similarly useful as it is selective in its action on insects and is being used increasingly for the control of scale insects and spider mites. Besides having no effect on the predators, the insect pests do not develop resistance to oil. Insecticides together with sterilization techniques, or sex attractants, are other practical forms of integrated control.

10. Control of rats by barn owls

Barn owls (*Tyto alba*) in Malaysia feed almost exclusively on the rat species which are pests of rubber and oil palm plantations. Barn owls are increasing in numbers and the provision of nesting facilities at appropriate locations is a cheap method of rat control (Lenton 1978). This should be worth trying in other crops such as cocoa and sugar cane.

Destroy the source of infection

Infected plant material is a frequent source of inoculum for future crops. All dead and diseased material should be pruned from the trees, infected fruits collected where possible, and burnt or buried. Frequent harvesting of cocoa and the removal of pods infected with black pod, reduces the incidence of this disease (Owen 1951). As *Phytophthora* starts to produce conidia within 2 days of the black spot starting to appear on the pod, harvesting every alternate day is the only way to ensure a minimum loss by this method. Such frequent harvesting is in general impracticable, but removal of infected husks from the plot means the removal of an important source of inoculum. Witches' broom has been kept in check on some West Indian estates by removing the brooms and 15 cm of bark twice a year and burning them. New plantings of tree crops, particularly rubber and tea, following the same crop or cleared

forest, are frequently attacked by root diseases which are very difficult to eradicate and cause a permanent loss of trees. These root fungi cannot live in the soil except in the woody root tissue and unless they can transfer from the old roots to the roots of the young trees they die out. Leach (1937) showed the importance of ring-barking trees some time before felling to reduce the food reserves in the roots to control *Armillaria mellea* which is a serious root disease of Africa. Adequate drainage helps to reduce this disease in the wetter parts of Nigeria. In Malaysia it is recommended to poison the trees or stumps when clearing, with sodium arsenite or 2,4,5-T ester in oil (Newsam 1964). Glyphosate may be used in the future. Poisoning hastens the death of the trees and the roots are rapidly invaded by saprophytic organisms. White root disease (*Fomes* spp.) and red root disease (*Ganoderma pseudoferreum*) are distributed by spores which fall on the cut stems of the old trees, germinate and spread down to the old roots which then infect the roots of the new trees by contact. Treatment of the freshly cut stumps with creosote is recommended to prevent these airborne spores germinating. Interplanting with legume cover crops reduces root diseases in young rubber (Newsam 1963).

Old decaying stumps of oil and coconut palms are breeding grounds for the rhinoceros beetle and *Ganoderma* which causes basal stem rot of oil palms. This disease can cause the loss of 40-50 per cent of the oil palms within 15 years of planting, where the land has previously been planted with coconuts (Turner 1965). Crop sanitation is the only method of controlling this disease. The decay of the oil palm bole can be accelerated by applying urea and covering with soil.

Stalk borers of maize, rice and sorghum are often in the stems when the crop is harvested. This straw should either be burned or carried away from the cropping area, particularly if the same cereal crop is to follow. *Sclerotium rolfsii*, an important disease of many crops, grows only in fresh debris. Therefore if the crop residues are dug in soon after harvest this disease is reduced.

The red or Sudan bollworm (*Diparopsis* spp.) has a 'diapause' (resting stage) as a pupa in a hard earthen cocoon in the soil, which carries the pest over to the next season. Emergence occurs at two periods in the year, and if ratoon crops of cotton are left and not uprooted the red bollworm will be carried over to the next crop. If the land is disc ploughed to 20 cm, 50 per cent of the diapause pupae in the soil are destroyed. In the Gash Delta of the Sudan, flooding the previous season's cotton land for 30 days killed all the diapause pupae in the soil (Tunstall 1958).

Neglected suckers are important breeding sites for coffee mealybugs

---

Plate 4.1 Cocoa farm in Ghana. (Courtesy Cadbury Bros. Ltd.)

and de-suckering is important both in maintaining the vigour of the yielding stems and keeping the mealybug population down.

Where clonal planting material is used, great care must be taken in selection to avoid planting virus infected material. Cassava mosaic has been distributed throughout Africa by the transport of infected planting material. So far, no important virus disease of rubber is known, but the risk of introducing and propagating swollen shoot of cacao or phloem necrosis of tea in clean areas is obvious and widely recognized.

Abandoned tea, particularly survivors in old nurseries, are important sources of new outbreaks of blister blight in India.

Before the advent of modern control techniques, chemicals, and spraying machinery, and crop prices which could carry the cost of spraying, 'cut and burn' was the most important 'weapon' in disease control, but it is no longer so.

Chemical attack on the parasite*

1. Soil treatment

The most important pest reservoir after infected plant material is the soil. Sterilization or partial sterilization of the soil is usually out of the question except in nurseries and seedbeds and for crops of high value. Root nematodes are now recognized as important causes of crop yield limitation in a wide range of tropical crops including bananas, tea, citrus and tobacco particularly on the more fondy soils. Nematodes are probably the major cause of 'banana sickness' on Central American soils. A summary of the more commonly available nematocides is given in Table 4.1, Tobacco nurseries are treated with ethylene dibromide or a proprietary fumigant to destroy the nematodes, or with methyl bromide to destroy both weeds, pathogenic soil fungi and nematodes (eelworms). This could become a standard practice for permanent sites of nurseries for other crops. The 'Rab' system of preparing rice seedbeds, where brush is collected and burned on the seedbed prior to sowing, may owe part of its success to the destruction of soil parasites and weeds.

Flood fallowing of banana land infected with Panama disease reduces the infection to under 10 per cent of its original level. The depth of flood water is up to 1.5 m and maintained for 6 months to create anaerobic conditions. The remaining infection is confined to the surface layer, and experiments have been carried out to try to eliminate the surviving parasites with a dithiocarbamate fungicide. This flood treatment is unsuccessful after cycles of control.

* For an authoritative treatment of crop protection, and details of the chemicals referred to in the text, see Martin (1973).

Table 4.1 Summary of main nematocides used in tropical crops (a) Soil fumigants

Trade name	Chemical	Control Properties	Remarks
D-D, Vidden D, Telone, Terr-O-Cide 30D, 15D	Include mixtures of 1,3 dichloropropene, 1,2 dichloropropane, 2,3 dichloropropene and related C_3 chlorinated hydrocarbons. Chloropicrin.	Root-knot nematode. Lesion nematode.	Place 20–25 cm below soil line. Phytotoxic. Waiting period 2–3 weeks, dependent upon soil and climate. Used for wide range of crops.
DBCP, Fumazone, Nemafume, Nemanax, Nemagon, Nemaset	1,2-dibromo-3, chloropropane.	Root-knot nematode. Lesion nematode.	As for D-D.
Bromofume, Celmide, Dowfume, KopFume, Nephis, Pestmaster, Soilbrom	Ethylene dibromide.	Root-knot nematode. Lesion nematode (poor).	As for D-D.
Metam-sodium, Metham, Carbam, Basamid-Fluid, Karbation, Metam Fluide, Sistan, Trimaton, Vapam, VPM, Mapasol	Sodium *N*-methyldithio carbamate.	Nematocide, fungicide and herbicide.	Most effective application is injection and covering. Cover can be omitted and as a herbicide it may be watered on. Phytotoxic. Waiting period, 5 days to 2 weeks depending on application method and local conditions.
A large number of trade formulations	Methyl bromide with or without additions of chloropicrin as a marker or for greater fungitoxicity.	Excellent general nematocide and herbicide. Good fungicidal properties.	Gas for seedbed use. Area must be gas-tight sealed until effects complete. Very wet or dry soil and low temperature give inadequate results.
Vorlex, Di-Trapex, Trapex	Chlorinated C_3 hydrocarbons 80% methyl isothiocyanate 20%.	Root-knot nematode. Lesion nematode. Black shank. Black root rot.	As for D-D but waiting period can be longer.
—	Chloropicrin.	Fumigant for soil pests including fungi and nematodes.	Highly phytotoxic. Often with other proprietary fumigants as a warning gas. Highly toxic and lachrymatory.

Table 4.1 (continued)

(b) Non-fumigants					
Common name	**Some trade names**	**Chemical type and toxicity**	**Acute Oral male rats LD 50**	**Control properties**	**Remarks**
Aldicarb	Temik	Carbamate	Oral 0.93 mg tech/kg. Very toxic, extra care in handling and attention to instructions. Only granular formulation available.	Root-knot nematode Numerous soil and above ground insect pests including aphids, white flies, and leaf miners. Acaricide.	Soil moisture needed to release active chemical from granules. Incorporation, pre-plant but good systemic activity effective from band or hole placement. Soil insecticide only. Waiting period 5 days for incorporation treatment. Incorporate within 30 minutes of application.
Carbofuran	Furadan	Carbamate	8–14 mg/kg	Root-knot nematode Acaricide Controls soil and fodder feeding insects.	Incorporation, pre-plant. Waiting period 2 weeks. Mainly soil insecticide. Achieves above ground effect by systemic properties. Incorporate within 30 minutes of application. Foliar application only translocated within leaf receiving spray.
Dazomet	Basamid, Cragg Fungicide 974, Mylone.	Tetrahydro-3 5-dimethyl-2H-thiadiazine-2-thione.	640 mg/kg	Nematocide (root-knot) soil fungi and herbicide	Acts by release of methylisothiocyanate Needs thorough incorporation in top 30 cm and a good watering. Normally used on seedbeds but has technical application to the field.

Common name	Some trade names	Chemical type and toxicity	Acute Oral male rats LD 50	Control properties	Remarks
Ethoprophos (Ethoprop)	Mocap. Prophos.	Organo-phosphate	62 mg/kg Dermal 26 mg/kg	Root-knot nematode. Wireworm. Soil insecticide.	Non-systemic. Granular formulation. Incorporation, pre-plant. Waiting period 5 days. Thorough soil mixing necessary within 30 minutes of application.
Fensul-fothion	Dasanit, Terracur P.	Organo-phosphate	7 mg/kg	Root-knot nematode. Wireworm. Also, cyst forming nematodes.	Use as Ethoprop. Long persistence, some systemic activity Insecticide and nematocide active against cyst forming and root-knot nematode.
Oxamyl	Vydate L – water-soluble liquid. Vydate G – granular	Carbamate	5.4 mg/kg	Root-knot nematode. Insects and mites.	Incorporation, pre-plant or in transplant water. Good systemic activity, application to foliage translocates to roots.
Fenamiphos	Nemacur.	Organo-phosphate	17 mg/kg	Root-knot nematode. Also controls cyst forming nematodes.	Incorporation pre-plant or drench post planting. Good systemic activity.

Note: For nematode populations above moderate, soil fumigants are the recommended chemical approach. (After Atehurst 1981).

Having obtained 'clean soil', great care must be taken to keep out the pathogen. Water infected with eelworms must not be applied to fumigated seedbeds. Infested soil on shoes and implements is a further source of reinfection. Until Panama disease became very serious in Central America, irrigation water was frequently treated with chlorine to destroy any *Fusarium* inoculum.

Trials in Cuttack, India, showed that the addition of 1 ppm of chlorine as bleaching powder to the irrigation water when the rice crop was 10 weeks old reduced bacterial leaf blight (*Xanthomonas oryzae*) to the same low level as the currently recommended chemical treatment, at a much lower cost (Padmanabhan & Jain 1966).

2. Seed dressings

Disease is often introduced with the seed. The infection may occur as hyphal threads within the seed as with loose smuts or as spores adhering to the seed coat. Sorghum smuts, *Sphacelotheca sorghi* and *S. cruenta* which were very serious in Bombay Province of India, are controlled by seed dressing (Butler & Jones 1949, p. 243). Copper sulphate solution or sulphur was used originally, but a dry mercurial or copper dust is better and more convenient. This treatment will not control soil-borne head smut of sorghum *Sphacelotheca reiliana* (Burns 1938, p. 148).

Cotton seed, with its fuzzy covering, often carries with it the bacterium *Xanthomonas malvacearum*, which causes bacterial blight, known in its various manifestations as angular leaf spot or blackarm. Crop residues also carry over the disease. If the seed is treated with strong sulphuric acid the bacteria are destroyed. Easier and more effective is to treat the delinted seed with an organo-mercurial seed dressing. After 5 years of routine seed treatment in Northern Nigeria the level of infection had fallen to such a low level that a reduced rate of dressing could be considered (King 1967). In East Africa copper based seed dressings were used, usually cuprous oxide, which is less toxic to man and cattle than mercurial dressings. The increased cash return from cotton seed treatment in Uganda in the 1958/59 season was estimated at £1¼ million. In the Chad Republic, where seed dressing is estimated to raise the yield of seed cotton 10–20 per cent, captan may be added where damping off is serious and aldrin where millipedes cause damage. Insecticides may also be incorporated in seed dressings as a protection against soil-borne insects. Experiments have been carried out with systemic insecticides as a seed dressing to protect the young seedlings.

Pineapple disease (*Ceratocystis paradoxa*) which causes poor germination of sugar cane setts under unfavourable growing conditions, such as drought or excessive moisture, can be controlled by dipping the cut ends of the setts in an organo-mercurial dressing or a dilute suspension of benomyl. In Queensland the setts are sprayed as they are cut in the cutter-planting machines.

Seed dressing is both cheap and easy and there is a good case for treating the seeds of all crops in areas where they may be attacked by fungi or insects before germination is complete, or carry a disease such as sorghum smut which develops to destroy the seed head.

At the National Vegetable Research Station in England it was shown that certain seed-borne diseases, even when the infection is deep in the seed, can be controlled by soaking the seed for 24 hours in 0.2 per cent thiram in water at 30 °C, then drying the seed in a cool air stream. The new systemic fungicides promise to control such infections.

3. Virus control

As there is no direct method of combating plant virus infections in the field, their spread must be restricted by controlling the vector which is frequently an aphid, but in other cases viruses may be spread by workers with their hands and tools as in the case of tobacco mosaic. 'Ratoon stunting' of sugar cane is spread on the knives or cutlasses used for harvesting. Hands, tools and farm implements must be disinfected whenever a risk of spreading any disease occurs.

'Clean' planting material can sometimes be prepared by heat treatment to kill the virus. Chlorotic streak virus carried in the sugar cane setts can be destroyed by a hot water treatment, 52 °C for 2 or 3 hours. Hot air treatment, 54 °C for 8 hours, is sometimes used. In order to prevent the spread of other cane diseases, such as gummosis and leaf scald (*Xanthomonas* spp.) during treatment, a fungicide should be added to the hot water, and its concentration maintained. Even where the planting material is to be heat treated, it should still be chosen from the healthiest areas available. Tissue culture of a small piece of virus-free meristematic tissue can now be used to produce in a laboratory virus-free plants for special purposes.

4. Crop treatment

Once the pathogen has invaded the plant it can be attacked with an eradicant chemical. Sulphur is used as an eradicant fungicide, being dusted in Sri Lanka and South India on to rubber trees to control *Oidium*, also on a number of other crops which are troubled with powdery mildew. Sulphur is also used to control red spiders but is being superseded by more effective organic acaricides. Organo-mercurials are used to eradicate paddy blast, but mercurial residues in the crop are causing concern and antibiotics are used in Japan. Certain of the newer fungicides act as eradicants and anti-sporulants, and so prevent the spread of the disease.

Banana leaf spot is now controlled in most countries by spraying about 4.5 litres of a highly refined mineral oil of low phytotoxicity, either from the air or by a motorized knapsack which mists the young banana leaves where the disease is developing. The oil does not kill the fungus but prevents it developing (fungistatic). Where the outbreak is severe and conditions very suitable for the spread of leaf spot zineb, maneb or a systemic fungicide such as benomyl or thiabendazole, may be added. The systemic fungicide in which the harvested bunches are dipped to protect them against storage rots should not be sprayed to control leaf spot otherwise resistant strains of the pathogen will develop more rapidly. To control speckle in Queensland it is usual to add a copper fungicide to the oil.

Use of insecticides

Insecticides should seldom be used as protectants but be applied as curative treatments as the pest starts to build up. Spraying should be carried out early, as small caterpillars are easier to kill and need less insecticide. Protectant use of an insecticide will often destroy predators and parasites and create a greater problem. When used as protectants, as in malaria control programmes, resistant strains develop rapidly.

There are four ways in which insecticides act:

1. They may enter through the insect cuticle - contact insecticides.
2. They may enter through the insect's breathing system - fumigants.
3. They may be eaten and act as a stomach poison.
4. They may move systemically through the plant sap and kill sucking insects.

Some insecticides have more than one mode of action.
Insecticides may also be classified by their chemical nature.

1. Botanical or natural insecticides

These are extracted from plants including pyrethrum, ryania, nicotine, derris, which are seldom used on tropical crops.

2. Inorganic compounds

Calcium and lead arsenate, which are rarely used now, and sulphur.

3. Chlorinated hydrocarbon insecticides

These include DDT, aldrin, dieldrin, endosulfan, lindane, and certain acaricides, chlorobenzilate, dicofol (Kelthane®) and tetradifon (Tedion®). Where still permitted aldrin being somewhat volatile is particularly useful for the control of soil insects.

4. Organophosphorus insecticides

At least 60 of these have been offered to growers, including malathion, parathion and dimethoate. All these insecticides have a P = S or P = O group in their structure. In the majority of cases they have two ethyl or two methyl groups joined to the phosphorus.

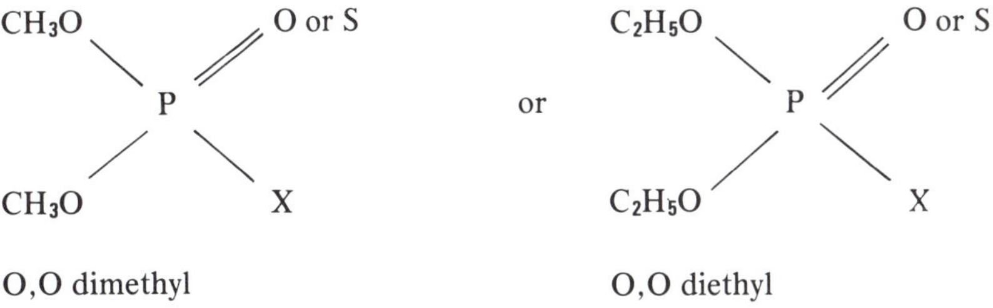

O,O dimethyl O,O diethyl

The X group varies considerably.

It is difficult to predict the activity or properties from the formula. Where the compounds are identical in all other respects the P = S compounds tend to be more selective than the P = O compounds and methyl compounds less toxic to humans. In general organophosphorus compounds are less persistent than the chlorinated hydrocarbons, and most of them have little action on caterpillars, azinphos-methyl (Guthion®) is an exception. Fenthion (Baytex®) which is a good flykiller, is very poisonous to birds but not to humans and has been used for the control of Quelea birds which are a serious pest in Africa. Most of these compounds act as contact killers. Dichlorvos (Vapona®) kills by fumigant action. Some are systemic.

5. Carbamate insecticides

These are a more recent group of insecticides. The ones introduced to date are based on the formula

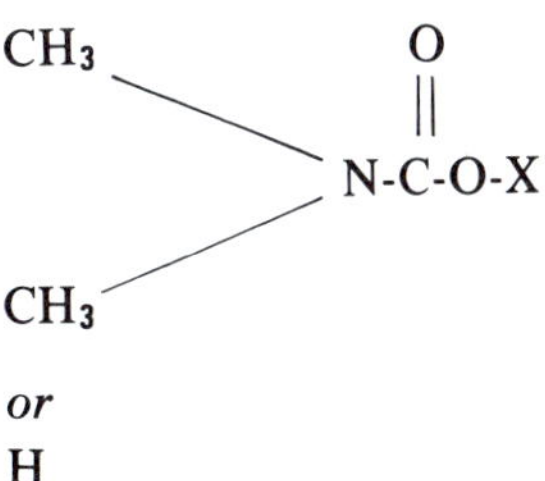

or

H

Where X is an organic group, generally a ring structure, Aldicarb is one exception. Carbaryl (Sevin®) the best known of this group is mainly used against caterpillars. Like DDT, carbaryl often leads to a build up of spider mites by destroying their predators. By formulating carbaryl with molasses, *Heliothis* is controlled much better.

6. Synthetic pyrethroids

Natural pyrethrum, prepared from the daisy-like flowers of *Chrysanthemum cinerariaefolium* has been used as an insecticide for over a century.

Elliot and his colleagues at Rothamsted studying esters chemically similar to the active ingredient in pyrethrum flowers discovered in 1967 a number of compounds based on resmethrin ester, but these were not stable in light. In 1973 they synthesized some halogenated esters which are light stable and can therefore be used to spray field crops and treat animals. The activity of these synthetic pyrethroid compounds is far greater than traditional insecticides; they are 10 to 100 times as active as DDT. They are specific to insects and degrade in the environment. Three compounds are currently available from the Rothamsted work, all esters of chrysanthemic acid:

NRDC 143 - permethrin - Ambush®, Permesect®
NRDC 149 - cypermethrin - Cymbush®
NRDC 161 - deltamethrin - Decis®

Chemical companies followed the lead of Rothamsted, and Sumito in Japan introduced fenvalerate. Other related compounds will undoubtedly follow, some with activity against different pests such as spider mites. The four compounds already developed act by contact and ingestion, they have no fumigant or systemic activity. Protection lasts from 7 to 21 days. They control a wide range of pests, particularly Lepidoptera and Homoptera, Diptera and some species of Coleoptera. They will not control mites, and certain pests that enter the plant rapidly; nor will they control pests that need a fumigant action such as cocoa capsid. They also control domestic insects, wood borers, pests of stored products and some ectoparasites other than ticks on animals. They have in general a low toxicity to mammals though deltamethrin is markedly more toxic than permethrin. With their wide range of insecticidal activity they destroy many insect predators and this could cause the upsurge of minor pests.

As so little active chemical is needed they are ideal for CDA drift spraying (see p. 337) on to crops such as cotton using 2 litres of spray per hectare, provided that a carrier of low volatility, such as mineral oil, is used to prevent the spray being lost by evaporation. These synthetic pyrethroids are liquids easily formulated for this use.

7. Systemic insecticides

The majority of insecticides are used as curative treatments, and are applied when infestation rises over a certain level, though certain of the insecticides such as aldrin and dieldrin have sufficient persistence to protect the crop from reinvasion.

In the early 1940s, Schrader and Kükenthal discovered that the organophosphorus insecticide octamethylpyrophosphoramide (OMPA), given the common name of schradan, was translocated through the plant tissues in sufficient amounts to be toxic to sucking insects. Since this discovery, the search has continued for new compounds having systemic activity, with a lower mammalian toxicity and activity against a wider range of pest species. There are now other systemic organophosphorus insecticides commercially available for use on crops including dimethoate, menazon, demeton-methyl and thiometon which are much less toxic to mammals than schradan. Dimefox which was found to be extremely efficient in reducing the population of cacao mealybugs, vectors of 'swollen shoot' on cacao (Hanna *et al.* 1955) is unfortunately highly toxic to humans.

Certain other organophosphorus insecticides, such as parathion, diazinon, malathion and azinphos-methyl, are soluble in the plant waxes and lipids and become absorbed into the leaves and fruits of certain plants but are not translocated.

Some of the safer systemic insecticides are suitable for use in peasant agriculture without hazard to the user, his animals, or fish unless grossly misused. For example dimethoate has been used very successfully against the cotton aphid, white fly and jassid; against citrus pests including red and other scales and aphids; on tobacco against white fly and aphis, the vectors of leaf curl and other virus complexes; and to control many fruit flies.

Systemic insecticides have a number of advantages:

(a) While they have some contact action they are not as persistently lethal to the natural predators as many contact insecticides.
(b) They protect new growth by translocation.
(c) They give overall protection despite inefficient application. Root absorption from soil application is often effective, and some can enter through the bark.
(d) They control insects protected from contact insecticides by rolled leaves, waxy repellant surface as possessed by certain mealybugs, or tents as constructed by the attendant ants over the cocoa mealybugs. This can often be improved by adding oil to the spray.

A further potential use of systemic insecticides is soil or seed treatment to give protection to the seedling. Phorate has been used successfully against cotton aphis, red spider, thrips, white fly, froghopper, leaf miners and lepidopterous larvae. Unfortunately phorate must be

Plate 4.2 Close-up showing entrance hole of maize stalk borer larva. Zimbabwe. (Courtesy Fisons Pest Control (Central Africa) Ltd.)

handled with great care. Disulfoton for soil application is also very toxic. No doubt a less toxic compound will be developed for seed treatment and soil application. There is still an outstanding need for a systemic insecticide to control lepidopterous larvae in the tropics, to prevent among other losses, the damage caused by the stalk borers of rice, maize (Plates 4.2 and 4.3), sorghum and sugar cane. Tetrachlorvinphos (Gardona®) acts against these pests but is not systemic.

Insecticides are the best method available at present for controlling insect pests and can increase the yields of many crops if used properly. The range of insecticides described above and in Table 4.2 will control practically all the economic pests in the tropics. The pests not controlled survive largely because the insect is protected from the insecticide. Many of the insecticides such as malathion, lindane, DDT, dimethoate and carbaryl and the new synthetic pyrethroids are very safe for the peasant farmers in the tropics to use, and with great care the more poisonous ones including parathion and endrin can be used safely. Local recommendations should be followed before deciding which insecticide to use, and if a choice is given the one with the lowest toxicity

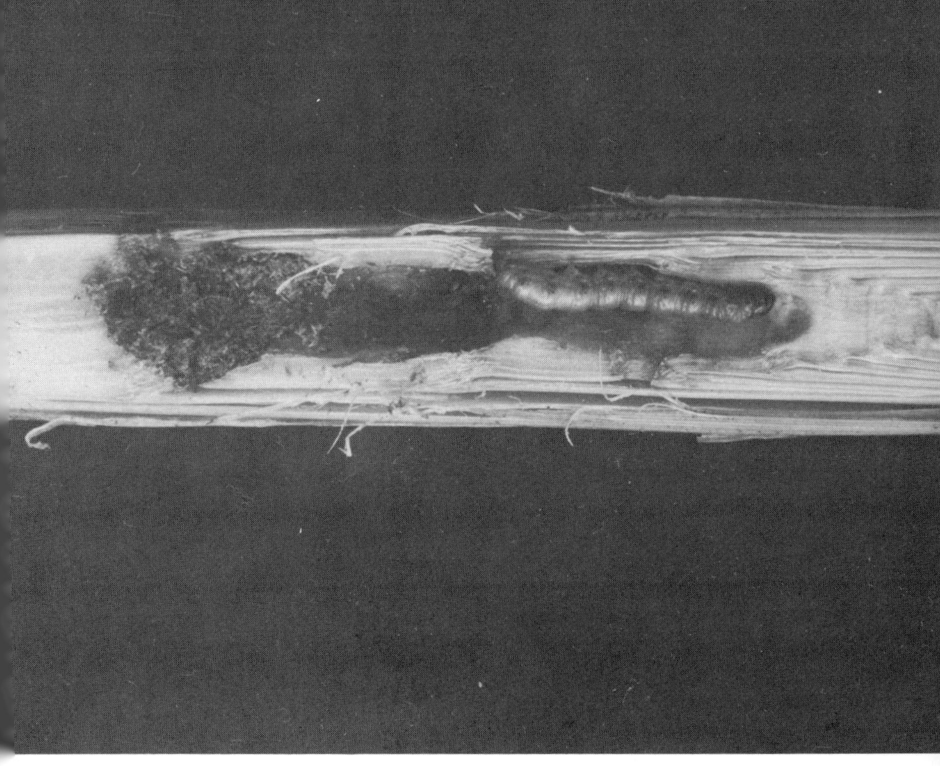

Plate 4.3 Close-up of maize stalk borer of larva in stem. Zimbabwe (Courtesy Fisons Pest Control (Central Africa) Ltd.)

chosen. The label and instructions should be read and followed in detail, and the empty containers destroyed.

In the Ivory Coast cotton is grown in standard sized cotton plots, for which growers can purchase standard bags of fertilizer and small packages of insecticide, which are just sufficient for a plot. Over 90 per cent of the growers use both fertilizers and insecticide but no irrigation, and yields are in the region of 2,800 kg of seed cotton per hectare. Without fertilizer and insecticide, yields would be around 670 kg per hectare. Good research facilities, grower education, extension service and marketing facilities encourage efficient cotton growing. Insecticides are being used with profit in many other parts of Africa.

In Thailand by sowing cotton at the correct time, timely thinning and weeding, and controlling the jassid with dimethoate and bollworms with carbaryl, growers can expect about 1,800 kg/ha of seed cotton worth in 1965 about £45, when the insecticides cost a little over £6 (Anthony & Jones 1966). Growing cotton in the traditional system growers harvest only 300 - 400 kg/ha of seed cotton. With new varieties, seed dressing for bacterial blight control, and the use of fertilizers, yields might be raised to 3,400 kg/ha.

Table 4.2 Summary of the main insecticides and acaricides used in the tropics

Aldrin. Acts by contact and ingestion. Particularly useful for soil insects. Persistent. Non-phytotoxic. Non-systemic. Non-tainting. Highly volatile: fumigant effect but not very persistent as a foliage spray. Controls bollweevils, cotton thrips, lygus, wireworms.

Amitraz (Mitac)®. Controls a wide range of mites at all stages and insects including *Bemisia tabaci*, and the eggs of many Lepidoptera including *Heliothis, Pectinophora* and *Spodoptera*. Also controls ticks and mites on cattle and sheep.

Azinphos-methyl (Gusathion®). Persistent contact insecticide with a wide range. Non-systemic. Cotton bollweevil and bollworm, aphids, cotton leafworms, lygus, fleahoppers, thrips.

BHC or **HCH** (benzene hexachloride). Mixture of isomers of which the gamma isomer is by far the most insecticidally active (Lindane). Usually contains 12–13% of this isomer. Stomach poison, persistent contact action, fumigant action. Active against aphids, bollworms, leaf-sucking plant bugs, e.g. cocoa capsid, grasshoppers, thrips, weevils. Used as a seed dressing. Protects against some soil pests. Can cause taint, particularly of root crops, largely due to associated isomers other than λ (gamma). More volatile than DDT, thus less persistent in the tropics but volatility sometimes valuable. Toxicity varies with isomers.

Bromophos. Contact and stomach organophosphorus insecticide of very low mammalian toxicity and a wide range of activity. Non-systemic. Used for control of flies, mosquitoes and cockroaches, in cattle sprays, and in food stores.

Camphechlor (Toxaphene®). A non-systemic contact and stomach insecticide. Controls the majority of cotton pests except spider mites and aphid epidemics; 2–4 weeks' residual effect for grasshopper control used as a cattle spray.

Carbaryl (Sevin®). A carbamate insecticide with a contact and slight systemic action. Good residual effect. Useful against caterpillars, froghoppers, bollweevil, bollworms, including pink bollworm, etc. Increasing importance as insects become resistant to chlorinated hydrocarbon, and organophosphorus insecticides.

Chlorpyrifos (Dursban®). Non-systemic organophosphorus insecticide acting by contact, ingestion and vapour action. Two to four months persistence as a soil insecticide. Effective against *Prodenia* on cotton, also mosquitoes, flies and ectoparasites of cattle and sheep. Protects stored products.

DDT. The first of the organic insecticides. Insoluble in water. A potent and persistent insecticide acting by contact and ingestion. Shows tendency to store in fat. *Para-para* isomer most active, usually 80% of this isomer in technical DDT. Wide range of usage. May increase aphids and spider mites.

Demeton methyl (Metasystox®). A contact and systemic organophosphorus insecticide. Used particularly against aphids.

Diazinon. Highly active insecticide. Little hazard to user. Used for control of coffee leaf miner, lepidopterous pests and DDT-resistant flies, bed bugs and locusts. Used for cattle sprays and sheep dips.

Dichlorvos (Vapona®). Quick acting volatile insecticide with knockdown action. Used against flying and crawling insects in houses, warehouses and greenhouses. Available in plastic strip for domestic and dairy, fly and mosquito control.

Table 4.2 (continued)

Dicofol (Kelthane®). A non-systemic acaricide particularly useful against strains resistant to organophosphorus insecticides. Needs good distribution.

Dicrotophos (Bidrin®). Contact and systemic insecticide and acaricide with a wide range of activity. Controls aphids, mites and thrips on cotton, coffee, vegetables and citrus. Also stem borers, leaf hoppers of rice and citrus scales. Two to three weeks' persistence.

Dieldrin. Persistent and highly active insecticide, acting as both a contact and stomach poison. Non-systemic. Widely used in public health and against boring beetles of trees and timber and some soil pests including Fidler beetle. Less volatile than aldrin therefore more persistent. Used as a foliage and surface spray. Controls bollweevils, but not bollworms, and cutworms. May increase aphids and spider mites.

Dimethoate (Rogor®, Roxion®). A safe contact and systemic insecticide active against aphids, jassids, whitefly, spider mites, mealybugs and other sucking insects on fruit crops, cotton, coffee, cereals, sugar cane, tobacco, pyrethrum and vegetables, also housefly control.

Disulphoton (Disyston®). A systemic insecticide and acaricide for protecting germinating seed against sucking insects and spider mites. Very toxic. Applied as a granular material at sowing.

Endosulfan (Thiodan®). A non-systemic contact and stomach insecticide. Controls many insects. Widely used on cotton in Africa as a replacement for DDT, applied by CDA. Highly toxic to fish.

Fenitrothion (Sumithion®). An organophosphorus contact insecticide with some acaricidal action, but poor against eggs. Effective against a wide range of rice pests, particularly stem borers. Controls cocoa capsid, and *Prodenia*. Possibly useful as a cattle spray. Low mammalian toxicity.

Gamma-HCH or **Gamma-BHC** or **Lindane**. Persistent insecticide by contact, stomach poison and fumigant action. Controls cocoa capsid, and rice stem borers (see BHC.). Less tainting than BHC.

Heptachlor. Related to Chlordane. Strong contact and some fumigant action. Useful soil and general insecticide - controls cutworms, white grubs and grasshoppers. Only moderately toxic.

Malathion. A contact insecticide with a low hazard to users, a short persistence, with a wide range of activity including aphids, scales, spider mites, mealybugs and flies. Can be used as concentrate spray.

Menazon. A highly selective systemic insecticide with low mammalian toxicity, used to control aphids. Useful as a seed dressing and soil drench.

Monocrotophos (Azodrin®, Nuvacron®). A quickly acting organophosphorus insecticide with both contact and systemic action against a wide range of pests including mites, beetles, bollworms.

Parathion or **ethyl parathion**. Extremely active by stomach and contact action. Non-systemic. Highly toxic to man. Wide spectrum of insects and spider mites controlled, including scale insects, caterpillars, aphids, leaf miners, weevils.

Phorate (Thimet®). A systemic insecticide with contact and fumigant action for the protection of germinating seedlings. Controls aphids, mites, leaf hoppers, thrips, leaf miners and certain nematodes on a wide range of crops. Very toxic. Applied as a seed dressing and granules.

Table 4.2 (continued)

Phosphamidon (Dimecron®). Systemic insecticide and acaricide with little contact action for control of aphids, jassids, mites, suckers, fruit flies and young caterpillars including rice and sugar cane stem borers.

Synthetic pyrethroids (see p. 316). Light stable compounds including permethrin, cypermethrin, fenvalerate and deltamethrin. Non-systemic stomach and contact insecticides, with no vapour action. Effective at a very low dosage rate against a wide range of pests resistant to other insecticides. Low mammalian toxicity. Useful in public health and against some cattle pests. Permethrin used to protect stored products. Promising for control of tsetse fly and mosquitoes. The differences in the range of biological activity between the compounds are not marked, but the effective dosage rate and toxicity varies between compounds.

Tetrachlorvinphos (Gardona®). An organophosphorus insecticide of low toxicity and selective action. Controls adults and larvae of lepidopterous and dipterous pests, and some Coleoptera. Due to rapid breakdown gives poor control of soil pests, sucking insects and mites. A fly killer. Used on stored products, also against pasture and forest pests.

Tetradifon (Tedion®). Kills the eggs and young stages of mites. Low mammalian toxicity. Long residual properties.

Triazophos (Hostathion®). Broad spectrum organophosphorus insecticide and acaricide with some nematicidal activity. Non-systemic. Active against aphids and some caterpillars, including *Agrotis*.

Note: The common names are given in bold.® = Registered trade name.

Insecticides or other crop protection chemicals cannot compensate for poor farming and must be used with the best practices and varieties to get the best yields.

Another example of many small farmers using insecticide profitably is the control of cocoa capsid throughout West Africa by mistblowing lindane. An important development in the Philippines is the application of granular insecticides to the paddy water rather than spraying the rice foliage. The insecticides appear to move not only in the water between the stem and leaf of the rice but also systemically within the plant. Two applications of granular lindane are more effective than 8 to 12 foliar sprays of endrin or parathion in controlling stem borers and maggots. Diazinon, highly effective against stem borers, leaf hoppers, plant hoppers and whorl maggot, degrades rapidly in paddy soils. Soaking the roots of rice seedlings in 0.04 to 0.1 per cent of the systemic insecticide carbofuron before transplanting protects the transplanted seedlings for 2 or 3 weeks for little cost (Athwal 1972).

A study of the habits of an insect pest can lead to improved control. For example the pink bollworm (*Pectinophora gossypiella*) hides under leaves and debris during the day and is only active at night. Therefore better control is obtained if the cotton is sprayed in the late evening or at night. Similarly, brown plant hoppers lie at the base of rice plants and if an insecticide is sprayed it should be applied to the base of the plant; a granular insecticide gives better control than a spray. The main insecticides used in the tropics are given in Table 4.2.

DEFEND THE HOST

Choice of planting date

If a crop can be grown the whole year round either because the rainfall is suitable or irrigation is practised, a continuous source of pests and diseases is provided. Many cotton growing countries legislate for a close season, by stipulating the dates when planting may commence, and when the crop must be uprooted and burnt. This may have to be extended to tobacco growing countries where irrigation is tending to eliminate the close season. During the close season the insect population will fall and the new crop will be able to get away before it builds up again to a serious level. Diseases without an alternate host rely upon volunteer plants and crop residues for their survival unless they are maintained in the soil. By spreading the planting dates, as occurs with peasant cotton, pests are presented with a succession of crops to breed and feed upon.

A crop grown at the time of the year most suitable to its continued active growth will normally be more resistant to pests and diseases than a crop struggling against adverse conditions.

Tobacco in Zimbabwe suffers from two virus complexes causing rosette and bushy top. Both are believed to be transmitted only by the aphid *Myzus persicae*. The behaviour of the aphids and the effect of virus infection on tobacco yield is shown in Figs. 4.1 and 4.2. The peak period for aphids alighting on the crop is the time when the crop becomes infected most readily. This explains the low virus infection found in early planted tobacco, though too early planting under irrigation is being discouraged as it allows infected aphids to build up and cause more severe infection in the main rain plantings. Late plantings are also discouraged as they build up the number of aphids which will pass on to the winter hosts (Legge 1960). Early planting also gives a higher yield, a better grade of leaf, and keeps *Cercospora* to a minimum (Stephen 1957).

Eleusine coracana escapes blast (*Piricularia* sp.) in Coimbatore if sown between October and April, but is heavily infected if planted between April and October (Butler & Jones 1949, p. 250).

The planting date must not be chosen to avoid a major pest, if by

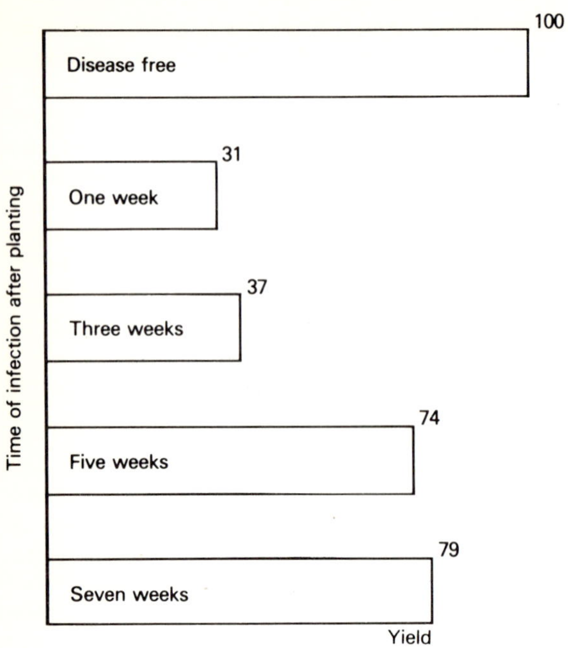

Fig. 4.1 Effect of rosette infection on yield. [Source Legge 1960]

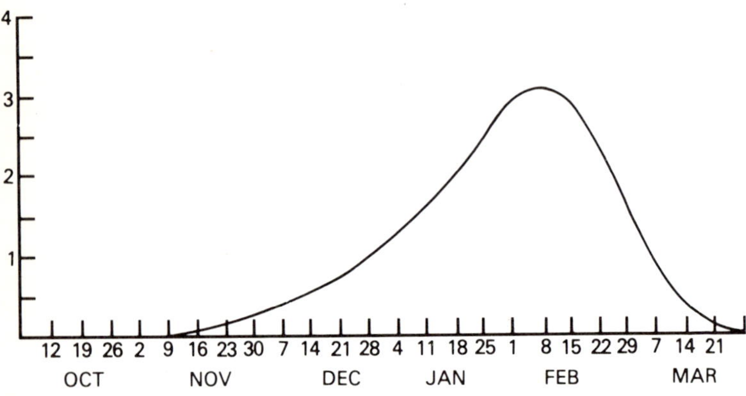

Fig. 4.2 Number of aphids alighting per plant. [Source Legge 1960]

delaying or advancing planting other factors cause a greater loss of crop. For example, in Thailand boll rotting can cause a serious loss of cotton, which can be reduced by delaying planting so that the bolls do not develop in the wettest part of the year. But early planting gives maximum yield, and it is better in many cases to use a wider spacing and thin early to one plant per hole rather than delay planting. Similarly in the Eastern Region of Tanganyika (Tanzania), up to 1954, March was the recommended sowing time for cotton. This was to have an interval between the maize crop flowering and the cotton bolls forming so that the American bollworm would be starved out. Unfortunately the rains normally finish in May, so March planted cotton yields only about half as well as cotton planted in February. Growers were also planting more late maize, which bridged the gap for the pest. Much better yields result from February planting and controlling the American bollworm with insecticides. In general, planting dates will be determined by factors other than the avoidance of pests and diseases.

Development of resistant varieties*

The selection or breeding of crop varieties resistant to the important pests and diseases is always part of a crop breeding programme, as when this is successful it is an ideal method of control profitable to the farmer. But resistance is only one factor and the breeder must combine adequate resistance with the other characters needed in an acceptable variety.

Except where the crop is totally immune, resistance is measured in relation to the infection suffered by the current varieties grown. Resistance may be due to the inability of the disease or pest to attack the plant, as frequently occurs when the disease spores from one host crop arrive on a completely different plant species, as most diseases are host specific, e.g. coffee rust will not attack maize. The physical nature of the host may prevent an attack, in the way the hairy leaves of cotton keep the jassid at a safe distance, and resistance to rice stem borers appears to be associated with a high silica content of the epidermis which hinders the caterpillar when boring and feeding. Alternatively resistance may follow from the reaction of the plant to the attack. The response may be hypersensitive, the areas around the infection being killed and so preventing further invasion, or the plant may allow the fungus to develop but prevent it sporulating, or the virus or insect may invade but not multiply.

A variety that is immune is good, provided that other essential characters are not sacrificed. In breeding for resistance to one disease, resistance to what had previously been a minor disease may be lost allowing it to develop into a major disease. Frequently breeders accept a compromise, a fair degree of resistance which ensures a worthwhile crop, which can be supplemented by chemical control on a reduced

* See Simmonds (1979), Chapter 7.

scale when conditions favour the pest, rather than immunity which may be vulnerable in an otherwise poor variety.

Van der Plank (1963, 1968) described two forms of crop resistance; vertical resistance, sometimes called race or pathotype specific resistence, which is often but not always associated with a major gene which is usually dominant and pathotype specific, conferring hypersensitive resistance, and horizontal resistance sometimes referred to as field resistance or tolerance, which is race or pathotype non-specific and dependent upon a number of genes each making a small contribution towards inhibiting the pest rather than producing hypersensitive resistance.

There are exceptions, a major gene may confer horizontal resistance, and vertical resistance may be dependent upon a number of genes. Selecting for horizontal resistance in the presence of vertical resistance is impossible.

Breeders have frequently discovered a source of major gene resistance, incorporated it in a variety, only to be defeated by a new pathotype often referred to as a physiological race, which may be a single phytotype or more often a population of phytotypes. An early example of this in the tropics was the arabica selection made by L. P. Kent, a Mysore coffee planter which Mayne (1932) showed to be resistant to strain 1 but highly susceptible to strain 2 of *Hemileia vastatrix*. When planted in Kenya, Kents showed little resistance to the local races of rust, but until the early 1950s it was the only cultivar distributed for planting in Bugishu, Uganda. With more than 25 physiological races of *Hemileia* known, horizontal or field resistance is essential and appears possible. A number of interspecific *arabica* hybrids show a high degree of resistance.

The evolution and distribution of new pathotypes is particularly serious with airborne diseases. In the case of pests and diseases of the soil, vertical resistance is much longer lasting as the pests are much less mobile and new pathotypes that may arise have slow and restricted distribution, and can be localized by not moving soil on plants, tools, boots, irrigation water, etc., from field to field.

It follows that when breeding for resistance, vertical resistance should be confined to soil pests and bacterial diseases that are only spread by violent storms (hurricanes) and splash. For the control of mobile airborne pests, reliance should be placed on horizontal resistance. Horizontal resistance is especially important in tree crops where a breakdown in resistance means a slow and costly replanting programme. Should a rust tolerant cultivar be developed, it would be an impossible task to replant all the coffee plantations of Brazil. The risk of planting too large an area with a single ‘resistant’ clone is well illustrated by the effect on the sugar cane industry of Mauritius when a more virulent form of gummosis (*Xanthomonas vasculorum*) and leaf scald (*X. albilineans*) appeared in 1964. The varieties planted on 60 per cent of the cane area were susceptible to this new form, and the dominant cane M.47/44 highly susceptible. Apart from the cost of replanting earlier than normal, the replacement resistant varieties were generally lower in yield, and the breeding programme required complete reorganization.

As we saw in Chapter 3 many of the economic crops of the tropics have developed from a very small initial importation, one tree imported from the Botanic Garden in Amsterdam is said to be the foundation of the South American *arabica* coffee industry. There is a great need to widen the genetic diversity of many of these crops, in the way cacao selections were moved from Trinidad to Ghana in the early 1940s to combat swollen shoot.

Many practical breeding successes have been achieved with tropical crops, including those listed in the table.

Crop	Disease/Pest
American cotton	Jassid
Arabica coffee var. 'Kents' (S. India)	Race No. 1. *Hemileia vastatrix* (Rust)
Banana variety 'Lacatan'	Panama disease
Cocoa (SCA 6 & SCA 12, crosses)	'Witches' broom' *Crinipellis perniciosa* (syn. *Marasmius perniciosus*)
Cocoa (some TSH clones)	'Witches' broom' *Phytophthora* and *Ceratostomella* (syn. *Ceratocystis*)
Cassava	Mosaic and brown streak
Maize	Streak
Malayan dwarf coconuts (Jamaica)	Lethal yellowing
Rice (IRRI varieties)	Bacterial blight, Tungro, hoppers
Sudan cotton	Leaf curl and blackarm
Sugar cane (Java)	Mosaic, streak and 'Sereh' viruses
Sugar cane	'Gumming' (*Xanthomonas vasculorum*)
Sugar cane	Leaf scald (*Xanthomonas albilineans*)
Sweet corn	*Xanthomonas stewartii*
Tobacco (Australia)	Blue mould (*Perenospora tabacina*)
Tobacco (South Africa)	Mosaic
Upland sea island cotton (USA)	Wilt. *Fusarium vasinfectum*

A source of resistance may occur in another cultivar or a related species. It may be found near at hand, as resistance to blackarm occurs in West Africa through years of natural selection of *Gossypium hir-*

sutum var. *punctatum*. It may occur in a related species which is not an economic crop, as resistance to blue mould was found in a wild *Nicotiana* in Queensland or as a sport which occurred in Hicks tobacco again in Northern Queensland. Or it may be necessary to search the original source of the plant in the manner that Pound found the Scavinia (Amazon) selection of cocoa varieties in the Amazon valley (Pound 1938).

The inheritance of resistance may be quite complicated. Knight (1957) found 10 genes of major importance in blackarm resistance, present in four species of *Gossypium* and transferred 9 of these to Sudan Sakel. Resistance, however, breaks down under heavy rainfall conditions and it has not been possible to introduce adequate resistance to blackarm in the American upland cotton grown in Uganda.

Rice breeders' attempts to develop varieties resistant to the main disease, blast (*Piricularia oryzae*), have been made more difficult by the existence of more than 200 different races of this disease. Resistance to one or more races probably occurs in all rice varieties and conversely, all or nearly all varieties are susceptible to one or more races and new races are arising all the time. Four other leaf diseases are serious in South-East Asia (Fig. 4.3) leafsmut and stackburn disease are rare. The current resistance position of the new IRRI varieties is given in Table 3.6.

In breeding for resistance, the aim may be achieved by varieties which have developed beyond the susceptible stage before the pest or disease strikes, or by introducing sufficient horizontal resistance that chemical control is easier and cheaper. For example, the collection of rice varieties at the International Rice Research Institute in Manila showed considerable variation in their resistance to stem borers. Some varieties are resistant to 'dead heart' and others to 'white heads' (the two effects produced by borers), and some to both types of infestation, but no variety escaped attack. This difference in susceptibility at different growth stages can be used in controlling borers, as a variety resistant to dead heart formation need not be protected earlier than the stage where the head is developing but has not shot (boot stage), and the 'white heads' can be prevented by a single application of gamma-HCH (see Table 4.2) rather than two (Anon. 1966c).

Resistant rootstock

The top grafting of desirable varieties of tree crops on to resistant rootstock is a well established practice. In South America it has been used to have a leaf canopy resistant to South American leaf disease, *Microcyclus* syn. *(Dothidella) ulei* on a high yielding trunk. Consideration has been given in Malaysia to top grafting a wind resistant canopy.

Citrus trees are generally grafted on a rootstock resistant to root rot (*Phytophthora* spp.). In Australia *Poncirus trifoliata* rootstock is resis-

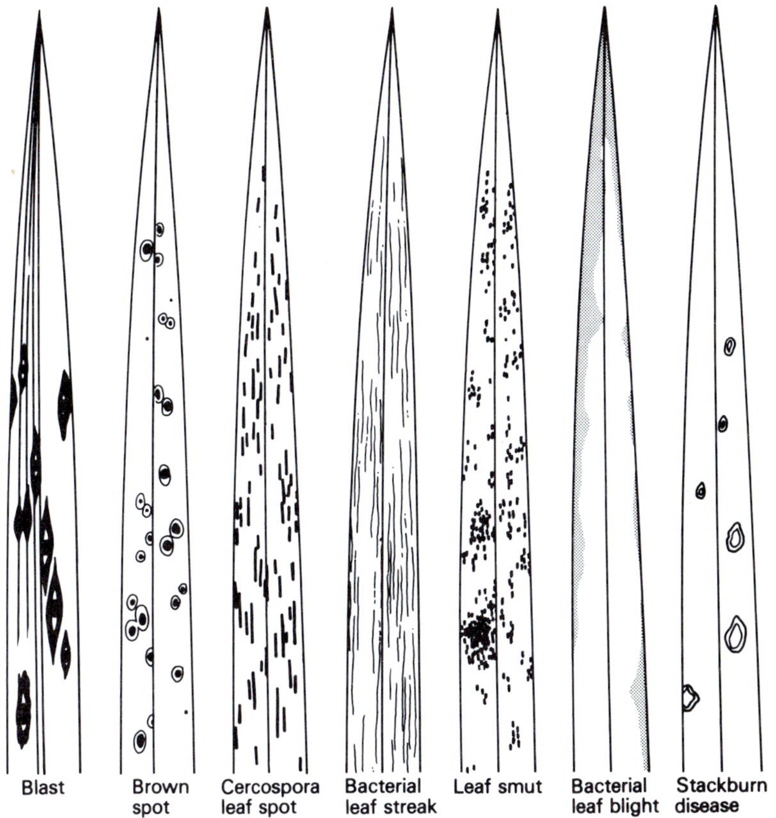

Fig. 4.3 Typical leaf lesions of seven rice diseases. All diseases are found throughout South-East Asia, although leaf smut and stackburn are rare. The diseases are caused by the following organisms:
Blast, or rotten neck: *Piricularia oryzae* – a fungus Brown spot, brown leaf spot, sesame leaf spot, *Helminthosporium* leaf blight or spot: *Cochliobolus miyabeanus* (*Helminthosporium oryzae*) – a fungus
Cercospora leaf spot or blight, narrow brown leaf spot, narrow brown spot: *Sphaerulina orizina* (*Cercospora oryzae*) – a fungus Bacterial leaf streak: *Xanthomonas translucens* f sp. *oryzae* – a bacterium
Leaf smut: *Entyloma oryzae* – a fungus
Bacterial leaf blight: *Xanthomonas oryzae* – a bacterium
Stackburn disease: *Trichoconis padwickii* – a fungus
(Courtesy The International Rice Research Institute)

tant to root rot and nematodes but susceptible to exocortis virus and often shows trace element deficiencies. Rough lemon, which produces vigorous trees, is very susceptible to root rot. Other rootstocks used include Rangpur lime, sour orange, Cleopatra mandarin, and the choice depends upon local experience, soil type including pH and salinity, tree vigour and fruit quality required, local virus diseases, as well as the need for resistance to foot rot and nematodes.

Legislation controlling the planting of resistant varieties

When a resistant variety is available, certain progressive farmers will be eager to try it. When the advantages are sufficiently great, government or a body controlling an irrigation area, or purchasing the crop, may insist on the resistant variety being planted. This may be done by direct legislation, as in Zimbabwe and many other tobacco growing countries where the permitted tobacco varieties are listed, or indirectly by restricting replanting subsidies to certain varieties. In a crop such as cotton where the seed must at one stage pass through a ginnery the variety to be planted can be regulated and the seed dressed with a protective chemical before issue.

Modify the crop environment

If all the factors relating to the balance between the crop and pest were known, it would be possible in many cases to avoid a large proportion of the crop loss by modifying the plant environment. Many of the bacterial diseases and root diseases are weak parasites and cause greater damage when the soil is cold and wet and air temperatures low, than under ideal growing conditions where the soil fertility and pH are suitable. *Armillaria mellea* which causes serious root rot in rubber in West Africa is much less serious when the land is well drained. In the Canary Islands, bananas may develop Panama disease if irrigation is excessive and the bases in the soil fall too low. With improved drainage and fertilization the diseased plants recover (Plate 4.4)

When a shade tree in a West African cocoa plot falls, damaging some of the trees and breaking the canopy, a severe attack of capsid often follows. Thus cocoa growers have realized the importance of maintaining good shade and canopy to restrict capsid damage and control weeds. Trials in Ghana (Cunningham & Lamb 1958) suggested that a better cash return is obtained by removing the shade, using a liberal dressing of fertilizer, and controlling the capsid attack by spraying with lindane. By this means, cocoa yields on small plots have been raised from 1,073 kg dry cocoa per hectare to 3,460 kg in the second year (Cunningham 1960).

Plate 4.4 Cut banana pseudo-stem showing dark patches due to Panama disease.

Birds-eye spot or *Cercosporella* disease of tea (*Calonectria theae*) was a problem in certain localized areas of Sri Lanka and India, until its association with *Acacia decurrens* shade trees was established and they were removed. Early planting of maize often avoids the serious crop loss from *Puccinia polysora* in West Africa. Close planting of groundnuts within the row reduces the incidence of rosette; early planting has a similar effect in some areas.

When blister blight appeared on tea in South India and Sri Lanka in 1946 many planters removed their shade trees with little effect on the disease. Under conditions ideal for an epidemic of blister blight, the removal of the shade did not reduce the humidity sufficiently to prevent spore germination.

Cassava bacterial blight, which is mainly spread by rain splash, was reduced when cassava was intercropped with maize or melons or mulched, which reduced the impact of raindrops (Ene 1977).

When planting a cover crop in young perennial tree crops, care has to be taken that the cover will not encourage eelworms. Conversely, in Sri Lanka *Tagetes* planted after pruning tea is thought to reduce the infection of nematodes. The rhinoceros beetle, a serious pest of newly established oil palms in Malaysia, can be effectively controlled at a low cost by maintaining a dense ground cover in the young plantations. This acts as a barrier and hampers the movement of adult beetles (Wood 1968).

The majority of crops growing under fertile conditions with adequate moisture will often survive a pest or disease outbreak and yield a fair crop. Balanced fertilizer application is more inclined to protect a crop from attack than make it more susceptible. Excess fertilizer, particularly nitrogen, produces much young growth susceptible to disease, and a lot of leaf which increases the humidity in the crop, aiding spore germination.

In Northern Nigeria the application of nitrogen to cotton without spraying insecticide increases bollworm attack and depresses yield (Hayward 1972).

Apply a crop protection chemical

(A summary of the main insecticides is given in Table 4.2, p. 320; herbicides in Table 4.10, p. 310; nematocides in Table 4.1, p. 310. See also Martin (1973) and Worthing (1979).) When an invasion by a fungus, weed or an insect is anticipated, the crop can sometimes be protected by spraying, and the invasion is prevented from getting into the crop. This applies to systemic insecticides and pre-emergent herbicides but more particularly to fungicides, where copper, the main fungicide used in the tropics, acts as a crop protectant.

Fungicides

The following are some of the crops protected in this way.

Coffee	Rust (*Hemileia vastatrix*) Coffee berry disease (*Colletotrichum coffeanum*)
Cocoa	Black pod (*Phytophthora palmivora*)
Banana	Sigatoka (*Mycosphaerella musicola*) (*Cercospora musae*)
Tea	Blister blight (*Exobasidium vexans*)
Rubber	Secondary leaf fall (*Phytophthora palmivora*)
Coconuts	Bud rot (*Phytophthora palmivora*)
Tobacco	Blue mould (*Peronospora tabacina*)

Where spraying was carried out Bordeaux mixture was originally the main fungicide, to be replaced later by other forms of copper fungicide particularly copper oxychloride and cuprous oxide which are much simpler to mix for spraying.

Fortunately, to control blister blight of tea very little copper fungicide is required and the bushes are small hence easy to spray, but spraying bananas with Bordeaux mixture is a much greater undertaking using large volumes of water. Guyot and Cuillé in the French West Indies developed a simpler means of spraying bananas using a motorised knapsack mistblower to apply copper fungicide in a much lower volume of spray. By mixing refined mineral oil with the copper fungicide they hoped for and obtained much better disease control. They then showed that without the copper fungicide, the oil alone controlled the banana leaf spot. This discovery ultimately revolutionized banana growing and probably saved the crop as a low value fruit. The oil can either be applied with mistblowers as a fine mist to protect the young leaves, or by aircraft at 7–15 litres/ha. When the disease is severe, rather than increase the amount of oil, a fungicide is added to the oil. Surprisingly, oil alone has not been shown to control any other disease economically, though it increases the effectiveness of many organic fungicides, particularly those with systemic action.

Apart from copper, sulphur and organic mercurials for treating seeds and sugar cane setts, were the main fungicides used in the tropics. The dithiocarbamates (zineb, maneb, mancozeb), since their discovery in 1934 have not become as widely used in the tropics as in temperate regions. Dinocap has been used to control powdery mildews on tobacco and valuable fruit crops where sulphur cannot be used.

The 1970s have seen the introduction of systemic fungicides (see Marsh 1977) with a low mammalian toxicity. These fungicides can enter the plant and protect it against parasitic fungi without damaging the plant. In the main they are very active, and only low dosage rates are required.

The extent of translocation of these new fungicides varies with the chemical. Some, if applied to the roots, move upwards to protect all the plant, but this is seldom economic except as seed dressings, or for nurseries and valuable fruit and vegetable crops. Most of these systemic fungicides only move outwards in the leaf and not from one leaf to another or to new leaves. Movement in woody plants is generally poor. One of the newer products is said to move both upwards and downwards in the plant. There are a number of systemic fungicides to control powdery mildews, some to control rusts and smuts, and the benzimidazole derivatives (e.g. benomyl, thiabendazole) which have been widely used against banana leaf spot, coffee berry disease and post-harvest fruit and tuber rots. They also control leaf spot of groundnuts and have replaced mercurials for treating sugar cane setts. More recently a number of new systemic fungicides with widely differing structures, including Curzate®, Ridomil® and Aliette®, have been shown to control phycomycete fungi, particularly *Phytophthora* and *Pythium* species. Ridomil has given very good control of tobacco blue mould

(*Peronospora tabacina*), Triadimefon (Bayleton®) shows promise for the control of coffee rust, and other new ones control tobacco blue mould *Peronospora* and *Ceratocystis* on cocoa. One of these could be useful in controlling cocoa black pod provided it does not affect the flavour of the chocolate. Ridomil is currently being used against blackpod in Cameroon.

Unfortunately after three or four seasons' use many fungi have developed tolerant strains, or resistance, to the early systemic fungicides. This has not happened with older fungicides, copper and sulphur, even after many years of use. As the benzimidazoles are the best chemicals for protecting harvested fruits they should not be used to protect the same crop in the field otherwise tolerant strains of post-harvest rots will develop more quickly. Resistance to benomyl in certain areas of East Africa has necessitated a change to other fungicides to control coffee berry disease.

Disease prediction

Farmers know by experience that certain climatic conditions favour the spread of the disease they are most concerned with. These observations can sometimes be explained scientifically and the information used to predict when outbreaks of a certain disease will occur. For example, tea blister blight spores require moisture to germinate, but 2.5 mm of rain in a day is quite adequate. Spores are very susceptible to sunlight, 30 to 60 minutes direct exposure will kill them, and an average of 225 minutes of sunshine per day over 5 days is enough to reduce the blister blight to an unimportant level. The Dutch workers in Indonesia (De Weille 1957) suggested a system of forecasting based on sunshine hours and a simple method was developed in Sri Lanka (Visser *et al*. 1961). Spraying is postponed by successive 5 day periods, until the average sunshine for the previous 5 days is less than 225 minutes. By the use of this system the protective spray applications normally applied at 10 day intervals in the wet (cloudy) season can be reduced by 20 to 50 per cent (Mulder and De Silva 1960) with a considerable saving in cost.

In India rice is attacked by paddy blast in the late seedling stage, 25-50 days after transplanting, or at flower emergence. When these susceptible conditions coincide with a minimum temperature of 20-24 °C (68 °C-75 °F) and a relative humidity over 90 per cent in the morning for 2 to 4 days, an outbreak of blast can occur (Padmanabhan 1967).

Disease prediction based on temperature, sunshine and humidity should be possible for most plant diseases and spraying in accordance with disease predictions is the most efficient and economical way of using a protective fungicide. In the same manner insecticide spraying based on pest scouting is the most economical way of using insecticides, provided that the area to be protected is not too large, and the pest does not build up very rapidly.

APPLICATION OF CROP PROTECTION CHEMICALS*

Applying crop protection chemicals is considerably more difficult than distributing plant nutrients, as a very small amount of active ingredient has to be applied to a large crop area, frequently to a particular part of the crop where it can be most effective.

These active materials are formulated for application in a variety of ways (Marrs & Middleton 1973).

Dusts

The finely ground active ingredient, either in the technical form as with a copper dust, or on an inert carrier in the form of a dust concentrate, is blended down with a local filler such as a talc or clay to give a dust, usually containing 2-4 per cent of active ingredient. Sulphur dust is an exception having only a minor amount of inert material such as kaolin to prevent aggregation. This can be applied to the crop by hand dusters, tractor-drawn dusters which blow the dust out through a boom, aircraft, or by drift dusting. The characteristics of the finished blended dust should be determined by the method of application, but often in the tropics there is an insufficient choice of fillers.

When choosing a filler, a number of characteristics are important.

1. Acidity or alkalinity. Certain insecticides break down very rapidly when formulated with an alkaline filler.
2. Bulk density. Usually between 480 and 800 kg/m^3.
3. Characteristics of the particle - size, shape, abrasiveness. If the particle sizes are very mixed, separation may occur.

The blended dust must retain its biological activity over a reasonable period. It must flow properly through the machine and not form balls; this can often be modified by the addition of small quantities of other materials. The blended dust must not reseparate during application. Separation of the active ingredient and inert filler has been overcome in the case of copper fungicides by preparing a coated dust but it is expensive to manufacture. Certain fillers have insecticidal properties of their own, in that they scratch the water-retaining layer of the insect, causing death under dry conditions.

Subject to meeting these characteristics, the cheapest filler available in a finely ground form is used.

* For a full treatment see Matthews (1979)

Table 4.3 A comparison of dusts and sprays for the distribution of crop protection chemicals

	Dust	**Spray**
Active ingredient	More required (as a dust).	More efficient (as a spray).
Formulation	Concentration should be blended locally or freight excessive.	Prepared in an industrialized country and exported formulated.
Filler or diluent	Local filler required.	Water generally used.
Stability	Active ingredient may react with local filler.	Generally good and known.
Mixing with other active ingredients	Difficult for farmer.	Simple if compatible.
Water	None required.	Required. May be difficult to obtain.
Dew	Improves retention.	Reduces effectiveness due to increased run-off.
Terrain	Suitable for hilly land, particularly drift dusting.	Ground machines and knapsack sprayers difficult to use on hilly land. Drift spraying possible.
Wind	Over 8 km/hour interferes with application.	Can be applied up to a velocity of 24 km/hour.
Deposit	Very visible.	Frequently difficult to see without close inspection.
Particle size	Fixed on manufacture. Usually too many fine particles.	Determined at spraying.
Supervision	Easy.	Difficult.
Equipment	Simple, cheap and fairly resistant to peasant labour.	Efficient, more complicated, suffers from mis-use.
Retention	Poor.	Good if well formulated.
Distribution	Poor.	Good.
Dilution	Fixed on blending.	Easily varied.
Adaptability	Non-variable.	Variable.
Cost	Expensive to formulate due to filler cost and more a.i. required.	Cheaper than dust, unless expensive solvents or formulants required.
Herbicides	Not very effective.	Usually applied as a spray.
Insecticides	Cotton in certain countries, peasant maize, and in sugar cane in Trinidad for froghopper control.	General. Efficient distribution of a small quantity of active ingredient.
Fungicides	Used on rubber and on tea in difficult country and where water is short.	Used on bananas, citrus and coffee. Preferred for tea.

Sprays

The majority of crop protection chemicals are applied as sprays and a comparison with dusts is given in Table 4.3. The active ingredient which is sold to the farmer either as a wettable powder, emulsifiable concentrate, solution, emulsion, soluble or flowable powder, is mixed with a quantity of water and applied to the crop with a spraying machine. Some chemicals are available in a variety of formulations, the insoluble materials only as wettable powders or flowable suspensions. Wettable powders are generally cheaper to make, package and handle; but liquid formulations are easier for the farmer to measure and mix and less liable to settle out in the sprayer. The newer herbicides and fungicides are mainly available as wettable powders, and insecticides as emulsifiable concentrates.

During manufacture wetting or spreading agents are usually added so that the sprayed crop is well covered. The farmer may add oil to increase the penetration of the spray into the leaf or insect cuticle, which besides making the product more active, reduces the effect of rain after spraying.

Sprays can be applied by aircraft (Plate 4.5) where there are large contiguous areas of a single crop, as in the irrigated cotton in the Sudan Gezira and banana plantations in South and Central America. Tractor mounted sprayers widely used in temperate regions are limited to field crops where tractors are available for other operations. Motorized knapsack sprayers and mistblowers have a place where small volumes of spray can be used effectively in valuable crops, as in the control of cocoa capsid in West Africa, banana leafspot in the Windward Isles, and tea in Sri Lanka.

Controlled droplet application (CDA) of sprays at ultra low volume (ULV)

Crop spraying developed in Western Europe and North America where water and the mechanical means of transporting it are freely available so the use of 200 or 2,000 litres of water to spray a hectare of crops is no impediment to treating large areas many times during the growing season. In the tropics, however, this is rarely the case especially on small holdings. Knapsack sprayers which are relatively cheap and easy to maintain, and need no fuel, are used on cotton in the Ivory Coast, Upper Volta and Malawi to apply insecticides; on tree crops, such as cocoa in Nigeria and tea in Sri Lanka, to protect them with fungicides; and to spray herbicides in sugar cane in Mauritius and the West Indies. In many other countries however, despite clear demonstrations that regular protective spraying increases the yield of certain crops, spraying has not been taken up. In Northern Nigeria for example, it

Plate 4.5 Helicopter spraying bananas against leaf spot disease. Central America. (Courtesy Fisohs Pest Control Ltd.)

was shown in the 1960s that four sprays of insecticide on cotton increases the yield from 300 to 400 kg/ha to nearer 2,000 kg, but this meant carrying nearly half a tonne of water per hectare, often from the swamp at the bottom of the hill some distance from the cotton. In many situations like this the development of controlled droplet application at ultra low volume using a simple low cost machine has made crop spraying a practical proposition for small farmers to carry out themselves.

Controlled droplet application (CDA) means controlling the size of the droplets in the spray to suit the purpose for which it is required. To spray weeds, and horizontal targets, a droplet of 175 to 250 μm in diameter* should be used to eliminate drift, and at the other extreme to control small flying insects such as mosquitoes, a very fine droplet with a diameter of around 35 μm is best. Between these extremes a 70 μm droplet is ideal for drifting fungicides and insecticides with the wind on to vertical targets. The small droplets used for controlling mosquitoes have not the momentum to impact readily on plant leaves or

* The droplet diameters referred to are the volume median diameters (VMD), i.e. half the volume of spray is of drops larger and half smaller.

Plate 4.6 Spraying oriental tobacco with a Micron Ulva sprayer at 2 litres/ha. Note that the wind takes the spray away from the operator.

house walls but drift away; however these can be collected on plant hairs and by flying insects.

Ultra low volume (ULV) spraying is the application of chemicals (Plate 4.6) in the minimum amount of spray liquid per sprayed unit area that will give acceptable control of the pest or disease. It is possible to spray at ULV using a wide range of droplets and conversely it is possible to spray at high volume using CDA but this is normally not necessary. By the use of a narrow range of appropriately sized pesticide droplets, many tropical pests, diseases and weeds can be efficiently controlled using spray volumes from 2 to 25 litres/ha; the higher volumes being required for weed control and tree crops. This eliminates the labour of carrying large amounts of water. Where the drift technique is used to apply 2 litres/ha, the spraying time is much reduced, 1 ha being sprayed on foot without a tractor in a little over half an hour. That this is possible is due to the development of the Micron® CDA sprayers by Edward Bals. These are light hand-held machines constructed mainly of plastic. The spray mixture, contained in a plastic bottle, is fed by gravity through a rigid plastic tube which meters the spray on to a spinning disc. This plastic disc is rotated rapidly by a small motor, powered by the torch batteries fitted into the handle of the machine. The spray

leaves the serrated edge of the disc in a fine mist of evenly sized droplets which drift downwind on to the crop. The size of the droplets produced depends mainly upon the speed the atomizer rotates, and the feed rate on to the disc. Where the disc is not over-fed the approximate formula is: droplet size in microns $= \frac{500{,}000}{\text{rpm}}$. The Micron Ulva machine which has been used for the application of insecticides and fungicides spins at approximately 7,000 rpm with new batteries and produces droplets mainly of 75 μm diameter. To judge the size, there are 200,000 such droplets in one drop produced by a dripping tap. The Micron Herbi used for applying herbicides spins at 2,000 rpm to produce droplets around 250 μm. The Mini Ulva which is really based on a rotating cup rather than a disc, can be used to produce droplets from 30 μm to 70 μm or larger by reducing the number of batteries in the handle of the Mini Ulva.

As the volume of a droplet is related to the cube of its diameter, if the diameter of the droplets is halved then eight times the number of droplets are produced from the same volume of spray.

The importance of this and the excellent crop cover possible with small droplets is shown in Table 4.4.

Table 4.4 Volume of a droplet = $4/3\pi r^3$ (1/μm = 0.0001 cm)

	Droplet size		
	30/μm (Mini-Ulva)	70/μm (Ulva)	250/μm (Herbi)
Number of droplets per cm^3 of spray	450×10^5	56×10^5	122×10^3
Droplets per cm^2 if 1 litre is applied evenly to a flat surface	450	56	1.2

When the droplets arrive on a surface they spread, so the distance between the edges of the droplets is even less and the chance of an insect or germinating fungus spore avoiding such a pattern of droplets is very slight.

Theoretically it is possible to give a very good cover to a crop using 70 μm droplets and 1 litre/ha of surface to cover; remembering that 1 ha of growing crops may have up to 4 ha of leaf area and it is often important to cover both surfaces which would mean 8 litres/ha; provided this can be evenly distributed by the machine. With a mobile insect or when using pesticides with a systemic or fumigant action, such good cover is not necessary. Small droplets start to evaporate from the moment they leave the disc and if the carrier is water they may rapidly become too small to impinge on the crop. Therefore with such low volumes it is important to use a pesticide in a non-volatile carrier or if larger spray volumes are required use an anti-evaporant oil with the water which will maintain a minimum droplet size. The correct carrier in CDA spraying may help the toxicant to spread and penetrate the wax cuticle of the plant or insect pest (Wrigley 1973), making the active ingredient more effective. With the correct formulation for CDA, less chemical is required to control the pest and ultra low dosage (ULD) rates of chemical, in some cases even 10 per cent of the normal rate or less, can be used which not only saves the farmer money but, increase the safety to the user, and reduce the contamination of the environment (Bals 1970). It is unnecessary to have high concentrate ULV formulations for CDA application.

The field application of CDA to cotton with hand-held machines was developed in Malawi (Matthews 1972a). The operator walks through the crop at right angles to the wind and allows the breeze to carry the mist of spray droplets away from him across the crop. Preferably application is made late afternoon when the land is cooling down (Johnstone 1972) as this avoids the upward convection currents and daytime thunderstorms. It also provides a fresh deposit for the moths and larvae which are particularly active at night. Matthews (1972a) gave typical yields in Malawi showing that CDA spraying was comparable with conventional low volume spraying using the specially designed tail boom cotton sprayer which is considerably heavier to carry through the crop when full of spray.

Yield seed cotton kg/ha	*1969/70*	*1970/71*
CDA single row swath	2,225	1,848
CDA 5 row swath	1,595	1,754
ULV 'Tail boom' sprayer	2,259	2,151
Control unsprayed	475	798

CDA has been widely used in peasant grown cotton in many parts of Africa. In Mozambique in 1971/72 50,000 ha of cotton were sprayed CDA with the yield increasing from 100 or 200 kg/ha unsprayed to 800 kg/ha from the sprayed cotton (Matthews 1972a), and about 100,000 farmers sprayed cotton for the first time in Tanzania using the Micron Ulva and endosulfan (Percy 1975). This technique is not confined in its usefulness to cotton or to insecticides. The Ulva has been used successfully on rice in the Far East, though with this method it is difficult to control the leaf hopper nymphs which feed at the base of the rice plants. Groundnuts can be protected at low cost against leaf spot (*Cercospora*). The Handy herbicide sprayer which is similar to the Herbi has been used by small farmers in Africa to apply wettable powders, particularly triazines in water suspension for weed control in cotton and maize, and this is particularly valuable in areas where the weeding of food crops creates an annual labour shortage about 6 weeks after the planting rains start and delays the planting of other crops. Glyphosate is particularly effective when applied with a Herbi to control perennial grasses and nutgrass. As the Herbi is a placement sprayer the disc is held 20 cm above the weeds.

So little spray liquid is required with CDA spraying that both the spray and the sprayer can be easily carried to the field on a bicycle and the protection of crops against insects and diseases can be extended into areas where water is not available (Tunstall *et al*. 1971).

The efficiency of this CDA system may be further improved by the use of electrostatically charged droplets which eliminates losses from drift and ground contamination. The Electrodyne system developed by ICI (Coffee 1979) uses very little electrical energy to produce and charge the droplets of a chosen size, and gives excellent crop cover using very little spray.

Cotton, the crop which is sprayed more than any other in the tropics, can stand a lot of damage, particularly by leaf eating caterpillars in the first 2 months. The fruiting points, particularly those which develop early in the season, however, need protection.

Regular spraying, even if no pest is present, costs money and may result in more damage as the parasites of the pests are destroyed. On large cotton schemes however there may not be time to protect all the crop when the pest reaches a dangerous level and routine spraying has to be carried out. Small cotton growers can base their spraying on scouting in their own cotton. For details of this and for information on spraying machinery see Matthews (1979).

Granules

The active ingredient is either adsorbed, coated on to, or mixed with a solid carrier, and released either by moisture or volatilization. The granules vary in size, 0.3-0.7 mm being common, but larger granules (1.5 mm) are used and microgranules (0.01-0.02 mm) are more effective where the active material is wanted on the leaves. The granules are applied at 15 to 25 kg/ha to the soil, flooded rice paddies, foliage or, as in the case of maize stalk borer, placed in the leaf funnel.

The advantages of granules are that no water is required for spraying; there is less drift; no measuring is needed for mixing; application can be made with a simple spreader or, for very safe chemicals, a gloved hand; there is good control over placement; greater persistence can be obtained if desired.

The disadvantage is that granules are expensive per kilogram of active material, due to the cost of the filler and granulation, and high transport costs for low concentrate material. If a fertilizer such as superphosphate could be used as a carrier this would partly offset the higher cost. At the IRRI, lindane and diazinon granules applied to rice at 30 day intervals were more effective than weekly spraying. This appears to result from a combination of fumigation as the active material volatilizes from the water surface, capillary action inside the leaf sheath, and diffusion into the rice plant. Granular herbicides have been effective in rice paddies at the IRRI and are commonly used in Japan. A dense granule that sinks rapidly is preferred.

Fumigants

These are used in the treatment of tobacco seedbeds against weeds and nematodes, and soil fungi; see Table 4.1. Fumigants are often used to disinfest stored crops, methyl bromide is commonly used for this purpose.

Aerosols

The active ingredient is in a solvent kept as a liquid by pressure and packed in a special container from which it is easily released as a spray. The aerosol is an expensive though convenient method of combining application with efficacy, but is only suitable for household and garden use.

Smokes/fogs

Warehouses are often treated with insecticidal smokes before grain is stored in them. HCH is commonly used in warehouses and azobenzene as a smoke in glasshouses. This can also be done efficiently with a fogging machine which can apply a wide range of insecticides. Misting with a CDA applicator would achieve the same effect and eliminate the fire risks.

Baits (Peregrine 1973)

An attractive food material or chemical attractant is combined with a toxic material as in the use of Warfarin® and coumachlor rat baits on rubber and oil palm estates; HCH (BHC) (see Table 4.2) with rice husks, etc., for cutworm and locust hopper control; Sevin® or HCH with coconut meal for the rhinoceros beetle; spent hops with malathion for fruit fly control. The Coconut Research Station in Jamaica has developed poisoned bait blocks for rat control where Warfarin®, maize meal and sugar are mixed into a paraffin wax block. These blocks are not damaged by rainfall or ants and are easily suspended in the palms.

Seed dressings

The active ingredient compounded with a filler and an oil sticker is applied to the seed prior to planting to destroy the fungi carried on the outside of the seed, and protect the seed during germination against soil fungi and insects. Mercurial compounds are the most effective fungicides, but the toxic hazard may be avoided by using sulphur, cuprous oxide, thiram or other organic fungicides which are often equally satisfactory. Aldrin, dieldrin and HCH were the most commonly used insecticides in seed dressings but the first two are not permitted in some countries due to their persistence which makes them so effective. Disyston® and Thimet® are under investigation as systemic insecticide dressings for the protection of both the seed and seedling; similarly the new systemic fungicides used as seed dressings protect the young plants. The classification of the fungi causing the major tropical plant diseases is outlined in Table 4.5.

VIRUS DISEASES OF TROPICAL CROPS

In 1884, a Japanese grower showed there was an association between 'Dwarf' or 'Stunt' disease of rice and leaf hoppers (*Nephotettix apicalis*), though it was not until the twentieth century that it was realized that the leaf hoppers were not the cause of the disease but merely the transmitters. This was the first virus disease shown to be insect transmitted. This vector is somewhat exceptional in being able to pass the virus to the next generation through the egg, the virus multiplying within the vector. Many viruses do not multiply in the vector.

Since 1884 12 different virus diseases of rice have been identified and classified, and as far as is known they are all transmitted by leaf and plant hoppers (*Homoptera*). The symptoms and control are described by Feakin (1970). Control of the vector by spraying will only protect the crop if a large contiguous area is sprayed at the same time. Left to individual farmers, a single unsprayed plot could provide a source of infection for all the surrounding crop. Where tungro is serious it is well worth while treating the nurseries with carbaryl to control the vector, particularly where virus-resistant varieties are being grown.

Table 4.5 Outline classification of some of the fungi causing serious diseases in tropical crops. The scheme proposed by Ainsworth (1966) is used.

LOWER FUNGI

PHYCOMYCETES or alga-like fungi. Their hyphae never unite into strands or tissues, and the mycelium is either filamentous or composed of isolated, rounded cells. Regular septation is never found, so that the living parts of the hyphae usually form a continuous cell.

Asexual reproduction by zoospores, conidia, sporangiospores, sporangia.

Order Peronosporales

Family	Pythiaceae – parasitic or saprophytic		
	Pythium	*P. arrhenomanes*	Sorghum and sugar cane root rot.
		P. complectens	Patch canker of rubber tapping panels.
	Phytophthora	*P. palmivora*	Bud rot of coconut palm.
		P. palmivora	Cocoa black pod.
		P. palmivora	Black stripe of rubber.
		P. palmivora	Secondary leaf fall of rubber.
		P. citrophthora	Citrus gummosis.
		P. nicotianae var. *parasitica*	Citrus gummosis.
		P. nicotianae var. *nicotiana*	Black shank of tobacco. *P. nicotianae* vars. also infect egg plant, peppers, castor beans, tomato and potato.
Family	Peronosporaceae (Downy Mildew)		
	Perenospora	*P. tabacina*	Downy mildew or blue mould of tobacco.
	Sclerospora	*S. graminicola*	Downy mildew of maize and sorghum.
		S. maydis	Downy mildew of maize and sorghum.
		S. sacchari	Downy mildew of sugar cane.

HIGHER FUNGI

Asexual reproduction by conidia.

The ASCOMYCETES all have spores developed within a mother-cell, the ascus. At some stage hyphae interwoven to form tissue.

Order Plectascales *Elsinoë fawcettii* – Citrus scab

Table 4.5 (continued)

Order Erysiphales Powdery mildews			
Family	Erysiphaceae - obligate parasites		
	Erysiphe	*E. cichoracearum*	Powdery mildew of tobacco.
	Sphaerotheca	*S. fuliginea*	Powdery mildew of sim-sim.
	Oidium	*Oidium heveae*	Rubber mildew. *Oidium* generally refers to conidial stage.
Order Hypocreales			
Family	Nectriaceae		
	Gibberella	*G. zeae*	Maize cob rot, scab of rice.
		G. fujikuroi	Seedling blight of rice, stem and ear rot of maize, a leaf disease of sugar cane. Conidial states - *Fusarium graminearum* (*G. zeae*) and *F. moniliforme* (*G. fujikuroi*)
	Calonectria	*C. rigidiuscula*	Cacao canker. (Follows *Phytophthora palmivora* or capsid damage.)
	Sphaerostilbe	*S. repens*	Violet root rot of tea. Red root of limes. Also infects rubber and arrowroot.
Order Sphaeriales			
Family	Sphaeriaceae		
	Rosellinia	*R. pepo*	Citrus root disease. Black root disease of cacao.
		R. bunodes	Root disease of citrus and tea.
		R. arcuata	Black root disease of tea.
		R. necatrix	Root disease of tea.
Family	Ceratostomataceae		
	Ceratocystis	*C. fimbriata*	Mouldy rot of tapped bark of rubber trees. Also disease of coffee, sweet potato, pigeon pea and cocoa. On cocoa associated with wounds, particularly *Xyleborus* attack.
		C. paradoxa	Black rot or pineapple disease of sugar cane, bleeding stem of coconut palm, pineapples, banana suckers.

Table 4.5 (continued)

Family	Mycosphaerellaceae		
	Mycosphaerella	*M. musicola*	Sigatoka disease of banana (conidial stage - *Cercospora musae*).
		M. berkeleyi	Late leaf spot of groundnuts. (Conidial state *Cercospora personata*).
		M. arachidis (*C. arachidicola*)	Early leaf spot of groundnuts.
Family	Pleosporaceae		
	Glomerella	*G. tucumanensis*	Red rot of sugar cane.
	Colletotrichum	*Glomerella cingulata* - perfect state	
	Imperfect stage	(*C. coffeanum*	Coffee berry disease.
		(*C. gloeosporioides*	Anthracnose of citrus, fruit rot of mangoes.
		C. gossypii	Anthracnose of cotton.
		C. falcatum	(Perfect Stage - Glomerella *tucumanensis*). Red rot of sugar cane.
	Magnoporthe	*M. salvinii*	A disease of ripening rice (Imperfect states *Sclerotium oryzae* and *Helminthosporium sigmoideum*).
Family	Valsaceae		
	Valsa	*V. eugeniae*	'Sudden death' of cloves.
	Diaporthe	*D. citri*	Melanose (black spotting) of citrus.
Family	Xylariaceae		
	Ustulina	*U. deusta*	Root disease of tea and rubber.
Order Dothideales			
	Microcyclus (syn. *Dothidella*)	*M. ulei*	South American leaf disease of rubber.

The BASIDIOMYCETES include a very large number of the higher fungi, representing the most diverse types.

The HEMIBASIDIAE include only one order, the *Ustilaginales* or smuts.

There are two families in the *Ustilaginales*:

Family	Ustilaginaceae - Basidia transversely sepated into four cells.		
	Ustilago	*U. maydis*	Maize smut.
		U. scitaminea	Sugar cane smut.
	Sphacelotheca	*S. sorghi*	Sorghum covered smut.
		S. cruenta	Sorghum loose smut.
Family	Tilletiaceae - Basidia continuous or one septate.		
	Tilletia	*T. barclayana*	Rice bunt.

The PROTOBASIDIAE - usually 4-celled basidium.

Table 4.5 (continued)

Order Uredinales The rusts			
Family Pucciniaceae – Teleutospores stalked.			
	Puccinia	*P. arachidis*	Groundnut rust.
		P. sorghi	Maize rust (brown).
		P. polysora	Maize rust (yellow).
		P. kuehnii	Sugar cane rust.
		P. psidii	Pimento rust.
	Hemileia	*H. vastatrix*	Coffee rust.

The EUBASIDIAE – The hyphae frequently unite to form large conspicuous masses. Causal agent of many root diseases.

Order Hymenomycetales			
Family	Thelephoraceae		
	Corticium	*C. salmonicolor*	Pink disease of cocoa, citrus, tea.
	(*Rhizoctonia* – imperfect stage)	*Koleroga noxius*	Threadblight of coffee, areca palm and tung oil.
		C. invisum	Black rot of tea.
		C. theae	Black rot of tea.
		Thanatephorus cucumeris syn.	'Sore shin' of cotton (*Rhizoctonia solani* – mycelium).
		R. solani	Disease of many crops.
		C. rolfsii	(Vegetative stage – *Sclerotium rolfsii*.) Root disease of groundnuts, damping-off of cotton, and infects sugar cane, sweet potato and black pepper. Seedling blight of rice.
Family	Exobasidiaceae		
	Exobasidium	*E. vexans*	Blister blight of tea.
Family	Agaricaceae		
	Armillaria	*A. mellea*	Root rot of tea, citrus. Collar crack of cocoa.
	Marasmius	*Crinipellis perniciosa M. equicrinis*	Witches' broom of cocoa. Horse hair blight of tea.
Family	Ganodermateceae		
	Ganoderma	*G. philippii*	Red or wet root disease of rubber and tea.
Family	Polyporaceae		
		G. lucidum	Basal stem rot of oil palm.
		Rigidoporus lignosus (*Fomes*)	White root disease of rubber, cacao.
Family	Hymenochaetaceae		
		Phellinus noxius (*Fomes*)	Brown root disease of rubber, cacao, tea.
	Poria	*P. hypolateritia*	Red root disease of tea.
		P. hypobruneae	Root disease of tea and rubber.

Table 4.5 (continued)

FUNGI IMPERFECTI

The DEUTEROMYCETES - Asexual forms. Some are conidial stage of perfect fungi.

Order Sphaeropsidales			
	Ascochyta	*A. gossypii*	Leaf spot of cotton.
		Phoma medicaginis	Black stem of lucerne.
	Diplodia	*Botryodiplodia theobromae*	Diplodia pod rot (wound parasite) of rubber, tea, cocoa.
		D. maydis	Dry rot of maize.
Order Melanconiales			
	Gloeosporium	*G. alborubrum*	Secondary leaf fall of rubber.
		G. limetticola	Withertip of limes.
Order Hyphales			
	Verticillium	*V. albo-atrum*	Cotton wilt.
		V. dahliae	Cotton wilt.
	Moniliophthora (*Monilia*)	*M. roreri*	Monilia pod rot of cacao.
	Thielaviopsis	*T. basicola*	Tobacco root rot.
	Pyricularia	*P. oryzae*	Paddy blast.
	Alternaria	*A. longipes*	Alternaria leaf spot of tobacco.
		A. citri	Alternaria rot of citrus.
Family	Cercosporidium		
	Cercospora	*C. henningsii*	Leaf spot of cassava.
		C. coffeicola	Coffee leaf disease.
		C. nicotianae	Frog eye spot of tobacco leaves.
		C. gossypina	Cotton leaf spot.
		C. elaeidis	Leaf spot of oil palm.
	Fusarium	*F. oxysporum* f.sp. *cubense*	Panama disease of banana.
		F. oxysporum f. sp. vasinfectum	Cotton wilt.
		F. oxysporum f. sp. *elaeidis*	Vascular wilt of oil palm.

Table 4.5 (continued)

Drechslera	*Drechslera turcica* (perfect state *Steosphaeria turcica*)	Maize leaf spot or leaf blight.
	D. oryzae – (perfect state *Cochliobolus miyabeanus*)	Brown spot of rice.
	D. sacchari	Eye spot of sugar cane.
Helminthosporium	*H. sigmoideum*	Sclerotial disease or irregular stem rot of rice.
Sclerotium	*S. rolfsii*	Root rot of groundnuts.
	S. oryzae	Sclerotium disease of rice.

For a description of the characters used in classification and details of these diseases see Butler and Jones (1949) or Brooks (1953). For descriptions Kranz *et al.* (1977).

Anyone working in the field of plant pathology should have a copy of The Plant Pathologist's Pocketbook compiled by the Commonwealth Mycological Institute, revised edition due 1982.

Transmission of plant viruses is mainly by sucking insects: aphids, leaf hoppers, whiteflies, mealybugs and tingids, of which the first two are the most important. Some viruses are transmitted mechanically by rubbing sap from a diseased plant on to a healthy one, and some of these viruses are transmitted by chewing rather than sucking insects, the virus probably being regurgitated. Tobacco mosaic is transmitted by the hands of the cultivators, entering the plants by small wounds, hence the importance of labour in tobacco fields disinfecting hands and tools and not smoking as the virus survives in the cured leaf.

Most of the mosaic diseases are aphid transmitted. *Myzus persicae*, the green peach aphid, can transmit over 50 virus diseases, and similarly some virus diseases have a large number of aphid species known to be vectors though they are not transmitted by other insects. The aphids vary in the length of time taken to acquire infection, the need for an incubation period, and the persistence of the virus.

Swollen shoot of cocoa first reported in 1936 in the Eastern Region of Ghana is economically one of the most important plant diseases in the world. Over 100 million diseased trees were cut out in this Region in an unsuccessful effort to control this virus disease to which the Amelonado trees were particularly sensitive. This virus is transmitted from tree to tree by mealybugs particularly *Planococcoides njalensis* and *Planococcus citri*, neither of which is ever present in sufficient numbers to be a pest in its own right. The planting of varieties with horizontal resistance appears to be the only practical method of control where this disease is serious.

In recent years swollen shoot has spread into the adjacent cocoa area in Togo where it is causing serious damage.

Leaf hoppers are responsible for transmitting streak and leaf necrosis virus of maize. They tend to inject and withdraw the virus from the phloem. The difference in ability to transmit virus between leaf hoppers is a genetic character in the case of corn streak virus where it is determined by a single sex-linked dominant gene.

As the nymphs of whiteflies are non-migratory they cannot transmit the virus they acquire until they become adult. The adults can acquire the virus independently.

Unfortunately many plant viruses have alternate hosts which makes the problem of eradication nearly insuperable, hence the control of virus diseases can only be done either by breeding resistant varieties, or by controlling the vector. Crop nutrition, particularly the soil nitrogen and phosphorus level, is important in determining the seriousness of the symptoms but cannot prevent the disease. The classification of the important pests of tropical crops is outlined in Table 4.6 and the toxicological risk associated with the insecticides used to control them in Table 4.7.

Table 4.6 Some important insect pests of tropical agriculture
Outline of the most important families
(Classification after Imms, Richards & Davies, 1977)

SUB-CLASS PTERYGOTA
(Winged, or secondary wingless, almost all with pronounced metamorphosis)

DIVISION EXOPTERYGOTA. The wings develop outside the body (metamorphosis simple and in some slight, mostly without a pupal stage).

ORDER ORTHOPTERA 'Straight wings' (Cockroaches, stick-insects, grasshoppers, locusts, crickets, etc.).

ORDER ISOPTERA 'Equal wings' (termites or 'white ants') - a few species destructive to living plants, e.g. tea, rubber and grasses; very destructive to wooden buildings, etc., in all tropical countries.

ORDER ANOPLURA 'Unarmed tails' (lice) - including parasites of domesticated mammals and birds.

ORDER THYSANOPTERA (thrips) - many serious crop pests in all countries, including a few vectors of virus diseases.
Caliothrips impurus - cotton thrips.
Selenothrips rubrocinctus - cacao thrips.

ORDER HEMIPTERA 'Half wings' or **Rhynchota** (bugs) - innumerable crop pests and some beneficial species, especially in the following families:

Sub-order **Heteroptera** 'Different wings'

Pentatomidae (shield bugs).
Antestiopsis orbitalis form *lineaticollis* - Antestia of Arabica coffee.
Scotinophara coarctata - black paddy bug.
Leptocorisa spp. - rice bugs.

Table 4.6 (continued)

Coreidae - e.g. squash bugs.
Pyrrhocoridae (red bugs) - e.g. *Dysdercus* spp. - cotton stainers.
Reduviidae (assassin bugs) - some beneficial predators.
Miridae (Capsidae) - many pests and a few beneficial predators.
Sahlbergella singularis - brown capsid (cocoa).
Distantiella theobroma - black capsid (cocoa).
Helopeltis spp. - cocoa, cotton, tea.
Taylorilygus vosseleri - cotton lygus.
Sub-order **Homoptera** 'Similar wings'
Cercopidae (froghoppers) - e.g. *Aeneolamia varia* - Trinidad sugar cane froghopper.
Cicadellidae (*Jassidae*) (leafhoppers) - many pests, including many vectors of virus diseases.
Jacobiasca lybica - cotton jassid.
Empoasca facialis - cotton jassid.
Delphacidae (leafhoppers) - some important pests, including vectors of virus diseases.
Peregrinus maidis - vector of virus stripe of maize.
Psyllidae
Phytolyma lata - galls of *Chlorophora excelsa*.
Aleyrodidae (whiteflies) - some pests of citrus, cotton, etc., including a few vectors of virus diseases.
Bemisia tabaci - cotton whitefly.
Vector cotton leaf curl virus, cassava mosaic, tobacco virus.
Aphididae (green-flies, plant-lice) - many pests, including the vectors of a majority of the plant virus diseases.
Toxoptera aurantii - tea aphis.
Toxoptera citricidus - black citrus aphis, vector of 'tristeza'.
Sipha flava - yellow sugar cane aphis.
Pentalonia nigronervosa - banana aphid, vector of 'bunchy top'.
Aphis gossypii - cotton aphis.
Rhopalosiphum maidis - maize aphis. Vector of sugar cane mosaic.
Myzus persicae - wide range of crops and important virus sector.
Coccoideae (scale insects and mealy-bugs) - one of the most destructive of all insect families, attacking almost all crop plants.
Planococcus citri - citrus mealybug } Vectors of swollen shoot.
Planococcoides njalensis - cocoa mealybug }
Lepidosaphes beckii - purple scale of citrus:
Aonidiella aurantii - California red scale.
Planococcus kenyae - coffee mealybug.

DIVISION ENDOPTERYGOTA. Wings develop inside the body (metamorphosis complete and pupal stage well defined, in all).

ORDER LEPIDOPTERA 'Scale wings' (butterflies and moths) - probably the most destructive order after the Coleoptera. Some outstanding families in the tropics are:
Gelechiidae, *Pectinophora* (*Platyedra*) *gossypiella*, pink bollworm of cotton, *Phthorimaea operculella* - potato tuberworm or potato tubermoth.
Psychidae - defoliators of tea, shrubs, etc., known as 'bag-worms'.

Table 4.6 (continued)

Tineidae - (clothes moths, etc.) *Lyonetiidae* - *Leucoptera coffeella* - coffee leaf miner.
Yponomeutidae - *Plutella xylostella* - diamond-back moth, on Cruciferae.
Cryptophasidae - *Nephantis serinopa* - coconut caterpillar.
Tortricidae (leaf-rollers) - a few crop pests. Tortrix caterpillar.
Cryptophlebia leucotreta - false codling moth of cirus and cotton.
Homona coffearia - tea tortrix.
Pyralidae - many serious pests, e.g. moth-borers of rice, sugar cane, maize, etc.
Chilo zonella - cane borer of sugar, rice, etc.
Etiella zinckenella - on cowpea, etc.
Diatraea saccharalis - sugar cane moth borer. Also attacks maize and rice.
Scirpophaga incertulas - yellow rice borer.
Chilo suppressalis - rice borer. Also attacks sugar cane.
Limacodidae - e.g. nettle-grubs of tea.
Castniidae - e.g. *Castnia licus*, giant moth-borer of cane.
Sphingidae (hawk moths) - many serious pests, e.g. hornworms of tobacco - *Manduca* spp. Tailed caterpillar of sweet potato - *Agrius convolvuli*.
Noctuidae - leafworms, armyworms, cutworms, stem borers, fruit borers.
Heliothis (Helicoverpa) armigera } American bollworm, corn ear worm.
H. zea } Wide host range.
Earias insulana - cotton spotted bollworm.
Earias biplaga - spiny bollworm.
Diparopsis castanea - Sudan or red cotton bollworm.
Diparopsis watersi - Sudan bollworm.
Busseola fusca - maize stalk borer.
Spodoptera frugiperda - Fall armyworm (maize, rice, etc.).
Spodoptera litura - cotton worm.
Agrotis ipsilon - black cutworm.

ORDER COLEOPTERA 'Sheath wings' (beetles) - over ¼ million described species. Largest order in animal kingdom. Probably the most destructive order. A few outstanding families in the tropics are:
Coccinellidae (ladybirds) - very few destructive species, many important beneficial predators on aphids, scale-insects, mealybugs, etc. Both adults and larvae beneficial predators.
Rodolia cardinalis - Vedalia beetle parasite of cottony cushion scale. First outstanding success of biological control.
Bruchidae (pea and bean 'weevils') - many pests, especially in leguminous seeds.
Chrysomelidae - very many pests, especially as leaf-eaters.
Coelaenomenodera elaeidis - oil palm leaf miner.
Dicladispa armigera - rice hispa.
Cerambycidae (longicorns) - many wood-boring pests, e.g.
Steirastoma breve - cocoa beetle.
Bixadus sierricola - coffee borer.
Anthores leuconotus - coffee stem borer.
Xylotrechus quadripes - white stem borer of coffee.
Curculionidae (true weevils) - the most destructive of any single family of insects, including pests of almost all crops.
Anthonomus grandis - Mexican boll weevil.
Cosmopolites sordidus - banana weevil.

Table 4.6 (continued)

Cylas formicarius - sweet potato weevil.
Rhynchophorus ferrugineus - red palm weevil.
Rhynchophorus phoenicis - palm weevil.
Sitophilus oryzae - rice weevil.
Scolytidae (bark-beetles) (often carry a fungus *Ceratocystis*) - many forest pests.
Hypothenemus (Stephanoderes) hampei - coffee berry borer, etc.
Xyleborus fornicatus - shot hole borer of tea.
Xylosandrus morigerus - coffee stem borers.
Scarabaeidae (chafers, etc.) - many pests of tropical crops, especially sugar cane and coconut. Larval stages root feeding white grubs; adults - chafers.
Dermolepida albohirtum - sugar cane root grub.
Oryctes rhinoceros - rhinoceros beetle of coconut and oil palms.
Heteroligus meles - yam beetle.

ORDER HYMENOPTERA 'Membrane wings' (bees, wasps, ants, etc.) - some destructive species, e.g. parasol ants, and innumerable important beneficial species, in many families, as parasites and predators of all stages of insects. Ants important in spreading and protecting mealybugs and aphids, e.g. *Planococcoides njalensis*, cocoa mealybug tended by *Crematogaster* ants.

ORDER DIPTERA 'Two wings' (two-winged or true flies) - includes leather jackets, mosquitoes, houseflies, tsetse-flies, blow-flies, *Simulium*, but comparatively few species of agricultural importance, e.g.
Cecidomyiidae (gall midges) - a few pests of tropical crops.
Contarinia sorghicola - Sorghum midge.
Syrphidae (hover-flies) - some beneficial predators, as larvae, on aphids, etc.
Muscidae
Stomoxys calcitrans - stable fly.
Musca domestica - housefly.
Glossina spp. tsetse flies.
Tephritidae (*Trypetidae*) (fruit-flies) - many serious pests.
Ceratitis capitata - Mediterranean fruit fly.
Dacus dorsalis - oriental fruit fly.
Tachinidae - many important parasites of serious pests.

For descriptions of many pests with photographs see Wyniger (1962); also Hill (1975); Kranz *et al.* (1977); Atwal (1976).

WEEDS

The control of weeds in tropical crops*

Tropical conditions are conducive to a rapid and heavy growth of weeds during the rainy season, and many weeds are adapted to survive a dry

* See Kasasian (1971); Deuse and Lavabre (1979).

Table 4.7 Hazards of insecticides

The table following lists in order of descending toxicity those insecticides commonly used in either agriculture or public health. Where any choice is available to the user, he should always select the least toxic insecticide which will control the particular pest in an efficient and economical manner.

The figures given refer to the active ingredient and not to the trade materials, the registered names of which are given. Formulation of active ingredient can reduce the hazard of toxic materials (Barnes 1959), for example Thimet granules on carbon.

Common name	Trade name	Class[a]	Acute LD 50 to rats (mg/kg)	
			Oral	Dermal
Phorate	Thimet	OP	2-3	70-300
Endrin		CH	3-6	60-120
Parathion		OP	3-6	4-200
Azinphos methyl	Guthion	OP	7-13	280
Carbophenothion	Trithion	OP	7-30	800
Monocrotophos	Azodrin, Nuvacron	OP	17-21	112
Phosphamidon	Dimecron	OP	—	1,500-2,500
Dichlorvos (DDVP)	Vapona	OP	25-30	75-900
Dicrotophos	Bidrin	OP	22-45	225 Rb
Endosulfan	Thiodan	CH	35	74-680
Dieldrin		CH	40	>100
Heptachlor		CH	40	200-250
Aldrin		CH	40-60	>200
Nicotine		—	70	140
Demeton methyl	Metasystox	OP	50-75	300-450
Vamidothion	Kilval	OP	64-100	1,160 Rb
Triazophos	Hostathion	OP	80	1,000
Chlorpyrifos	Dursban	OP	82-163	202
Thiometon	Ekatin	OP	100	>200
Deltamethrin	Decis	Py	135	>2,000 Rb
Dimethoate	Rogor and Roxion	OP	200-300	700-1,150
HCH [BHC]		CH	200	500-1,000
Fenitrothion	Sumithion	OP	250-673	1,500-3,000
Camphechlor	Toxaphene	—	283	1,000
Chlordane		CH	283	1,600
DDT		CH	300-500	2,500
Fenvalerate	Belmark	Py	300-630	—
Diazinon		OP	300-600	500
				1,200
Cypermethrin	Cymbush	Py	303-4,123	2,400 Rb
Carbaryl	Sevin	CAR	400	500
Trichlorphon	Dipterex	OP	650	2,800
Fenchlorphos	Ronnel	OP	1,000-3,000	5,000
Menazon	Sayphos	OP	1,200-1,600	500
Permethrin	Ambush, Permesect	Py	430->4,000	—
Malathion		OP	1,400-1,900	>4,000
Bromophos	Nexion	OP	3,750-5,180	>1,000 Rb
Tetradifon	Tedion	CH	5,000-14,700	>10,000 Rb

[a] *Note*: OP = Organophosphorus type. CH = Chlorinated hydrocarbon type. Rb = Rabbit. CAR = Carbamate type. Py = Synthetic pyrethroid.

Source: Jones *et al*. 1968 and others

season and an annual burn. The first nine weeds listed by Holm *et al.*, 1977 in their survey of the world's worst weeds are common in the tropics, namely

1. *Cyperus rotundus* - nutgrass
2. *Cynodon dactylon* - Bermuda grass
3. *Echinochloa crus-galli* - barnyard grass
4. *Echinochloa colona* - jungle rice
5. *Eleusine indica* - goosegrass, or fowl foot grass
6. *Sorghum halepense* - Johnson grass
7. *Imperata cylindrica* - lalang
8. *Eichhornia crassipes* - water hyacinth
9. *Portulaca oleracea* - purslane

Others, including *Digitaria sanguinalis* and *D. scalarum* in Central Africa, *Salvinia* in Sri Lanka, *Rottboellia exaltata* in West and Southern Africa are challenging for the 'top ten'.

The wide range of broad-leaved weeds are generally easily controlled by hand pulling, slashing with a cutlass, hoeing or mechanical cultivation, the traditional methods of weeding. Most areas, however, have their own particular problem weeds which have developed by a combination of climatic factors and cultural practices.

Perennial grasses such as *Axonopus compressus, Cynodon dactylon, Digitaria scalarum, Imperata cylindrica, Panicum repens, Paspalum conjugatum* and *Pennisetum clandestinum* are serious in many areas, and in a variety of crops largely due to the rapidity with which they can regrow from the stolons or rhizomes which survive weeding.

The annual grasses *Eleusine africana* and *Brachiaria eruciformis* rapidly colonize maize fields, seriously affecting the nitrogen supply available to the crop. Other annual grasses including *Setaria pallidefusca* and *Digitaria timorensis* are serious in sugar cane in Mauritius.

Rice paddies have many broad-leaved weeds, and serious grass weeds such as *Echinochloa crus-galli* which is difficult to differentiate from rice seedlings in the early growth stages, and may be transplanted as rice from nurseries. Flooding is the traditional way of killing weeds in paddies. Few seeds of *Echinochloa* germinate if well covered with water.

Cyperus rotundus and the related sedge species which are referred to as 'nutgrass' are a problem in most tropical areas. As the underground 'nuts' are efficient sources of regrowth, if the top growth is destroyed normal cultivation just spreads and encourages them. Fortunately they can now be controlled chemically where there is no annual crop.

Deep cultivation of the cane fields of Trinidad during certain parts of the dry season reduces the invasion of perennial grass weeds (Blackburn *et al.* 1952), but due to the differences in the seasons, this method is not so effective in Jamaica.

Parasitic weeds can be serious pests, particularly on soils of low fertility and where host crops have been grown too frequently. As the seeds can remain dormant in the soil for many years until stimulated to germinate by the roots of host plants, they are difficult pests to control by rotation.

Orobanche spp. (broom rape) (Pieterse 1979) are most important in hot dry areas, and can parasitize broad-leaved crops including tobacco, beans, cotton and certain vegetables. As *Orobanche* occurs mainly on poor soils improving the soil fertility decreases the attacks.

Striga spp. which are serious in Asia and Africa, live mainly on grasses, maize, sorghum and millet, though some species can parasitize broad-leaved crops such as tobacco and beans. In Northern Nigeria in a heavy infestation there may be two million *Striga* plants per hectare weighing over 3 t, which gives some indication of the seriousness of the competition to the crop. Some varieties of sorghum and pearl millet are resistant to *Striga* but there is always a risk of this resistance breaking down. Short Kaura 5912 sorghum is resistant to a light attack of *Striga* but not a heavy infestation. The International Crops Research Institute for the Semi-Arid Tropics (ICRISAT) has been selecting sorghum for resistance to *S. asiatica* in India, and it is hoped that some of these selections will also be resistant to *S. asiatica* elsewhere and *S. hermonthica* in Africa.

There are two forms of resistance in sorghum, some varieties exude only small amounts of stimulant, others stimulate the germination of *Striga* but are not parasitized.

Hand pulling of *Striga* as recommended in many parts of Africa is not practical but 80 kg/ha of nitrogen applied 4 weeks after sowing to sorghum in Northern Nigeria suppressed the parasite. Very low rates of MCPA, 2,4-D, mecoprop and ametryne as spot applications gave very effective control of *Striga* (Ogborn 1970).

Crop loss

Weed competition seriously reduces crop yields in the tropics; Ashby and Pfeiffer (1956) estimated that such losses are two to three times as great as in temperate zones. Prentice (1957) quoted a loss of 50 per cent in the yield of a cotton crop due to *Digitaria*, with a greater loss in a bean crop. Ogborn has pointed out that in Northern Nigeria weed competition in June–July largely determines eventual crop yields. The farmers' ability to weed at this critical period is the bottleneck in agricultural production and delays cotton planting.

In Central Luzon, Philippines, there are on average 280 plants per square metre of *Echinochloa colona* in direct seeded rice areas. A density of 80 per square metre in the first 40 days after sowing can reduce the yield by a quarter as the maximum tillering of the weed coincides with the critical period of competition for the rice. It produces more tillers than rice and about 42,000 seeds per plant, which are shed from the seventh week and remain dormant for a time (Mercado & Talatula 1977).

A loss of half the potential crop must be quite common where a heavy infestation of *Echinochloa* occurs in rice, and this is confirmed in many yield trials. This grass weed appears to be much more efficient in taking up nitrogen and phosphate than the crop as shown in Table 4.8.

Table 4.8 Nitrogen and phosphate uptake by rice and barnyard grass (*Echinochloa crus-galli*) in Australian flooded rice

Treatment	Nitrogen uptake (kg/ha)		Phosphate uptake (kg/ha)	
	Rice	Grass	Rice	Rice
Stam F-34, 4.4 kg a.i. per ha Water after 4 days	100	—	18.5	—
Control	36.8	56.3	7.3	14.5
Stam F-34, 4.4 kg a.i. per ha Water after 6 days	111.8	1.14	—	—
Control	15.5	94	—	—

Source: French and Gay (1963).

The importance of early weeding in cassava is illustrated by Doll (1978) who estimates that weed competition in the first month could reduce the crop by a quarter, and in the first two months by a half. He considers four or more weedings in the first 3 or 4 critical months after planting worth while, and recommends intercropping, closer spacing, and chemical treatment but surprisingly not the development of varieties with denser foliage in their early growth to suppress the weeds.

Crops without weeds

Maximum yields will be obtained when the crop is grown without any weed competition. If this ideal cannot be achieved, weed control in the early growth stages of the crop is essential. Lamusse (1965) showed that weed infestation of sugar cane that started 12 weeks after planting had no significant effect on yield. The most important period for controlling weeds in sugar cane was from the time the primary shoots just started to appear above ground, to the beginning of stalk elongation. Inadequate weed control during the first 6 weeks of sugar cane growth can reduce the yield by as much as 45 per cent even if clean weeding is carried out for the rest of the growing period. No weeding at all caused an almost complete loss of crop.

This importance of preventing weed competition from the earliest stages is well illustrated where atrazine is sprayed just after planting maize. Given adequate rainfall, the atrazine treated areas show a marked 'stimulation effect' which is attributed to the complete absence of weed competition rather than a positive growth stimulating factor.

Bunting (1959) stressed the crucial nature of weed control, particularly in the early stages of the groundnut crop. As this work was done at Kongwa (Tanzania), it is probable that competition for moisture was the major factor. On the other hand the weeds in the irrigated cane of Guyana are competitors for oxygen and nutrients.

Competition to perennial tree crops in their early establishment years is particularly serious and may not only influence the time taken to come into bearing, but also the ultimate crop yield.

Jones and Maliphant (1958) found a highly significant correlation between cocoa yields in the sixth, seventh and eight years and the girth of the trees at 3½ years, indicating that the yielding capacity was associated with the rapidity of growth in the years of establishment when weed competition, lack of moisture or too heavy shade could have a serious retarding effect.

The importance of weed prevention in young vegetatively propagated tea was seen at Dessford Estate 2,000 m above sea level in Sri Lanka. To get a good spread of the tea bush, all the young plants were pegged out. The lygus bugs from the grass weeds caused a lot of damage to the young tea, and to hand weed all the pegs had to be taken out and replaced. By spraying these grasses and other weeds with simazine, taking care not to spray within 20 cm of the young plant during the first 6 months, it was only necessary to pull out the few resistant grasses once a year. Removing the grass controlled the lygus bugs, and without the weed and pest competition about a year was saved from planting to plucking compared with the old method.

The vogue for clean weeding on the steep hillsides of Sri Lanka led to considerable erosion and a decline in yield from loss of fertility. 'Clean weeding' was replaced in many countries for tea, coffee and other tree crops, by selective weeding and slashing. The weeds known to be detrimental were removed and the rest slashed to form a mulch and give protection against soil erosion. A cover of *Oxalis* in Sri Lanka has no detrimental effect on the tea but helps to bind the soil against erosion, and shows the soil pH is sufficiently low. The prevention of erosion, however, cannot justify allowing young tree crops to struggle against tall weeds, particularly grasses. Removal of the grasses and tall weeds and slashing of 'soft weeds' is frequently used on steep slopes to reduce erosion. Clean weeding and mulching of alternate lines along the contours, planting cover crops or constructing soil erosion tied bunds are better ways of preventing erosion and crop competition.

Non-chemical weed control

The use of hand tools, forks, hoes and cutlasses, can damage roots, trunk and young shoots of the crop. Continuous hoeing of the soil damages the soil structure, assists erosion, and aids the decomposition of the soil organic matter. Hand weeding of young rubber in West Africa often creates a 'dish' round the young tree as weeding is always away from the trunk to prevent cutlass damage. This 'dish' collects water during the rains and is an ideal breeding ground for bark fungi. In Trinidad and neighbouring cocoa areas of South America, the loss of trees from *Ceratocystis fimbriata* transmitted by the shot hole borers (*Xyleborus* spp.) is serious. Some planters believe the borer enters through the slight cuts in the bark of the cocoa tree made when cutlassing. The removal of *Digitaria scalarum* from tea and coffee plots in East Africa requires special forks, which do severe damage to the tree roots, from which some trees may never recover. Mechanical cultivation of sugar cane has often been overdone so that the cane has been set back more by hoeing than by the weeds.

Root damage may also be important in spreading root diseases and nematode infections.

A further disadvantage of manual or mechanical methods of weed control is that the weeds are not destroyed until they are quite large, by which time they have already had a detrimental effect on the crop.

In Thailand *Imperata cylindrica* ('lalang') is being controlled with large cultivator discs which bury the grass about 250 mm deep. Small farmers are able to hire the services of contractors for this tractor work. Good control of Johnson grass can be achieved by using spiked harrows to bring the rhizomes to the soil surface where they dry out.

Chemical weed control

The big advance in chemical weed control in temperate regions came at the end of the war following Templeman's observation of the selectivity of alpha-naphthyl acetic acid between charlock and wheat which led to the introduction of MCPA and 2,4-D as weedkillers. These hormone herbicides, which remove the broad-leaved weeds from cereals, have had relatively little impact on tropical agriculture, except in those restricted regions of high altitude growing temperate cereals, in sugar cane, in sisal, and to a small extent in maize and rice. The reasons for this are various.

1. Most of the smallholders' plots can be hand weeded at no cost in cash, and in subsistence farming this is important.
2. Grasses and sedges which are not controlled by 2,4-D and MCPA are the important weeds in the tropics.
3. Apart from cash crops, intercropping is the general practice and it is difficult to have a herbicide which will take the weeds out of a mixture of widely different plant species.
4. Until recently no suitable spraying equipment was available.
5. The unavailability of suitable herbicides in small packs.
6. The general lack of an extension service which can advise on weed control.
7. Vegetables, cotton and tobacco are very susceptible to hormone herbicides and the risk to nearby crops would be too great in many farming systems.

However with the introduction of the Herbi (and Handy) sprayers, and chemicals capable of controlling the 'world's worst weeds' (p. 356) there are an increasing number of situations where chemical weed control is practical and economic. The introduction of herbicides in peasant farming in the tropics requires careful consideration, as there is the risk of following the experience of the temperate cereal farmer, who has eliminated his easily controlled broad-leaved weeds only to have his fields infested with previously rare grass weeds, which are both more competitive to the crops and more costly to control.

In some crops grown in the humid tropics half the labour devoted to producing a crop is used in weeding. Herbicides should be used to supplement hand weeding and to deal with weeds such as nutgrass and Johnson grass which thrive on hand weeding; and to eliminate work demand peaks.

Most estates have introduced chemical weed control, and there is often a place for herbicides in smallholders' tree crops, tea, coffee, cocoa, etc., particularly in the early stages before the canopy is thick enough to shade out the weeds. Maize and sugar cane are sprayed pre-emergence with a triazine such as atrazine, and diuron is useful on sugar cane provided that the soil is not deficient in phosphate. Both these herbicides, which depend upon soil moisture for their activity, have been used to keep weeds out of tea, coffee, sisal, pineapples, oil palms, citrus and other tree crops. To remove established non-woody weeds paraquat has been widely used. This chemical kills the top growth of most green plants on which it is sprayed. It is absorbed very quickly and its effect is visible in 2 or 3 days. As it will kill young crop plants if it falls on green tissue, it can only be used if these crop seedlings are protected from the spray; or it is sprayed on the land before the seedlings are planted out. As paraquat leaves no soil residue, planting can follow as soon as convenient. Glyphosate acts in a similar way, but as it is translocated into the rhizomes it gives a longer period of control of perennial grasses and nutgrass. It is particularly effective at low rates when sprayed with a Herbi, and is often activated by adding sulphate of ammonia or oil to the spray. These herbicides have in many cases replaced dalapon which is a very effective grass killer if used when the grasses are growing actively. Used carefully and kept off the crop, low rates of dalapon have given good control of annual grasses in sugar cane in Mauritius. Dalapon has been used successfully in lucerne, tea, coffee and many other tree crops. In East Africa it was the first chemical that was used to clear tea plantations of *Digitaria*. Aminotriazole controls certain grasses resistant to dalapon such as *Paspalum conjugatum*, a creeping grass common in Malaysian rubber plantations, as well as many other grasses and broad-leaved weeds. A trace of this chemical on a green leaf causes severe chlorosis so it must not be sprayed over young seedlings.

Fleming (1958) gave the responses of the major grass weeds in East Africa to dalapon and aminotriazole (Table 4.9).

Asulam has been widely used to spray established sugar cane to kill Johnson grass and Para grass and would probably be very effective against other grass weeds in sugar cane.

To combine knockdown effect and persistence of weed control, mixtures of herbicides are frequently sprayed. In Malaysia mixtures of MCPA, diuron, and aminotriazole sometimes with paraquat; or paraquat, diuron and 2,4-D amine, are sprayed along the rubber tree rows and in oil palm circles, and harvesters' paths (Teoh *et al*. 1978). Aminotriazole is often used with diuron or a triazine herbicide in established perennial tree crops to kill the established weeds and keep the crop weed free as in Cameroon.

A number of herbicides have been developed for annual crops including cotton, soya beans and tobacco but their use depends upon the local cultivation system, particularly irrigation.

Weed control in rice

Weeds are spread in rice crops with the seeds, in the irrigation water, and by the animals and tools used to cultivate the paddies. Certain grass weeds, particularly *Echinochloa crus-galli*, are easily mistaken for rice in the seedling stage, and transplanted from the nursery. The seeds of canary grass (*Phalaris minor*) in India are distributed in irrigation water from other farmers' fields and after germination this grass is particularly difficult to weed out by hand. For centuries weeds have been controlled by 'drowning' them with irrigation water. Alternative ways of controlling them could free valuable water to irrigate an extended area. Hand weeding of rice is difficult and takes a lot of time and the fertiliz-

Table 4.9 Reaction of some grasses to dalapon and aminotriazole in East Africa

Name	Annual perennial	Habit	Reaction to dalapon	Reaction to aminotriazole
Chloris gayana	Perennial	Shallow stoloniferous	2 applications at 11 kg/ha kills.	Requires 2 applications at 17 kg/ha to kill.
Cynodon dactylon	Perennial	Shallow stoloniferous	Total of 17 kg/ha kills.	Resistant to 11 kg/ha. Strongly checked by 22 kg.
Dactyloctenium aegyptium	Annual	Tufted	Killed at 5.5 kg/ha but rapidly comes in as seed again (5 weeks).	Checked by 11 kg/ha.
Digitaria scalarum	Perennial	Rhizomatous	Requires a total of 11 kg/ha to give 90% control. Lower roots not affected by initial treatments.	Resistant to rates up to 33 kg/ha. (All trials in E.A. are without cultivation.) Checked at rates above 33 kg/ha.
Digitaria ternata	Annual	Tufted	Susceptible to 5.5 kg/ha. Comes in as seed (3–4 weeks).	Checked at 11 kg/ha.
Digitaria velutina	Annual	Tall tufted	Susceptible to 5.5 kg/ha. Comes in as seed (3 weeks).	
Panicum maximum	Perennial	Short rhizome	Requires at least 22 kg (2 applications)	Severe chlorosis at 11 kg. At 22 kg checked not killed.
Pennisetum clandestinum	Perennial	Rhizomes	Killed at 22–33 kg/ha (2 applications).	Resistant to 22 kg/ha. Strong check at 33 kg/ha.
Pennisetum purpureum	Perennial	Tall tufted	Susceptible to to 22 kg/ha. Strongly reduced 11 kg/ha treated young.	
Setaria verticillata	Annual	Tall, shortly creeping	Killed at 11 kg/ha.	Killed at 22 kg/ha.

ed high yielding varieties are even more difficult to hand weed. Thus the use of herbicides in rice-growing areas is likely to increase output more than displace labour. In most tropical rice-producing areas growers can only afford cheap herbicides such as 2,4-D. In the Philippines 2,4-D or MCPA applied 4 days after transplanting not only controls broad-leaved weeds and sedges but also some germinating grasses. Weed control in direct seeded rice is more difficult, as the crop and weeds start to grow at the same time and the crop suffers serious competition at this critical early stage. Propanil and molinate give excellent control of grass weeds when applied in the correct way, but they are usually too expensive for field use in the tropics. Where *Echinochloa* is serious in the nurseries it is usually worth while spraying the nursery beds with propanil (Stam F-34®) to prevent the weed being transplanted into the paddies. Benthiocarb (Saturn®) and butachlor (Machete®) are two promising herbicides for both direct seeded and transplanted rice, but at present their cost is about the same as hand weeding. 2,4,5-TP added to these herbicides will improve the control of Cyperaceae.

Paraquat (Gramoxone®) is being used in Malaysia and Sri Lanka to clear up the weeds in rice paddies before planting. This saves many cultivations, including puddling, and cuts down the time taken to prepare the paddy from 2 months to 2 weeks, making double cropping possible. possible.

Glyphosate (Roundup®) can also be used for this purpose giving good control of perennial species including nutgrass (*Cyperus rotundus*).

Red rice, which is a serious problem in Guyana, cannot be controlled chemically. Great care should be taken to rogue out this weed at harvest and make certain it is not spread by sowing seeds of red rice in the nursery beds. The best control in the field is achieved by deep ploughing the soil to bury the dropped seed at least 20 cm when the soil is saturated with water but firm. The expected yield increase is worth nearly twice the cost of ploughing (Giglioli 1956). Rai (1973) has suggested controlling red rice by growing a short strawed variety (70–90 cm), which grows vigorously in the seedling stage and matures in 90–110 days before the red rice has shattered. A change to sowing pre-germinated seed would also help.

Control of water weeds

Water storage tanks and dams, and the channels used for distributing irrigation water are ideal for water weeds. These use the water, and block the flow.

In a large irrigation scheme in India, submerged aquatic weeds have cut the flow of the main canal by 80 per cent. This impeded flow has increased seepage from the canals and contributed to waterlogging and increased salinity. Of the 32,000 ha put into irrigation between 1961 and 1965, the water table by 1966 was within 1.5 m of the surface on half the area, and 2,000 ha are already out of production through waterlogging and salinity (Holm 1966). About 320 km^2 (200 square miles) of the surface of Lake Kariba are now covered with *Salvinia molesta*, a water fern, though something appears to have halted its spread in recent years. The water lettuce (*Pistia stratiotes*) is increasing rapidly in Lake Volta, Ghana. The water hyacinth (*Eichhornia crassipes*), one of the 'world's worst weeds' (see review Pieterse, 1978), increases the surface evaporation of water sometimes as much as four times the rate of open water (Little 1967). Water hyacinth only appeared in the Congo River in 1952, but by 1955 it covered 1,500 km (938 miles). In 1958 it appeared in the Nile, and apart from threatening the irrigation scheme of the Gezira it is travelling towards the new Nasser Lake. It is widespread throughout Asia. In Sri Lanka *Salvinia molesta* in 1960 occupied at least 20,000 ha of land, mainly paddy fields, particularly where drainage was poor (Dias 1967).

Control of many of these weeds has been attempted in the past, but recent successes with paraquat at Kariba and in Sri Lanka offer a better control of water hyacinth, water lettuce, *Salvinia* and duckweeds, but the cost of reclaiming the areas already infested will be enormous.

Water hyacinth is very susceptible to many herbicides including 2,4-D, MCPA, paraquat, glyphosate and aminotriazole. However, the use of certain of these is restricted where the treated water might be used to irrigate susceptible crops like cotton or tobacco. The control achieved so far with these herbicides has in any case not been long-lasting. This weed has been used in mixtures to feed cattle.

The Commonwealth Scientific and Industrial Research Organisation (CSIRO) in Brisbane are controlling *Salvinia* with a kerosene/wetting agent mixture which floats on the water and sinks the *Salvinia* which then dies. This approach could have a wider use with no risk to irrigated crops.

Yield stimulation

Old rubber trees due for replanting within 6 years were often treated with 2,4-D or, more commonly, 2,4,5-T applied in oil to the scraped bark near the tapping cut. A 1 per cent 2,4,5-T ester can increase the yield of latex 10–30 per cent. The action of these chemicals is believed to be due to the production of ethylene in the tree which delays the collapse of the vessels and the coagulation of the latex at the ends of the cut vessels. These chemicals have now been largely replaced by Ethrel® (2-chloroethyl-phosphonic acid) which breaks down in the rubber tree to produce ethylene. This is used in the same way but gives about twice the yield increase of 2,4,5-T and is used on trees not due for uprooting and replanting.

More recently Dickenson working in the UK has developed Ethad, a novel way of introducing gaseous ethylene into the tree. Field trials in Malaysia indicate that Ethad can be applied to the bark without scraping and will raise the yield of poor trees to that of good clones, and the yield of good clones to 5,000 kg/ha and maintain the increased yield longer than other stimulants. Coming at a time when rubber prices are high, this promises to introduce revolutionary changes both in rubber production and the breeding programme.

Table 4.10 Summary of the main herbicides used in the tropics[(a)]

Alachlor (Lasso®). A selective pre-emergence herbicide with 10–12 weeks' persistence. Controls annual grasses and many broad-leaved weeds in maize, soya beans, cotton, sugar cane and brassicas.

Ametryne. Pre- and post-emergence selective herbicide used to control both grasses and broad-leaved weeds in sugar cane, bananas, citrus, coffee and fruit orchards, and as a directed spray in maize. Low solubility.

Aminotriazole (Amitrol) (ATA) - 3 amino-1,2,4-triazole. Readily absorbed by roots and aerial plant parts causing severe chlorosis. Actively translocated. Activated with oil. Kills both broad-leaved and grass weeds. Little persistence in the soil. Cotton defoliant.

Asulam (Asulox®). A post-emergence herbicide absorbed by leaves and roots. Used against grasses in sugar cane. Particularly *Sorghum halepense* and *Rottboelia exaltata*.

Atrazine - 2-chloro-4-ethylamino-6-isopropylamino-*S* triazine. A pre-emergent herbicide, slightly more soluble than simazine. Active through the leaves as well as the roots. Field of use similar to simazine but more effective under drier conditions. Foliar activity increased by adding oil.

2,4-D (as amine or ester) - 2,4-dichlorophenoxyacetic acid. The most used herbicide in the world. Selectively removes broad-leaved weeds from cereals and sugar cane. Used pre-emergence on groundnuts. Used alone or in mixtues as a directional spray in other broad-leaved crops.

Dalapon - 2,2-dichloropropionic acid. A water-soluble grass killer. Absorbed by the foliage and roots. Most effective when grass actively growing. Can be used in many tree crops if carefully applied. Also used for removing grass from irrigation and drainage ditches.

Diuron - N'-(3,4 dichlorophenyl) *N, N*-dimethylurea. A selective persistent herbicide of low solubility used on citrus, pineapples, sugar cane, rubber, oil palms, tea and other perennial crops. Also used at high rates as a soil sterilant.

DSMA (Ansar®). A selective post-emergence contact herbicide with some power of translocation. Used to control grass weeds in cotton, and in mixtures.

EPTC (Eptam®). A herbicide used pre-planting, which kills germinating seeds and inhibits bud development from underground portions of some perennial weeds. Incorporation in the soil mechanically or by irrigation is necessary to avoid loss by volatilization. Useful for the control of perennial *Cyperus* spp.

Fluometuron (Cotoran®). A herbicide mainly root-absorbed but with foliar activity. Suitable for the control of broad-leaved weeds and grass weeds in cotton. Intermediate persistence with a half-life of 60–75 days according to soil conditions.

Glyphosate (Roundup®) – *N*-phosphonomethyl glycine. A non-selective post-emergence foliar-absorbed herbicide slowly translocated through the plant system thus giving good control of deep-rooted perennial weeds including *Imperata cylindrica, Paspalum* spp., *Sorghum halepense, Cynodon dactylon, Cyperus* spp. and *Eupatorium odoratum* as well as shallow-rooted weeds. More effective when sprayed in a Herbi than as a dilute spray. Negligible soil persistence. Low mammalian toxicity. Used in rubber, coffee, tea, oil palms, citrus, coconuts and other tree crops. Useful for weeding ground prior to planting. Apply 2–3 weeks before planting tobacco to control *Cyperus* and *Cynodon*.

MCPA – 4-chloro-2-methylphenoxyacetic acid. Selectively removes broad-leaved weeds from cereal crops. Used also in sisal.

Molinate (Ordram®) – *S*-ethylhexahydro-1 H-azepine-1-carbothiolate. A selective herbicide used on rice fields prior to seeding or transplanting to prevent the growth of a wide range of weeds. The herbicide must be worked into the soil.

Nitrofen (TOK E®). A pre-emergence herbicide, toxic to a number of broad-leaved and grassy weeds in cereals when left as a thin layer on the surface. Activity rapidly lost if incorporated with the soil.

Paraquat (Gramoxone®) – 1,1'-dimethyl-4,4'-dipyridylium-2A. A rapid acting non-selective weedkiller, inactivated on contact with the soil. Used to control a wide range of weeds, either before planting, after crop emergence by a directed spray, and in perennial crops. Good for controlling floating water weeds.

Propachlor (Ramrod®). A pre-emergence herbicide effective against annual grasses and certain broad-leaved weeds in maize, cotton, soya beans, sugar cane, peanuts. Persistence of 4–6 weeks.

Propanil (Stam F-34®) – 3',4'-dichloropropionanalide. A selective post-emergence or post-transplanting herbicide used to control many broad-leaved and grass weeds including *Echinochloa* in rice. Can be used in rice nurseries.

Simazine – 2-chloro-4,6-bisethylamino-*S* triazine. A nearly insoluble pre-emergent, selective herbicide for use in maize, sorghum, citrus, sugar cane, rubber, tea, oil palms, etc. Also as a soil sterilant.

2,4,5-T – 2,4,5-trichlorophenoxyacetic acid. Used as ester in oil or water to kill trees and scrub. Also used as a latex stimulant in rubber and to kill volunteer rubber seedlings.

2,4,5-TP Mixed with other rice herbicides to control Cyperaceae. (Propanil, Molinate, Benthiocarb.)

Oil. Certain oils marketed as Lalang Oil, are used to eliminate small remaining patches of lalang (*Imperata cylindrica*) in Malaysian rubber estates. The oil is wiped on the leaves with a rag or foam rubber sponge. The oil is non-toxic to humans and animals, but toxic to rubber seedlings and other plants. Also used as adjunct to atrazine, glyphosate, aminotriazole and asulam to improve post-emergence action.

(a) There are more than 130 herbicides commercially available. Details of their chemical structure and properties are given in the *Weed Control Handbook*, edited by J.D. Fryer and R. Makepeace. Though this refers to the United Kingdom, the principles are universally applicable. Local recommendations should be followed.

Note: Where the acid is referred to, the material is generally used as the sodium or amine salt, or ester.

5 THE PLACE OF CATTLE IN TROPICAL AGRICULTURE

Cattle keeping in the tropics covers nearly as wide a range of systems as cropping. There are the large nomadic herds of the Hamitic tribes of East Africa and the Fulani of West Africa; the religion-protected underfed cows of India which roam unculled in excessive numbers; the ranches of the Venezuelan Llanos and other parts of South America and North Australia; the more intensively managed large herds of Queensland grazing improved pastures sown with legumes and fertilized; the small intensive dairy herds developed in the West Indian Islands grazing Pangola grass paddocks.

Mixed farming has developed to a limited extent mainly in the medium rainfall areas of East and West Africa, but cattle owners in the tropics, even in Australia, show a reluctance to combine crop and animal production. Even the well-informed grazier of Australia is generally unwilling to cultivate part of his land to sow lucerne or an improved pasture mixture.

Possession of cattle in the tropics is not only a sign of wealth and social position; cattle are money. This applies to peoples as widely different as the Masai of Africa and the Australian grazier. The lives of the nomadic Fulani are controlled by the needs of their cattle. In the dry season they find water and grazing in the river valleys of the south, but once the rains start they move north to the drier healthier parts. In the dry season, their cattle graze the crop stubbles of the Hausa who value the cattle manure and may offer the herdsman food to encourage him to graze his cattle on their land. In Rwanda and Burundi, cattle keeping was the prerogative of the conquering Batutsi, who restricted the Bahutu to cultivating.

While herd size remains the sole aim, improvement is unnecessary. Years of natural selection have produced animals well adapted to this extensive system. The herdsmen have their traditional skills and medicines, and only seek assistance for inoculation against epidemics such as rinderpest. The numbers form an uneasy balance with the available grazing and water.

The bullock is an important source of power both for cultivating the land and pulling carts. In Asia and the Far East draught cattle and buffaloes provide over half the total energy employed in farming.

The interest in ox ploughing where conditions make it practical is shown in Teso (Uganda) where the number of ploughs increased from 48 in 1914 to 8,300 in 1932, 30,740 by 1949 and over 40,000 in 1968. The four-oxen plough was introduced originally but this was rapidly replaced by the lighter two-oxen plough. Teso is in the short grass area with few large trees and has a terrain well suited to ploughing, and furthermore the people are traditionally cattle owners.

There is no doubt that both milk and meat production can be considerably increased if a market is provided for the increased production at a satisfactory price. This is illustrated in Jamaica where a factory to produce condensed milk was completed in 1941. In that year 290 suppliers delivered 792,716 gallons (3,604 klitres) of milk, but by 1953 suppliers had risen to 2,325 and production to 2,888,598 gallons (13,132 klitres) (Arnold 1955). The new condensary offered a market at a fixed price and arranged collection, even in some areas where production was initially low. Stud bulls and later an artificial insemination service were available which enabled the small cattle owner to keep an extra cow and improve his herd with sires of a quality far beyond his own resources. Factories are established in Barbados and Trinidad, and in response to a guaranteed price and collection service the production of quality liquid milk is increasing rapidly. Government control of the imports of dried and condensed milk is often necessary to get a local dairy industry started. Government aid in the introduction of quality cattle is also essential.

In Kenya a fermented milk product *Maziwa lala*, produced from skim milk by the Cooperative Creameries, is in big demand in the urban areas. The product is rich in proteins and minerals and is viscous and tasty. It is popular with both children and adults (Shalo & Hansen 1973). This is a traditional drink of certain nomadic tribes in East Africa who frequently make it in gourds. It offers an attractive way of improving the diet in other countries where milk is produced, and a good market for milk producers.

The large-scale movement of families from the country to the towns as in Nigeria and Thailand, has created a demand for milk and meat in urban areas at prices that encourage increased production by the surrounding farmers.

Increased production will result from the improved management of animals suited to the local conditions. As standards of management are raised, the quality of cattle (or pigs and poultry) can be improved which in practice means greater amount of European (*Bos taurus*) blood in the cattle and less of the Zebu (*Bos indicus*) strain. The lack of impression on the local Zebu type cattle that regular importations of British Friesian and Jersey cattle had on the general herds of the West Indies until recent times when management was improved, is a warning that the two must go together.

Zebu and European cattle have a number of physical differences (Plates 5.1 and 5.2). Zebus have their characteristic hump behind the head, which persists even after three or four crosses with European bulls. The skin which is thinner and usually pigmented, is much looser over the body and hangs down below the neck (dewlap and brisket). In a mixed herd the skins of the Zebu and grade animals collect fewer ticks than the pure European cattle, which suggests that the Zebu skin is either less attractive or even slightly repellent to ticks. Zebus are smaller animals slower to mature, with a longer head and pointed ears, longer legs and a less well developed rounded udder. These differences may be a result of long selection among the European cattle. The Zebu tolerates higher temperatures and poorer grazing conditions.

IMPROVED MANAGEMENT

An indigenous herd of cattle is the result of natural selection over generations with a little influence from the owner or herdsmen. The genetic make up of the herd is in harmony with, or a compromise with, the environment. Such cattle can live or maintain their numbers under the local conditions, where other breeds, particularly newly imported European cattle, would not even survive. The Criollo cattle of South America, as an example, are cattle introduced by the Spaniards selected over the centuries by their ability to survive. To the visitor they appear to be degenerate European stock.

Production of milk or beef by unimproved herds is at present generally limited by management rather than genetics. There are variations between animals and a certain amount of improvement can be achieved by selection but a greater response is more rapidly achieved by better management, and with better management, the blood of higher yielding animals can be introduced, followed by selection.

Better management may involve:

1. Housing.
2. Fly control.
3. The provision of shade in grazing areas.
4. Better feeding including pasture improvement and controlled grazing.
5. Greater attention to calf rearing.
6. Provision of mineral licks and salt.
7. Better watering.
8. Disease control including control of ticks.

Plate 5.1 Sahiwal-Pusa, the outstanding Sahiwal bull used in early cross-breeding, which led to the development of the Jamaica Hope. (Courtesy Ministry of Agriculture and Lands, Jamaica.)

Housing

It is impracticable to house the large herds, though they may be kraaled for safety at night when predatory animals or tribes are about.

Housing is frequently provided for a small herd kept near the homestead for safe-keeping at night, shelter from the heat of the day, or to give a respite against biting flies.

Housing is normally made in the local building materials such as mud and wattle with thatch which keeps the animals cool. More prosperous owners may use bricks, murram or concrete blocks or similar more permanent materials, with a corrugated iron roof. A corrugated iron roof is more expensive and much less efficient as a heat insulator, but it is permanent and gives protection against fire. An aluminium roof by reflecting heat will keep the building cooler than corrugated iron. Insulation could be valuable for reducing the temperature in the buildings.

The entrance to the house may be closed with poles supported in horizontal position from which they may be easily slipped. Poles may also be used round the inside walls at the height of the animal's body, to protect the wall from the animals.

Plate 5.2 Katarega, Zebu bull, founder of the Serera herd, Uganda. (Courtesy Ugandan Department of Information; photograph Ron Ward).

Housing of calves is of particular importance. Individual pens should be provided; these should be easy to clean out and the entrance darkened to make a fly trap. If possible the windows should be covered with gauze. Where blood lapping bats are a problem as in Venezuela the houses should be screened (Kverno & Mitchell 1976). A mineral lick, in addition to water and a feeding trough, is ideal.

Separate isolation quarters should be available for sick animals and as a quarantine for any animals brought into the herd.

Fly control

In addition to the tsetse fly (*Glossina* spp.), biting flies of the *Stomoxys* spp. are sometimes a serious problem. Not only are their bites unpleasant but they can carry diseases such as 'surra' and anthrax. The onset of the rains sees a rapid increase in the number of *Stomoxys*, and this is frequently reflected by a drop in milk production and a loss in condition of the herd. Traditionally cattle have been housed in the middle of the day in buildings which are kept dark by covering the entrance with mats or making the entrance in the form of a dark passage. At the beginning of this passage strings may be hanging down to disturb the flies on the cattle. The flies prefer to remain in the light. The houses in these areas are constructed without windows, though a hole may be left just below the ridge to attract any flies which get into the house. At the height of the fly season, smudge fires are kept burning in the houses. Insecticide spraying is uneconomic unless the flies are a risk to public health, but a cheap method of control is to use strings impregnated with insecticide stretched below the roof, or having Vapona® (Dichlorvos or DDVP) strips (see Table 4.2) in the buildings. Fly breeding areas around the buildings such as manure heaps and long grass should be treated or eliminated. Coaker and Passmore (1958) in Uganda showed that *Stomoxys* are most active between 8 am and 9 am and 3 pm and 4 pm. Midday is a period of minimum activity. Where these flies are troublesome it would appear to be advisable to water the cattle in the early morning, house until 9 am, re-house between 3 and 4 pm and water again in the evening, or better still graze by night provided that the fencing is adequate. Night paddocking avoids both the flies and the heat of the day and the cattle spend more time grazing at night than during the day.

The common housefly is attracted to the dairy and can bring many human diseases for distribution in the milk. The room where the milk is handled should be screened with double doors, kept very clean and the flies controlled by insecticides applied when there is no milk in the dairy.

Shade trees for pastures

The higher the proportion of European blood in the cattle, the greater is their tendency to seek shelter in the heat of the day. With pure bred European cattle such as Holsteins or high grade cattle, they are often brought into buildings in the middle of the day to avoid the heat. Where cattle are outside throughout the day shade trees or thatched shelters should be provided in the pastures. The choice of trees is important as they must not be poisonous or have poisonous fruit (e.g. *Albizia versicolor* with poisonous pods and seeds which has caused many deaths in Zimbabwe) as they are often browsed by cattle, and must not interfere with the growth of grass under their canopy. The shade trees may simply be those that have survived the annual burn, part of the hedging or fencing, browse trees or be protected trees such as the Shea butter nut (*Butyrospermum parkii* = *B. paradoxum*) or valuable timber trees see Plate 5.3. Suitable shade trees include the wide spreading Saman or raintree (*Samanea saman*), the flamboyante or flame tree (*Delonix regia*), the tamarind (*Tamarindus indica*) and species of *Ficus, Acacia,* and *Albizia* such as *Albizia lebbeck*. When clearing forested land for

Plate 5.3 Groundnut crop in *Butyrospermum*. (Courtesy Rose Innes)

pasture it is often preferable to leave clumps of trees for shade rather than single species which may not survive the change of micro-climate.

Better feeding*

If a high yielding pedigree cow is sent from Europe to India or tropical Africa, even if it avoids the many new diseases to which it becomes exposed, it will yield very little milk because the feed from the local pastures is too low. Higher production understandably demands a higher level of feeding. Cattle keepers in the tropics are handicapped by a lack of information on the feed requirements of their animals for maintenance and production, and frequently inadequate information on the nutritional value of the locally available feeds. Consequently recommendations are based on experience from temperate countries, the southern states of the USA, and more recently on Australian experience.

Tropical cattle appear to be better adapted to the utilization of coarse fibrous herbage (roughage), and to make more efficient use of drinking

* See FAO (1978b).

water, than temperate cattle. Maintenance of body temperature requires less food in the tropics.

Farm animals can be divided into two classes, the non-ruminants such as the horse and pig which have a simple stomach, and the ruminants which include cattle, water buffalo, sheep and goats which have a compound stomach. Micro-organisms (bacteria, fungi and protozoa) present in the rumen can break down cellulose to products of nutritional value to the ruminant. Plant cell walls are broken down making the cell contents available to the ruminant. The fermenting organisms can utilize non-protein nitrogen compounds such as urea to synthesize amino acids.

Cattle require for their nutrition, adequate amounts of proteins for tissue building and milk; carbohydrate or fats as a source of energy; minerals (see p. 21) of which calcium and phosphorus are the most important for bone building and milk; and vitamins which are naturally available to animals feeding on green matter and exposed to sunlight for part of the day. Members of the vitamin B group can also be synthesized in the rumen. Common salt is important to tropical cattle. Proteins are built up in the animal's body from amino acids. There are nine different amino acids considered essential in animal nutrition and five others are essential to some animals but only partly essential to others. In the case of cattle, these five are normally synthesized in the rumen or replaceable by other amino acids.

Proteins fed to animals are broken down to amino acids during digestion and then rebuilt into body structures or milk and wool. Each protein has a different biological value which is largely a reflection of the total amino acids and the balance between different essential amino acids, though certain protein feeds are considered by some to also have beneficial growth factors. Animal proteins have a higher biological value than vegetable proteins as they are better balanced to the needs of animals. The biological value of a protein feed is very important with pigs, but not very important in cattle where the amino acids are synthesized or changed in the rumen. Conversely, food protein, particularly soluble protein such as casein and grass protein, may be wasted by conversion into ammonia. Excess protein is used by animals to provide energy, which is a waste as carbohydrate can usually be provided in a cheaper form than protein.

Calculation of feed requirements

Nutritive value of feed. There are a number of systems of expressing the nutritive value of feeding stuffs, none of which is perfect. The aim is to assess total amounts of carbohydrate and protein in the feed which the animal can utilize.

1. Carbohydrate. Human nutritionalists quote the value of different foods in calories, which is the gross energy available in the food, but this is rarely used with livestock. The net energy value, which allows for the energy required to digest the particular feed is more valuable. For practical purposes the net energy of a food is related to the net energy of pure starch for fattening bullocks. This is the Starch Equivalent (SE) which can be defined as the number of kilograms of pure digestible starch equal in value for fattening purposes to 100 kg of the feeding stuff. One kilogram of starch equivalent has an energy value of 2,361 kilocalories (kcal). While by definition starch equivalent refers to the fattening value of food fed to a bullock together with a maintenance ration, experience has shown that it is applicable to both work and milk production. Thus, if the starch equivalent of coconut meal is given as 75, then 100 kg of this feed will be equal to 75 kg of starch for production. The starch equivalent is calculated by chemical analysis. The starch equivalent of any food, such as coconut meal, will vary for different sources, qualities and grades. It will also vary in its value, from one cow to another, and within the same cow at different planes of nutrition or production. Despite these variations, starch equivalent is a useful practical value for comparing different feeds and calculating feed mixtures.

2. *Protein*. Proteins contain about 16 per cent of nitrogen and the crude protein content of a feed is calculated by multiplying its nitrogen content by 100/16. As Kjeldhal nitrogen determinations include non-protein compounds, the true protein in the feed will be less than the Kjeldhal figure, but this is not very important for ruminants where the non-protein nitrogen compounds are converted to protein. American workers often use the crude protein percentage, or the total digestible nutrients (1 kg TDN is approximately 0.91 kg SE or 2,140 kilocalories) and digestible crude protein. Digestible crude protein (DCP) is the digestible true protein (DTP) together with the other nitrogen compounds. In the early days of animal nutrition studies, these other nitrogen compounds were thought to have half the value of true protein and the term protein equivalent (PE) was used for

$$\mathrm{DTP} + \frac{\mathrm{DCP} - \mathrm{DTP}}{2} = \frac{\mathrm{DTP} + \mathrm{DCP}}{2} = \mathrm{PE}$$

This generalization on the value of non-protein nitrogen compounds is now known to be invalid, but as these compounds generally occur only in small quantities the error is of no practical importance. The greatest error occurs with green foods, root crops and silage. For cattle, DCP is more accurate; and true protein for pigs. Protein equivalent has passed out of favour with animal nutritionalists. The ratio of protein to starch equivalent (SE) is important in balancing feeds. Some typical figures are given in Table 5.6 for a number of feeds available in the tropics, but these should only be taken as an indication and local determinations should be used.

Nutritive requirements of cattle. In 1924 a committee in England recommended that for maintenance a cow requires 6 lb or starch equivalent including 0.6 lb or protein equivalent per 1,000 lb, and for production 2.5 lb SE including 0.6 lb PE per 10 lb of milk with 3.7 per cent fat. More recent work suggests that 0.5 lb PE per 10 lb of milk is adequate but the SE for maintenance should be raised. For milk with a higher fat content, say 5.2 per cent, the SE should be raised to 3.25 lb and PE 0.65 lb for each gallon of milk produced (or 0.32 kg SE and 0.065 kg PE per litre).

In-calf cows need a production ration in addition to their maintenance ration.

For beef cattle, in addition to the maintenance ration, for each kilogram of liveweight gain a calf needs 1.3 kg SE rising with increasing weight of the animal to 4 kg SE for a fat bullock. The total PE required for both maintenance and liveweight gain is 1.25 to 1.5 kg. For rapid liveweight gain much more starch equivalent (2 kg for every kilogram liveweight gain required from 1 year or 250 kg) must be included in the ration.

Bullocks working a full day require about twice as much starch and protein as bullocks idling. The approximate food requirements of cattle in the tropics are given in Table 5.1.

This clearly illustrates how higher producing animals make more economical use of the available food, and this applies to both pasture and concentrate feed. It is therefore advisable to have the animals with the highest production potential which the standard of management can support. With increasing availability of veterinary attention and quality feed, a greater amount of European blood (*B. taurus*) can be present in the cattle.

The total bulk of the feed is important in feeding cattle. Too rich a feed will not satisfy the animals' hunger and too poor a feed may not contain adequate nutrients in the amount of food the animal can consume. The range is about 7 to 14 kg of dry matter per day for every 450 kg liveweight of the animal. Young calves take in less in proportion to their weight and need more concentrated food easily digested.

Unsatisfactory though the basis of these levels of feed requirement may be, they form a guide to the amount of feed required for cattle, particularly where productivity is being increased.

Grazing

Animal production in the tropics and sub-tropics relies on native and naturalized pastures, not sown pastures. Even in Queensland less than 2 per cent of the grazing is planted to improved pastures.

Table 5.1 Approximate food requirements for cattle (all units in kilograms)
Calves

Liveweight	Dry matter	SE	DCP	PE
50	1	0.75	0.16	—
100	2	1.5	0.32	—
150	3	2.5	0.48	—
200	4	3.5	0.59	—
250	5.5	4.1	0.70	—
500	12	6.4	1.00	—

Recommendations for India are lower, particularly on protein, but it is false economy to underfeed young stock.

Newly born calves need colostrum the first 4 or 5 days after birth. Young calves have a poorly developed rumen and need more easily digested feed, also a high level of calcium and phosphate, 15 to 20 g of each element daily.

Working bullocks
The lower figure is for small bullocks, the higher for large animals.

	SE	DCP
Basic feed not working	1.7 -2	0.14-0.18
Additional feed per hour worked	0.14-0.2	0.02-0.023

If the bullock is still growing the basic ration should be increased by half.

Beef cattle

Liveweight	Dry matter	SE	DCP
400-600	10-12.5	4.5-7.5	0.6-0.95

Gaining 0.5 kg per day
For every additional 0.5 kg liveweight gain required feed 1 kg rising to 2 kg extra SE.

Dairy cattle All units in pounds*.

Type of animal	Approx. weight lb.	Maintenance		Production per gallon at 3% butterfat		Extra butterfat	
		SE	DCP	SE	DCP	SE	DCP
Small Zebu	600	4.5	0.5	2.5	0.6	0.5	0.15
Jersey	700	4.5	0.5	2.5	0.6	0.5	0.15
Large Zebu	900	5.5	0.7	2.5	0.6	0.5	0.15
Holstein	1,250	7.2	1.0	2.5	0.6	0.5	0.15

Table 5.1 (continued)

Thus the requirements of three typical animals in the tropics could be:

	Zebu		Jersey		Holstein	
Size in lb.	900		700		1,250	
Yield in gal./day	1		2		3	
Butterfat %	5.75		5.0		3.5	
	DCP	**SE**	**DCP**	**SE**	**DCP**	**SE**
Maintenance	0.7	5.5	0.5	4.5	1.0	7.2
Milk production	0.6	2.5	1.2	5.0	1.8	7.5
Extra butterfat	0.4	1.4	0.3	1.0	0.07	0.25
Total	1.7	9.4	2.0	10.5	2.87	14.95
Per gallon produced	1.7	9.4	1.0	5.25	0.96	4.98

* One pound is 0.454 kg and one gallon 4.55 litres.

Williamson and Payne (1978) classify natural grazing into five main types, (3) and (4) below being grouped together.

1. Rain-forest grazing. Annual rainfall usually over 1,500 mm (1,000 mm in some monsoonal regions). Grasses are usually absent in virgin rain forest, but following repeated 'slash and burn' management grasses are established. Some of the dominant ones are:

South and South-East Asia	Africa	Central and South America
Imperata cylindrica lalang	*Imperata cylindrica*	*Axonopus* spp.
Themeda trianda red oat	*Pennisetum purpureum*	*Bouteloua* spp.
Dicanthium spp.	*Digitaria* spp.	*Paspalum* spp.
Sehima spp.	*Hyparrhenia* spp.	*Hyparrhenia* spp.
Ischaemum spp.	*Andropogon* spp.	*Imperata* spp.
	Ctenium newtonii	*Panicum maximum*

Though the original rain forest had a number of leguminous trees, herbaceous legumes are slow to establish when the forest is cleared. As moisture conditions are suitable, legumes such as *Alysicarpus, Calapogonium, Centrosema, Crotalaria, Desmodium, Indigofera, Pueraria* and *Stylosanthes* can ultimately be established. Where forests are currently being cleared to establish grazing this process could be accelerated by the introduction of seeds of these legumes, in the way that 'lalang' pastures of South-East Asia have been improved by interplanting with Stylo (*Stylosanthes guianensis* syn *S. gracilis*).

2. *Dry woodland grazing*. Annual rainfall 640-1,400 mm and 4-7 months dry season.

Though occurring throughout the tropics the largest area of dry woodland grazing is in East and Central Africa (*miombo* in Tanzania which covers half the old Tanganyika land area) where more rainfall is expected and a high grass (*Hyparrhenia*) with low tree (*Combretum*) vegetation is dominant. Where there is less rainfall, the trees include *Brachystegia, Julbernardia* and *Isoberlinia* and the grasses a mixture of creeping grasses with tussocks, including *Hyparrhenia, Aristida, Eragrostis, Cymbopogon*. These grasses are palatable when young but rapidly their quality and palatability decline. Thinning the trees and coarse grasses improves these pastures, but as this is done by burning in the dry season the grazing is often converted to savanna.

3. *Savanna*. Annual rainfall 500-1,500 mm. This name (Savanna) is said to be of Carib origin, describing a grassy plain with scattered trees adapted to the long hot dry season with their leathery small leaves, thick corky bark, and extensive root system. The majority of the tropical rangeland is savanna, of which there are many types and associations. Ground legumes are scarce, but the carrying capacity of many could be improved by the introduction of species such as *Stylosanthes humilis* - Townsville stylo.

The coarse grasses are often dominant and these are only of value when young. If overgrazed, as is common, bush spreads and in Africa this can encourage tsetse fly. Careful use of fire is necessary to maintain the balance between grass and scrub. Where the rainfall is reasonable and reliable, sown pastures can be established.

4. *Steppe and semi-arid thornbush*. Annual rainfall 350-750 mm. In East Africa tropical steppe occurs roughly between 1,500 and 2,000 m and semi-arid thornbush below 1,200 m.

The perennial and annual grasses, which vary in their contribution to the grazing according to the management, are medium or short and give a sparse cover. Dominant grasses include *Aristida, Cenchrus, Eragrostis, Themeda*, but there are few herbaceous legumes. Many of the thornbushes are legumes, such as *Acacias* and *Cassias*, and in South and Central America *Prosopis*, which provide useful browse.

In the very dry areas the semi-arid thornbush merges into desert scrub, and the ground cover is there for a short period after any rain.

5. *Montane grazings*. Annual rainfall 1,000-5,000 mm; often only a short dry season. Altitude 1,000-3,000 m. Characteristically they support short grasses, few legumes, but many low-growing herbs. The soil is often acid but the productivity of the pastures high.

Montane grazings may be surrounded by forest destroyed by logging or fire; or the drainage may be inadequate for trees. These grazings are common in South-East Asia. They are the wet *patanas* of Sri Lanka usually dominated by *Chrysopogon* and the *paranas* of South America.

In the Kenya highlands Kikuyu grass (*Pennisetum clandestinum*) with the Kenya wild white clover (*Trifolium semipilosum*) is common. In Zimbabwe *Loudetia simplex* is a common grass. Temperate grass/legume mixtures can be grown in many of these areas. If the fertility can be improved and the land limed as is frequently necessary, productivity can be high.

6. *Seasonally flooded and permanent swamp grazing.* Grazing is usually abundant as the floods dry out but the palatability rapidly falls off and the areas are usually burnt to get a more valuable regrowth. Consequently the plant species must be resistant to flooding, fire and drought. The *llanos* of Venezuela and Colombia are a good example, where the important grasses are *Paspalum, Trachypogon* and *Leersia hexandra* which is well grazed by cattle when kept low. On a much smaller scale swamp grazings are common around Lake Victoria and Lake Chad and in the Nile and Niger valleys. They also occur frequently in Indonesia. The margins of permanent swamps are important for dry season grazing.

The extensive nature of the grazing system is shown by the calculation that if in Masailand each beast requires 6 ha and a basic herd is 13–16 cattle, then the critical population density would be 2–3 people per 2.59 km^2 (square mile).

In Kenya Masai at 4.3 people per 2.59 km^2 (square mile) it is overstocked and at 2.0 in Tanzania generally understocked with local overstocking (Allan 1965).

Management of natural grasslands

With few exceptions these grazings are completely unfenced and in most areas are grazed by the cattle of many owners, hence any management practice or improvement must depend upon the agreement of a group of owners and this generally excludes expensive operations such as fencing and fertilizer application. The tools of management are the hoe, cutlass and most important, fire.

Cattle in a pasture with a wide choice of herbage, graze the plants selectively, but they also select different parts of the plant. Tropical grazings offer a wider variation in herbage than temperate pastures. In East Africa Bredon *et al.* (1967) found that the forage actually eaten by Zebu steers contained 66.4 per cent more crude protein and 7.7 per cent less crude fibre than the average for the clipped pasture, and indeed more crude protein than if the leafy portion of the grass only had been selectively grazed. If the grazing is limited, apart from the plants that are avoided, all the grasses are uniformly defoliated. Soil conditions can influence the attraction of grasses to cattle; Rose Innes (1977) has observed that *Bothriochloa bladhii* and *Heteropogon contortus* which in the savanna grazings of Ghana are normally avoided by cattle, are readily taken up when growing on termite mounds. On the Western Accra plains he found that *Schizachyrium schweinfurthii* was the first grass chosen by cattle with *Setaria sphacelata* a close second, and lists others in order of preference.

Savanna grasses yield a small volume of nutritiously valuable herbage early in the grazing season, and as the volume increases the quality falls, and the needs of the cattle are only fully met halfway through this cycle. In the dry season the available herbage is often inadequate and the cattle lose weight. In temperate regions the surpluses are preserved as hay or silage for the periods of shortage but grass conservation is virtually unknown in the tropics. The other major difference between the two regions is that in the tropics the herbaceous plants other than grasses make little or no contribution to the value of the grazing. The indigenous legumes are of little value and indeed some of them are toxic. On the other hand, woody browse plants, particularly leguminous species, are very valuable in the tree savanna grazings. They are well adapted to the environment, including seasonal burning, and either remain green through the dry season or start to grow just before the rains when the grasses are dry and of little value. The fruits and foliage of these browse species are very nutritious often high in phosphate. Rose Innes (1977) gives an indication from studies in Ghana that these browse plants in the dry season can supply 20–40 per cent of the total herbage intake, and as the crude protein of the browse plants is far higher than in the grasses, without the browse many cattle would not survive a severe dry season. The species he studied with Mabey included *Baphia nitida* (Plate 5.4), *Grewia carpinifolia, Bandeiraea (Griffonia) simplicifolia* and *Millettia thonningii* from the coastal savanna and *Acacia albida, Afzelia africana, Bauhinia rufescens, Ficus graphalocarpa* and *Pterocarpus erinaceous* from the interior savannas, also the introduced legume *Leucaena leucocephala*. The contribution of browse has been underestimated in many tropical countries. In Australia certain succulents are grazed, including bluebushes (*Maireana* spp.) and saltbushes (*Atriplex* spp.) and in the Karoo of South Africa *Pentzia* spp. are valuable dry season feed.

For many years the balance between the grazing land and the numbers of cattle was maintained by the loss of cattle to diseases and epidemics such as rinderpest. Improved veterinary services with inoculation against certain diseases has led to an increase of livestock numbers grazing a restricted area. The effects of overgrazing are that the most sought after species are grazed out, leaving gaps in the cover which are colonized by species of lower acceptability not so severely grazed. Continued overgrazing removes most of the acceptable species which are lost and the pasture is reduced to bare ground with annual species, and

Plate 5.4 Browsed *Baphia*. (Courtesy Rose Innes)

tufts of unacceptable coarse grass such as *Sporobolus pyramidalis* which is common in overgrazed areas of Teso, Uganda and *Cymbopogon afronardus*. Scrub trees which are not browsed may also increase if there is sufficient rainfall, and the carrying capacity which originally was very low by temperate standards, becomes negligible.

Improvement of natural grazing

The traditional method of managing tropical grazing has been by burning. Burning destroys the dry unpalatable grass and this allows new palatable shoots to grow. The burn destroys the insect pests, parasitic worms, and the cattle ticks that are in the grass. During the burn the nitrogen and sulphur are lost but the phosphate and bases are recycled in the ash to feed the new shoots. Many useful grasses including species of *Andropogon, Hyparrhenia, Themeda, Trachypogon, Paspalum,* and *Panicum maximum* are well adapted to survive a burn. Where grazing has been protected from burning some of the valuable species are lost, smothered out by the dead vegetation.

Fire can be used at different seasons with different effects on the grazing. (see Plate 5.5)

Plate 5.5 Total degradation of parkland. Semi-starving cattle grazing a poor annual grassland, heavily grazed and burnt. Near Bawbiu, a northern boundary of Ghana.

1. *Early dry season burn*. The grasses are not fully dried out and the fire is not too hot. The damage to the grasses and browse plants is not too severe. Enough vegetation and ash is left after the fire to protect the soil from the sun and rain. After the fire the trees and grasses often shoot again to provide a little grazing in the dry season. The *Themeda triandra* pastures on the northern boundary of Uganda and Kenya (Karamoja-Turkana) form the dry season grazings and are burnt at the start of the dry season.
2. *Late dry season burn*. By this time all the old grass is very dry, the fire is very hot and difficult to control, and the damage to the shrubs which may include valuable browse plants is severe. Little is left to protect the soil. Following the fire the grass may flush which provides a valuable bite before the rains start.
3. *Early rains burn*. As the shrubs and woody vegetation shoot just before the rains start, a fire at this time will set them back by destroying this new soft growth, and as the grasses have not quite started to grow they are undamaged and benefit from the reduced competition from the trees and woody shrubs, provided the grass is not grazed too early after the fire.

Fire can therefore be used to encourage either trees and shrubs or grasses as desired. Exclusion of fire leads to senescence in grasses, stimulation of woody growth, and the fairly rapid development of closed canopy woodland or deciduous forest (Rose Innes 1977). Conversely it was shown in Northern Nigeria that by burning early in the rainy season, uncleared tree savanna could be converted to grassland in 4 or 5 years (Blair Rains 1963). Such burning can reduce the prevalence of tsetse flies (*Glossina morsitans*) which is in dry upland tree savanna.

Unfortunately in addition to cattle owners, hunters and shifting cultivators also use fire and their interests in the season to burn may conflict with the interests of cattle owners.

In dry areas, such as the northern parts of Kenya and Australia, grazing should only be fired every other year. Cattle owners have sound ideas about the practical effects of burning, the difficulty is that with the existing system of land tenure, particularly in Africa, it is difficult to get general agreement and common action. In many areas all the land of the tribe is open grazing for the members of the tribe, or a particular clan or section of the tribe, indeed access to grazing is generally more open then access to water. Seldom can an individual in tropical Africa claim an exclusive right to a particular area of grazing.

Provided there is some cooperation between the cattle owners and herdsmen, natural grazings can be maintained and improved by agreement to rest certain areas under a system of rotational grazing, agreement on burning, the use of swamps for dry season grazing, the closing of certain over-used cattle tracks that are causing gulley erosion, and the chopping out of undesirable shrubs which alone can double the carrying capacity of a grazing area. An agreement on stocking rates is usually impractical to enforce. The construction of dams or boreholes to increase the number of watering points will reduce the overgrazing around the existing water supplies, but it also tends to encourage the increase in livestock numbers feeding on the same grazing. Improved watering facilities rapidly improve the condition of cattle.

The quality of the natural pastures can be increased by sowing seeds of legumes which tolerate grazing and burning and are adapted to the area. Stylo (*Stylosanthes guianensis syn. S. gracilis*) has been used successfully in many areas. The soil surface is broken before seeding, and once established in an area it soon spreads, particularly if phosphate is not deficient. Large areas of grazing in Queensland have been improved by seeding Stylo in the natural grasses (Tothill and Peart 1972). However it is much easier to introduce a grass into a legume sward than a legume into a grass sward. Similarly the planting of leguminous browse plants will increase the value of grazings. These young shrubs must be protected from grazing until they are established. This can be done by planting the shrubs in a thicket, or surrounding them with cut useless scrub.

Fencing

As enclosures were the start of agricultural improvement in England, and the improvement of ranching in the USA started with the introduction of barbed wire in 1874, fencing is the key to major pasture improvement and is essential to efficient tick control and controlled grazing.

No effective live hedges, that are cattle proof, have been devised for the tropics. Such live hedges need to be fire proof where the pastures are burned regularly.

Fencing is unfortunately very expensive by peasant standards, but where there is a good economic incentive to improve production and loans for agricultural development are available, small paddocks of about 2 ha can be enclosed and improved. Ordinary fence posts have a short life in the tropics due to termites and basal rotting. Hardwood posts such as teak or greenheart can be used if available, or the ordinary timber posts pressure treated with a pentachlorphenol/insecticide wood preservative. Live fencing posts, possible browse shrubs, or concrete posts are alternatives. The top wire of the fence should be plain wire and at least one of the middle ones barbed to prevent the cattle pushing through. When stapled to posts, the wire should pass on the same side of the post as the cattle otherwise the wire may be pushed off the post. Electric fences have not been very successful in the tropics [see Bishop 1979].

Sown pastures

Grass is the cheapest of all cattle feed and its management should be directed to providing high quality feed throughout the year. In the humid tropics, sown pastures, adequately fertilized and competently managed, can carry five animals per hectare throughout the year and elephant grass cut for forage ten animals per hectare. The technique has been developed in Puerto Rico (Vicente-Chandler *et al.* 1964) but could be applied throughout the humid tropics where cattle can be kept. This is a far cry from the extensive cattle raising systems of South America, Northern Australia, and much of Africa where 10–20 or more hectares may support only a single animal. In many cases lack of rainfall or drinking water is the limiting factor, but there are still large areas where the stocking rate could be considerably increased if economically justified, as in many Caribbean islands, in prosperous oil countries like Venezuela, and around urban areas in countries such as Nigeria. As yet, however, the use of sown grass, or grass/legume pastures in the tropics is of only minor importance even in Queensland.

Re-seeding

The best results are achieved when the original pasture is destroyed and replaced by chosen species of grass and legumes. As the fertility of these areas is low and largely in the top few centimetres, ploughing by burying the topsoil means the new pasture has to grow on very poor subsoil. It is therefore much better to rip out the old pasture with a tine cultivator which opens up the subsoil, or if this is not possible disc the old pasture. A summary of the characteristics of the more important tropical pasture grasses is given in Table 5.2 (see also Bogdan (1977) and Skerman (1977).) As wide variations occur in response to local conditions, not only between species of tropical grasses but also between strains (cultivars) of the same species, it is necessary to use the indications from local trials carried out with a wide range of introductions.

The feeding value of grasses falls off with growth, the percentage dry matter increasing but the protein content on a dry matter basis decreasing, often fairly rapidly as is shown (Table 5.3) with a Pangola grass (*Digitaria decumbens*) pasture in Trinidad.
There are large differences in feeding value between species, the leafy grasses being higher in protein than a stemmy grass such as *Chloris gayana*. This change is minor with a pasture legume where the fibre content hardly increases and the protein content remains high. This is fortunate as grazing to perfection is impossible. Pastures get out of hand in the wet season and make inadequate growth in the dry season.

A good pasture should have the following qualities:

1. Ease of establishment.
2. Harmony with soil and climatic conditions.
3. Resistance of the desirable species to grazing.
4. Even production throughout the year.
5. Even feeding value through the year.
6. Resistance to fire if necessary.
7. Ability to crowd out undersirable species.

Provided that rainfall is adequate and there are no trace element or phosphate deficiencies, high rates of nitrogen, 400 kg/ha or more, applied to quality grasses such as Pangola, Kikuyu or Rhodes (*Chloris gayana*), give high stocking rates and high production by increasing the dry matter and protein yield of the sward (Olsen 1972).

However, with the high cost of nitrogen fertilizer, mixed grass/legume pastures will become more important. Not only can legumes replace purchased nitrogen, but their nutritional value is higher and they are used more efficiently than grass of the same digestibility rating. Legumes reduce the decline in a digestibility of a pasture and remove the protein limitation over much of the year. In mixed farming systems the following crop benefits; grain yields are significantly higher after a legume-based pasture than following a grass sward.

The International Centre for Tropical Agriculture in Colombia

Table 5.2 Some of the more important pasture grasses in the tropics

Tribe	Species	Common name	Natural habitat	Distribution[a]	Habit of growth
Andropogoneae	*Andropogon gayanus* Kunth	Gamba grass	Wide range of soil types; long dry season.	*Tropical Africa,* Brazil, Queensland.	Erect tufted to 1.8 m.
	Bothriochloa insculpta (A. Rich.) A. Camus	Sweet-pitted grass.	Low to medium rainfall; 900 to 1,800 m	*East Africa*, South Africa.	Tufted to 1 m weakly stoloniferous; turf-forming when grazed.
	Dichanthium caricosum (L.) A. Camus	Nadi blue grass.	Moderate summer rainfall. See level to 900 m	*India, Burma, S.E. Asia,* Fiji, West Indies, Hawaii.	Turf-forming; stoloniferous culms to 0.6 m.
	Hyparrhenia rufa (Nees) Stapf.	Jaragua grass.	Medium to heavy rainfall with moderate dry season; well-drained heavy soil. Sea level to 1,500 m	*Tropical Africa,* Brazil, Central Americas.	Tussock-forming with flowering culms to 3 m.
	Ischaemum indicum (Houtt.) Merril (= I. *aristatum*, L.).	Batiki blue-grass (Fiji); Toco grass (Trinidad)	Medium to heavy, fairly well-distributed rainfall; low medium altitudes.	*India, S.E. Asia*, W. tropical Africa, Polynesia, West Indies.	Turf-forming weakly stoloniferous culms to 1 m.
	Sorghum sudanense (Piper) Stapf	Sudan grass.	Moderate summer rainfall; wide range of soils.	*Sudan*, USA, S. Africa, Australia, S. America.	Tufted erect, heavy tillering; annual to 3.5 m.

[a] Indigenous area are in italics.

Palatability	Nutritive value	Persistence	Planting method	Yield[b] (tonnes)	Remarks
Good when young, but flowering culms unpalatable.	Moderate	Perennial, with good grazing qualities.	Normally from seed but supply limited.	30	Very drought-resistant. Favoured in N. Nigeria and N. Ghana. Used for hay.
Good	Moderate	Perennial under close grazing.	Seed; supply limited.	10-20	Promising for leys at medium altitude in E. Africa.
Foliage moderate, culms avoided.	Fair. Good as hay.	Perennial under close grazing.	Root divisions and self-sown seed.	Up to 10	Particularly suitable for sheep in West Indies. Favoured in Fiji.
Moderate for young foliage only.	Fair	Perennial under close grazing.	Root divisions and self-sown seed.	30	Regular mowing required to control flowering during short days.
Good	Fair	Perennial under close grazing.	Root divisons; limited amount of self-sown seed.	10	Important for milk production in Fiji. Several strains available.
Good; sugar content high.	Good, but some danger of prussic acid poisoning when very young.	For short-term leys only.	From seed.	20-40	Subject to breeding and selection in USA. Suitable for irrigated areas on fringe of tropics. Used for hay, silage and grazing.

[b] Figures are expressed as weight of fresh material in t/ha per year; and are intended for comparison only; yield being a complex function of soil type, rainfall and management.

Table 5.2 Some of the more important pasture grasses in the tropics (continued)

Tribe	Species	Common name	Natural habitat	Distribution[a]	Habit of growth
	Sorghum almum Parodi	Columbus grass.	Moderate summer rainfall; wide range of soils; sub-tropics.	*Argentina*, S. Africa, Austrália	Tufted, erect to 3.5 m short thick rhizomes.
	Themeda triandra Forsk.	Red oat-grass.	Drier, low fertility areas.	*Tropical Africa*, S. Africa.	Tufted to 1 m weakly rhizomatous.
Chlorideae	*Chloris gayana* Kunth	Rhodes grass.	Over 600 m Up to 1,250 mm summer rainfall.	*S. and E. Africa*, California, Queensland, India, S.E. Asia.	Stoloniferous, turf-forming; erect to 1 m
	Cynodon dactylon (L.) Pers.	Bermuda grass, Bahama grass, Star grass, Doob grass, Giant Star grass.	Sea level to 1,500 m Dry to medium rainfall. Prefers a high pH.	Pantropic	Rhizomatous and stoloniferous; open sward to 30 cm to 1 m
Eragrosteae	*Eragrostis curvula* Nees (Schrad.)	Weeping love grass.	Semi-arid tropics and sub-tropics; summer rainfall.	*Zimbabwe S. Africa*, USA, Brazil, Australia.	Tufted; long spreading narrow leaves to 1 m

[a] Indigenous areas are in italics.

Palatability	Nutritive value	Persistence	Planting method	Yield[(b)] (tonnes)	Remarks
Good	Good, but some danger of prussic acid poisoning.	Perennial under controlled grazing.	From seed which is readily available.	20-40	Increasing importance in sub-tropics in recent years as a multipurpose grass; very drought-resistant.
Moderate	Fair	Maintained as constituent of natural grassland by annual burning; discouraged by overgrazing.	Self-sown seed.	20	A useful grass under poor conditions
Very good even when mature.	Moderate; leaf/stem ratio falls rapidly with age age.	Annual to perennial according to variety. Withstands grazing well.	Normally by seed which is available commercially.	30-50	Of major importance in Central Africa. Named varieties include Mpwapwa, Nzoia, Katambora.
Good when young.	Fair. Great range in leaf/stem ratio according to selections.	Perennial and withstands heavy grazing.	Seed available for lawns; sprigs and runners for pasture.	Up to 10 for common strain. 10-30 for improved selections.	An extremely variable species. High yielding, non-rhizomatous strains exist. Nomenclature confused with *C. plectostachys*
Poor to fair.	Low	Perennial, stands grazing; can regenerate from self-sown seed.	From seed, commercially available in S. Africa.	10-20	Very drought-resistant. Suitable for range improvement.

(b) Figures are expressed as weight of fresh material in t/ha per year; and are intended for comparison only; yield being a complex function of soil type, rainfall and management.

Table 5.2 Some of the more important pasture grasses in the tropics (continued)

Tribe	Species	Common name	Natural habitat	Distribution[a]	Habit of growth
Paniceae	*Acroceras macrum* Stapf	Nile grass.	Medium rainfall, poor drainage.	*South Central S. Africa*, East and South to. Transvaal.	Erect, somewhat sparse stand from slender rhizomes.
	Axonopus compressus (Swartz) Beauv.	Savanna grass Carpet grass.	Well distributed medium to heavy rainfall; alluvial soils. Sea level to 600 m.	*Caribbean and Central America*: probably now Pantropical.	Short, dense turf to 4.5 cm rapidly spreading stolons.
	Brachiaria brizantha (Hochst. ex A. Rich.) Stapf	Palisade grass, Signal grass.	Humid savanna.	*Central and South Africa*, Sri Lanka Queensland.	Weakly rhizomatous, straggling decumbent stems up to 1.5 m
	Brachiaria decumbens Stapf	Surinam grass (in Jamaica). Signal grass	Humid tropics, 4-5 months dry season; tolerates shallow calcaerous soil.	Tropical Africa, Queensland, Hawaii, Venezuela, W. Indies.	Trailing perennial, stolons rooting at nodes. Stiff, erect leaves.
	Brachiaria mutica (Forsk.) Stapf	Para grass.	Sea level to 900 m humid poorly drained areas.	*Tropical Africa and S. America* now widespread throughout low altitude tropics. Probably introduced into Brazil	Trailing, decumbent stems, to 2 m

[a] Indigenous areas are in italics.

Palatability	Nutritive value	Persistence	Planting method	Yield[b] (tonnes)	Remarks
Good	Fairly high.	Withstands moderate grazing under moist conditions.	Root divisions or rhizomes	20-30	No drought resistance. Dual purpose hay or grazing. Difficult to eradicate.
Good when young.	Fairly high when young.	Withstands heavy grazing in rainy season. No growth during drought.	Root divisions.	10-20	Shallow rooted. Tolerates shade, e.g. leguminous trees.
Fair when young.	Moderate when young.	Perennial, stands controlled grazing well.	Root divisions.	30-50	The major planted pasture grass in Ceylon (milk production).
Good at all stages in wet season.	Good, slightly less than Pangola.	Good when rotationally grazed, good recovery.	Root divisions, stem cuttings Limited seed production.	20-40	One of the best pasture in Congo. Possible alternative to Pangola in Jamaica particularly for beef.
Apical growth very good, rapidly reducing in lower stems.	Moderately high when young.	Perennial, but requires careful management for grazing.	Vegetatively from root divisions or lower stem lengths.	40-50	More commonly used used as soilage. Little drought resistance. Promising for legume mixtures.

[b] Figures are expressed as weight of fresh material in t/ha per year; and are intended for comparison only; yield being a complex function of soil type, rainfall and management.

Table 5.2 Some of the more important pasture grasses in the tropics (continued)

Tribe	Species	Common name	Natural habitat	Distribution[a]	Habit of growth
Paniceae	*Cenchrus ciliaris* L.	African foxtail, Buffel grass.	Light soil low seasonal rainfall.	*South-central Africa*, North-East Central Africa, India, Queensland.	Tufted to decumbent; to 1 m weakly rhizomatous
	Digitaria decumbens Stent	Pangola grass.	Dry, alluvial sub-tropical areas to moist, well-drained full tropics. Sea level to 1,500 m.	*S. Africa*, Caribbean and Central America, Hawaii.	Dense trailing mat to 0.6 m vigorously stoloniferous
	Echinochloa pyramidalis (Lam.) Hitchc. and Chase	Antelope grass.	Heavy soil, seasonal swamp, low to medium altitude.	*Tropical Africa*.	Tall tufted robust to 4–5 m rhizomatous.
	Eriochloa punctata (L.) Desv. ex Hamilt	Carib grass, Malojilla grass.	Sea level to 900 m moist medium-heavy soil.	*Tropical South and Central America*, Brazil, Florida, Caribbean.	Fine stemmed: trailing from decumbent stems to 1.2 m
	Melinis minutiflora Beauv.	Molasses grass	Well drained soil, moderate rainfall. Sea level to 2,000 m	*Tropical Africa*, South America. Sri Lanka, Philippines, Queensland.	Fairly dense cover of trailing to decumbent stems to 1.2 m

[a] Indigenous areas are in italics.

Palatability	Nutrative value	Persistence	Planting method	Yield[b] tonnes	Remarks
Fairly good young.	Moderate when young.	A good grazing species.	Mostly by seed.	30-50	Many ecotypes available. Valuable species in drier areas for ley or permanent pasture and for legume mixtures.
Very good for considerable period of growth.	Generally high when young, falling somewhat at maturity.	Good when well managed; true perennial.	Root divisions or stem lengths disced-in.	20-40	An outstanding grazing species in humid Central America. Liable to virus decline.
Good	Fair	Good when well managed.	Root divisions; limited amount of viable seed produced.	Over 50	Suitable for hay or soilage under irrigation. Uganda and Transvaal. Good regeneration after burning.
Good	Good	Short-term perennial; requires good management. Not drought-resistant.	Root divisions disced-in stem lengths.	10-20	Best managed as for silage; restricted usage.
Very good when cattle are accustomed to it.	Good	Perennial when well Not very fire-resistant but fairly drought-resistant.	Seed, which is available commercially.	30-50	Characterized by leaf odour. Promising for legume mixtures.

[b] Figures are expressed as weight of fresh material in t/ha per year; and are intended for comparison only; yield being a complex function of soil type, rainfall and management.

Table 5.2 Some of the more important pasture grasses in the tropics (continued)

Tribe	Species	Common name	Natural habitat	Distribution[a]	Habit of growth
	Panicum antidotale Retz.	Blue panic grass.	Low altitude, light soils, low-medium summer rainfall.	*N.W. India*, Queensland, Southern U.S.A.	Densely tufted to 1.8 m rhizomatous.
	Panicum coloratum L. var. *makarikariense* (Goossens) Van Rensb.	Makarikari grass. Coloured guinea grass.	Low summer rainfall; deep alluvial soil.	*Tropical South Africa*. Central and East Africa.	Variable; erect to spreading from decumbent stolons.
	Panicum maximum Jacq.	Guinea grass.	Low to medium rainfall; well drained soils. Sea level to 1,200 m	*Africa*, wide-spread in humid tropics.	Tufted to 3.5–4 m
Paniceae	*Paspalum dilatatum* Poir.	Dallis grass.	Humid sub-tropics; heavier soils; low altitude.	*Tropical and sub-tropical, S. America* S.E. United States, Australasia. S. Africa.	Strongly tufted, deeply rooted, slow lateral spread; culms to 1.5 m
	Paspalum notatum Flügge	Babia grass	Light well-drained medium to high rainfall. Sea level to 1,500 m	*Mexico to Argentina*, East and West Africa, Florida.	Low growing, stiff dense cover from short shallow rhizomes; up to 0.5 m
	Pennisetum clandestinum Hochst. ex Chiov.	Kikuyu grass.	Fertile well-drained soil, 1,500–3,000 m in tropics and low altitude sub-tropics; medium rain-fall.	*East and Central Africa,* California, Brazil, Australasia, S. Africa, Hawaii.	Low growing, deep rooting, rhizomatous and stoloniferous.

(a) Indigenous area are in italics.

Palatability	Nutritive value	Persistence	Planting method	Yield[b] (tonnes)	Remarks
Fair, only when young.	Fair only.	Very persistent and drought-resistant. resistant.	Rhizomes or seed in Queensland.	50	Rapid growth in rains or under irrigation, soon developing woody culms.
Good	Good	Persistent and drought-resistant.	Stem cuttings or seed, quality of which varies with the ecotype.	20-40	Selection and improvement current in S. Africa and Zimbabwe
Good, particularly when young.	Good but falls rapidly with age.	Perennial, but requires rotational grazing for long-term production.	Root divisons or encouragement of self-seeding. Seed of some ecotypes available commercially.	Over 50	One of the best tropical fodder grasses. Very variable species.
Very good.	Moderately good.	Very persistent under grazing.	Commercial seed available but establishment slow.	20-30	Used for grazing or hay in maritime sub-tropics. Associates well with sub-tropical legumes.
Rather low, improved by breeding.	Moderately good.	Very persistent under grazing and in time of drought.	Root divisions for soil conservation work; seed of improved varieties for pasturage.	10-20	Slow-growing and low-yielding. Variety Pensacola used for beef production in USA.
Good	Good	Very persistent; excellent for permanent pasture; difficult eradicate.	Vegetatively from cuttings or runners.	10-20	Associates well with white clover. Excellent drought-resistance. Weed hazard.

(b) Figures are expressed as weight of fresh material in t/ha per year; and are intended for comparison only; yield being a complex function of soil type, rainfall and management.

Table 5.2 Some of the more important pasture grasses in the tropics[(a)] (continued)

Tribe	Species	Common name	Natural habitat	Distribution[(a)]	Habit of growth
	Pennisetum purpureum Schumach.	Elephant grass, Napier grass. Merker, grass.	Medium rainfall; heavier well-drained soil. Sea level to 1,500 m	*Tropical Africa*, wide-spread throughout the tropics.	Giant tufted grass up to 6 m; limited spread from short rhizomes; deep-rooted.
	Setaria sphacelata (Schumach). Stapf and Hubbard. [S. anceps]	Golden timothy grass.	Low to medium rainfall. Sea level to 2,000 m	*Tropical and South Africa*. Australasia	Tufted to decumbent and stoloniferous; very variable.

(a) Indigenous area are in italics.

Source: Commonwealth Bureau of Pasture and Field Crops.

For further details see Bogdan (1977).

Table 5.3 Decline in food value of Pangola grass with age

Age in days	10	15	21	42
% Dry matter	14.8	20.3	21.4	21.1
% Constituents on a dry matter basis				
Crude protein	14.9	13.8	9.2	4.8
Crude fibre	31.0	29.6	35.3	36.3
Ash	11.4	10.9	12.2	6.9

Source: Butterworth (1961).

(CIAT) have been studying the problems of the 850 million ha of arid, phosphate deficient, aluminium soils. Much of the area (40 per cent) is well drained savanna where beef production is limited by the lack of feed in the dry season; 25 per cent of the area is at the other extreme. The land is poorly drained and flooding restricts wet season grazing. In the worst conditions of extreme acidity the best legumes are species of *Stylosanthes, Desmodium* and *Zornia* and the most promising grass Gamba grass - *Andropogon gayanus*. Under better conditions they are trying a *Centrosema* spp. One of their improved pastures is carrying 2.5 beef steers per hectare with a daily gain of over half a kilogram. The loss of weight in the dry season was prevented by feeding urea and

Palatability	Nutritive value	Persistence	Planting method	Yield[b] (tonnes)	Remarks
Very good when young and leafy.	Good	Up to 15 years under controlled cutting; 3-5 years rotational grazing.	Stem cuttings, or whole lengths buried in furrow.	200 (as soilage).	Probably the highest-yielding grass known under soilage management. Borderline for grazing.
Good	Good	Perennial; longevity reduced by grazing.	Seed readily available and establishment fairly rapid.	40-50	Multipurpose hay, silage, soilage or grazing. Varieties: Nandi, Kazangula.

(b) Figures are expressed as weight of fresh material in t/ha per year; and are intended for comparison only; yield being a complex function of soil type, rainfall and management.

cassava with a little molasses. In a comparison of natural savanna grazing with a molasses grass (*Melinis minutiflora*) pasture the cattle on the molasses grass ate 70 per cent more of the mineral supplement possibly because the burning of the savanna provides the minerals. The minerals helped the body weight increase, improved the calving rate - 29 per cent increase on native savanna grazing, and there was a 45 per cent reduction in pre-weaning mortality largely due to the phosphorus.

Management of sown pastures

Having sown a pasture it is essential to maintain it by rotational grazing with adequate rest periods, application of fertilizer, particularly nitrogen if there are no legumes, mowing or brush cutting, and spraying out the undesirable weeds. The possibilities of production from pastures in the humid tropics is best illustrated by Pangola grass which has been studied probably more than any other grass in the tropics.

Pangola grass pastures (Digitaria decumbens)* This low growing leafy grass was found originally in the Pongola river valley in the Eastern Transvaal (South Africa) and introduced into the USA in 1935. In the

* See Bogdan (1977) for fuller details.

1940s it was released in Florida, became very important in Puerto Rico from where it spread through the Caribbean, South America, Asia, Australia and back to Africa. It is palatable to all domestic livestock, stands heavy grazing and grows well over a wide range of rainfall and soil conditions. Pangola crowds out less valuable grasses and weeds, though some can survive, and is tolerant to drought. It responds well to nitrogen fertilizer and irrigation. As the seed is infertile it must be propagated vegetatively. If the runners are spread on the surface and immediately disced in, this will give a good stand provided that rain falls soon after discing. If the land is flat planting should be on broad based cambered beds for drainage as Pangola dislikes waterlogging.

In Australia, this grass makes little growth when maximum temperatures are below 24 °C, and average temperatures under 16 °C, growth resumes at 27 °C. In the warmer months in Queensland production has reached 190 kg/ha of dry matter a day and nearly 1,270 kg/ha of dry matter in January (mid-summer) with 560 kg of nitrogen applied per hectare per year (Anon 1966a). The response of Pangola in these trials to nitrogen was up to 336 kg/ha, 71 kg extra dry matter for each kilogram of nitrogen. At 560 kg nitrogen the increase was still 31 kg dry matter for each kilogram of nitrogen with diminishing response up to 1,800 kg/ha.

About 45 per cent of the nitrogen applied as fertilizer finds its way into the herbage, the return decreasing with higher rates and more frequent cutting. Less frequent cutting resulted in higher annual yields of dry matter and plant nitrogen, at all rates of nitrogen application. Similar large responses in yield and protein content of Pangola to nitrogen are reported from Mexico, Costa Rica, Columbia, Taiwan and British Honduras.

If all this dry matter could be utilized it should in theory support 7.5 beasts per hectare, compared with 2.5 animals per hectare for grass legume pastures.

Properly managed improved pasture, such as Pangola grass in Trinidad with 1,270 mm of rain per year, is said to maintain 2.5 cows per hectare in good condition and produce 9 litres of milk per cow. To achieve this requires adequate fertilizer, brush cutting, and spraying to control weeds and ticks. This productivity has not been possible with pedigree Holsteins at Los Bajos in South Trinidad with excellent Pangola grass rotationally grazed (Plate 5.6). Starch rather than protein is thought to be deficient.

Other legumes have been grown with Pangola including Stylo, Townsville stylo, Siratro, *Pueraria, Calopogonium, Centrosema*, Alyce clover and species of *Desmodium*, though they seldom survived more than 2 or 3 years even when nitrogen was not applied to the Pangola.

In Trinidad, stocking at the rate of 5 animals per hectare of Pangola

Plate 5.6 Pangola pasture on cambered beds. Six months previously area was old cocoa. Some of the shade trees, including Immortelles, remain. Los Bajos, Trinidad. (Courtesy BP Ltd.)

showed a liveweight gain of 2,783 kg/ha in 2 years, but with 7.5 animals per hectare the gain was only 988 kg/ha in 2 years. Grazing for 6 days or less and resting for 18 days is recommended (Wilson and Osbourn 1963). The digestible crude protein falls off between 2 and 3 weeks of new growth and tends to be lower in the dry season. The response to fertilizer is such that the highest rate used, 270 kg/ha, gave the greatest net financial return. In Puerto Rico grazing heavily fertilized Pangola grass to around 15 cm from the ground every 2 weeks in the wet season and every 3 weeks in the dry season, when growth is slow, is advised.

Good Pangola grass pasture in a good rainfall area of Jamaica will give 780 kg liveweight gain for fattening steers, 450 kg liveweight gain in a cow/calf enterprise, or 3,570 litres of milk per hectare. A small dairy farm with 10 ha of pasture can support 18 cows in milk.

Where stunting virus of Pangola is serious, *Cynodon nlemfuensis* (Star grass) and *Brachiaria decumbens* are being studied as replacements for Pangola.

Legumes for sown pastures*

Norris (1966) says 'Now the world can no longer afford the luxury of blindness to one of its greatest potentials, the fixing of nitrogen by legumes in its vast tropical lands.' He considers that the ingrained association of 'lime' with 'legume' (clovers, medics, peas and vetches) from temperate regions, supplemented in the tropics by the lime hungry species *Phaseolus vulgaris* has caused the erroneous conclusion that liming is necessary for all legumes, and must therefore make the successful cultivation of legumes in the mainly acid soils of the tropics hardly worth while. Norris points out that legumes are tropical in origin, and tropical in their principal distribution at the present day. The leguminous tree of the rain forest has a low nutrient demand and at the other extreme the clover has a high nutrient demand. Between these extremes are hundreds of species in the genera *Desmodium* or *Phaseolus* that can derive their nutrients and nodulate on poor acid soils, and from these it should be possible to derive many useful species for grass legume mixtures.

The place of legumes in tropical grazing is likely to be decided by economics. In the high cost high return farming system in Puerto Rico legumes were only advised for use with molasses grass (*Melinis minutiflora*) on the poorer hilly areas. On the better land, heavily fertilized Pangola grass supplies a greater quantity of protein per hectare than the available legumes. Without fertilizer the grass yields less protein than legumes and the fibre content of the grass is higher. The more important pasture legumes are summarized in Table 5.4. In Australia, legumes are an integral part of the improved pastures of Queensland. In the wet tropical areas of Queensland, *Centrosema pubescens* or *Stylosanthes guianensis* is grown with Para or Pangola grass. At Coolum Research Station in Queensland, Pangola with *Lotononis bainesii* and *Trifolium repens* has been successfully maintained as pasture for about 5 years, though Pangola is aggressive and crowds out legumes.

Horrel (1958), after examining over 50 legume species grown without fertilizer at Serere in the short grass area of Uganda, suggested that *Stylosanthes guianensis, Calopogonium orthocarpum, Centrosema pubescens, Desmodium* spp., *Pueraria phaseoloides* and *Glycine javanica* = *G wightii* were the most promising species for a mixed grass and legume grazing. The genus *Desmodium* which has over 300 species he considered worthy of further exploration. Many of these legumes have been mentioned earlier as soil covers for perennial crops. The *Trifolium* species were all disappointing but apparently *T. semipilosum* and *T. repens* (Louisiana white clover) are growing satisfactorily in the

* See Skerman (1977) for a detailed review.

long grass area where the dry season is shorter and the temperature lower.

The characteristics of the main legumes suitable for grazing in the tropics are given in Table 5.4. While there is a good choice of legumes suitable for pastures in the humid tropics, there are none suitable for areas with less than 550 mm of annual rainfall. Caribbean Stylo, *Stylosanthes hamata* cv. Verano as the Spanish name Verano implies, can stand a harsh tropical dry season and may prove useful in areas with a short growing season (Anon. 1973). The range of cultivars is continually increasing, for example Harding and Cameron (1972) describe four new ones showing promise in Queensland. *Centrosema pubescens* cv. Belalto combines well with *Panicum maximum*, and two cultivars of *Stylosanthes guyanensis* cvs Cook and Endeavour and *Desmodium heterophyllum* cv. hetero can stand heavy grazing when grown with stoloniferous grasses such as Pangola and *Brachiaria decumbens*.

Where sufficient water is available and suitable calcareous soil conditions occur, alfalfa (*Medicago sativa*) will outyield most other legumes. In soils new to lucerne, the seed must be inoculated. The useful life of a lucerne field can be extended by chemical weed control. Previously the weeds tended to become dominant in the second year.

Rhizobium

Thus the use of tropical legumes in pastures depends upon the correct choice of legume for the particular environment, associated grasses and the system of pasture management. The legume must have an efficient *Rhizobium* strain and adequate fertility, particularly phosphate for nodulation (Jones 1972).

Tropical legumes will generally nodulate without inoculation but the amount of nitrogen fixed by the legume may be considerably greater if inoculated with the correct strain of *Rhizobium*, which should be obtained from a reliable source.

The *Rhizobium* bacteria which fix the nitrogen in the nodules of the legumes may produce an alkaline or an acid reaction when grown in a culture. The alkali-producing types are associated with tropical legumes on acid soil and the acid-producing *Rhizobium* strains with the legumes of fertile soils such as lucerne, clover, peas.

Associated everywhere with tropical legumes is the 'cowpea type' of *Rhizobium* which cross-inoculates from one host species to another, and from one genus, tribe or family to another. This type of *Rhizobium* is very common in the soil and may be very effective with sown legumes. In contrast the *Rhizobium* associated with certain legumes is highly specific and will not cross-inoculate except to closely related species and may even be specific to a cultivar of one species.

Uninoculated seed planted in a new environment may be compatible

Table 5.4 Some of the more important pasture legumes in the tropics[a]

Species	Common name(s)	Habit	Ecological requirements
Alysicarpus vaginalis (L.) D.C.	Alyce clover.	Laterally branching, erect to 1 m Perennial or self-sowing annual.	Fertile, well-drained soils. Tolerates dry seasons.
Cajanus cajan (L.) Millsp.	Pigeon pea.	Short-lived perennial branching shrub.	Light to medium well-drained soils.
Calopogonium mucunoides Desv.	Calopo	Poorly-branching, hairy climber. Annual to short-lived perennial.	High rainfall, wide range of soils. Survives dry season by profuse seeding.
Calopogonium orthocarpum Urb.	Calopo (E. Africa).	Trailing perennial rooting at nodes.	More drought-resistant than *C. mucunoides*.
Centrosema pubescens Benth.	Centro	Trailing perennial.	Wide range of soils. Drought-resistant. Tolerates poor drainage and short-term flooding. At least 1,000 mm rainfall.
Desmodium canum (Gmel.) Schinz and Thell.	Kaimi clover.	Perennial, stems erect and prostrate, dimorphic leaves.	Humid climate, well-drained, acid soils.
Desmodium intortum (Mill.) Urb.	Kuru vine Greenleaf (Australia) Intortum (Hawaii)	Coarse, spreading to trailing perennial, rooting at nodes.	Wide range of soils. Responsive to P and K. Tolerates water-logging and drought. At least 1,000 mm annual rainfall.

[a] Excludes shrubby species used only as browse.

Distribution	Admixture with grasses	Utilization	Palatability	Remarks
Tropical Asia, Southern U.S.A., Sudan, etc.	Temporary leys with *Brachiaria brizantha* in Sri Lanka	Hay or grazing.	Good as hay.	Pods shatter readily. Somewhat similar but less widespread is *A. rugosus*. 600,000 seeds/kg.
Tropical Africa, Asia, S. America, Australia, Hawaii, Fiji	In alternate rows, 1.2 m apart, with erect grasses, e.g. Guinea, Rhodes.	Grazing, hay.	Good.	Good dry-season fodder reserve, drought-resistant. Careful grazing management critical for persistence.
S. America. Widespread in tropics.	Good with *Melinis minutiflora* and *Chloris gayana*.	Grazed, in natural or sown mixtures. Good cover crop.	Low	Widely used as a green manure and cover crop in plantation agriculture. Indifferent fodder value. 70,000 seeds/kg
India, Uganda	Good with Guinea grass.	Grazed in mixtures.	Fair	Aggressive, profuse seeder.
S. America. Widespread in tropics.	Good with many species; preferably those of erect habit	Grazed in mixtures.	Moderate	One of the best and most widely adapted legumes for growing with grasses, particularly in Queensland. Slow to establish. 35,000 seeds/kg
West Indies, Hawaii.	Good addition to native range in Hawaii.	Grazing	Moderate	Persistent but low-yielding. Tends to become woody with age. Profuse seeder.
Central and South America, Hawaii, Queensland, Zimbabwe, Kenya.	Good with *Digitaria decumbens*, *Branchiaria mutica*, etc. if if well-managed	Grazed alone or in mixture; hay, soilage.	Good	One of the most promising grazing-type legumes investigated to date. Does not induce bloat. 280,000 seeds/kg

Table 5.4 (continued)

Species	**Common name(s)**	**Habit**	**Ecological requirements**
Desmodium uncinatum (Jacq.) D.C.	Spanish clover Silverleaf (Australia).	Trailing perennial.	Less tolerant of soil-moisture extremes than *D. intortum*.
Glycine wightii (R. Grah. ex Wight & Arn.) Verdc. syn. *Glycine javanica* (syn. Neonotonia wightii (Arn.) Lackey)	Glycine (Australia) Perennial soyabean.	Trailing, deep-rooted perennial.	Fertile, neutral to slightly acid well-drained soils. 760 - 1,750 mm.
Indigofera hirsuta L.	Hairy indigo.	Prostrate to erect annual.	Poor, sandy soils.
Indigofera spicata Forsk. (*I. endecaphylla*	Trailing indigo.	Prostrate, creeping perennial	Wide range of freely-draining soils. 1,250 - 2,500 mm
Lablab purpureus (L.) Sweet = *Lablab niger* Medic. (= *Dolichos lablab* L.)	Lablab Dolichos (Australia) Bonavist bean. Hyacinth bean.	Semi-erect, trailing stems, large leaves, deep-rooted, annual to biennial.	Humid climate, well-drained soils, preferably cultivated.
Leucaena leucocephala (Lam.) de Wit (= *L. glauca* (L.) Benth.)	Koa haole (Hawaii) Lamtoro.	Arborescent shrub or small tree if not maintained leafy by management.	Wide range of freely-draining soils. 750 - 1,500 mm rainfall. Low altitudes.

Distribution	Admixture with grasses	Utilization	Palatability	Remarks
Brazil, Hawaii, Queensland, Kenya.	Good with *Digitaria decumbens*, *Brachiaria mutica*, etc., if well-managed	Grazed alone or in mixture; hay, soilage.	Good	Very promising. Good recovery from grazing. 200,000 seeds/kg
Old-world tropics, Australia, S. America, USA.	Good with erect-habit grasses when well-managed. May be over-sown into *Hyparrhenia* grassland, Zambia.	Grazing, pure stand or mixture, silage.	Moderate to good.	Very promising tropical legume, good seed-setting but productivity rather low. Some tolerance of soil-salinity. Varieties: Tinaroo, Clarence, Cooper. 150,000 - 200,000 seeds/kg
Africa, S.E. Asia, S. USA, S. America.	Natural constituent of grasslands, or sown in pure stand.	Grazing, hay.	Low to moderate.	Productive but rapidly becomes coarse and unpalatable. 440,000 seeds/kg.
Pantropic	A good in-filler with erect tussock-grasses, e.g. *Paspalum plicatulum* in Queensland.	Grazing.	Good to moderate	Drought-resistant. Low production. Variable toxicity to livestock depending on ecotype. 440,000 seeds/kg
Tropics and Southern sub-tropics.	Good with soliage grasses.	Rotational grazing of variety Rongai.	Good	Provides fodder well into the dry season. 3,000 seeds/kg
Pantropic, Hawaii, Greensland.	In alternate rows, 1.2 apart, with erect grasses, e.g. Guinea, Rhodes.	Grazing (cattle only) or soilage.	Good	Very drought-resistant. High moisture content. Becomes woody unless topped frequently. 26,000 seeds/kg

Table 5.4 (continued)

Species	Common name(s)	Habit	Ecological requirements
Lotononis bainesii Bak.	Lotononis (Australia).	Creeping, twining, perennial, rooting at nodes, sward-forming.	Moist, sandy soils, tolerates water-logging.
Macroptilium atropurpureum (DC) Urb. (= *Phaseolus atropurpureus* Moc. and Sesse ex D.C.)	Siratro	Deep-rooting, creeping, stoloniferous perennial.	Wide range of well-drained soils, responds to improved fertility. 900 - 1,800 mm rainfall.
Macroptilium lathyroides (L.) Urb. (= *Phaseolus lathyroides* L.)	Phasey bean. Phasemy bean.	Erect with twining stems to 1.2 m annual to short-term perennial.	Heavy, well-drained soils, but fairly tolerant of water-logging. At least 750 mm. rainfall.
Pueraria phaseoloides Benth.	Tropical kudzu Puero.	Vigorous twining perennial forming dense mat.	Wide range of soils, some tolerance of water-logging. At least 1,250 m rainfall.
Stylosanthes guianensis (Aubl.) (S. gracilis Kunth.)	Stylo	Erect, shrubby perential, to 1.5 m	Tolerant of acid, poorly-drained soils. 900 to <1,250 mm rainfall.
Stylosanthes humilis HBK (= *S. sundaica* Taub.)	Townsville stylo	Annual to biennial, prostrate to ascending, to 750 mm. Self-regenerating from seed.	Free-draining light, poor soils. 625 - 1,250 mm rainfall.

Prepared by R.H. Forster.

Distribution	Admixture with grasses	Utilization	Palatability	Remarks
S Central Africa, Queensland.	Forms stable association with trailing and erect grasses, e.g. Pangola, Guinea.	Grazing	Very good.	The most palatable and frost-resistant tropical legume known. *Rhizobium* strain critical. Re-seeds freely. 3,500,000 seeds/kg
Central and South America, Queensland, etc.	Compatible with many grasses, e.g. Rhodes, Guinea, Pangola.	Grazing, hay.	Good	Persistent under grazing. The most important pasture plant bred in Australia. 80,000 seeds/kg
India, Queensland, Sudan, Zimbabwe	Good association with *Brachiaria mutica, Paspalum scrobiculatum.*	Grazing	Good	Productivity variable owing to pest and nematode attack. Self re-seeding. 120,000 seeds/kg
Indonesia, Puerto Rico, Pantropic.	Good with tall soiling grasses under high rainfall.	Grazing, or soilage.	Fair to good (according to hairiness).	Careful management necessary. Useful for rapid cover of newly opened land. 85,000 seeds/kg
Brazil, Africa, Australia.	Good association with Pangola and *Brachiaria decumbens.*	Grazing, hay.	Low to moderate.	Slow to establish. Persistent and drought-resistant Very tolerant of hormone-herbicides. Important in Central Africa. 350,000 seeds/kg
S. America, USA, Queensland.	Associates very well with Pangola in Florida, *Cenchrus ciliaris* in Queensland.	Grazing, good for hay.	Increasing with age.	Easily established from seed. An important constituent of pastures in dry areas. 450,000 seeds/kg.

Notes: The following are adapted to the sub-tropics and tend to behave as winter-annuals: *Trifolium alexandrinum* (Berseem), *T. resupinatum* (Persian clover), *T. subterraneum* (subterranean clover). Drought-tolerant varieties of *Medicago sativa* (lucerne) are available which persist under irrigation and soilage management. At high altitudes in the tropics, these and other temperate pasture legumes can be grown, e.g. *Trifolium pratense* in Colombia, *T. repens* (and *T. rueppellianum* and *T. semipilosum*) in Kenya. See also Skerman (1977).

with the *Rhizobium* in the soil or it may select a compatible strain slowly, and the legume only benefits the second year after planting. Or the legume may not find a compatible strain of *Rhizobium*.

Even with a legume that develops a symbiotic relationship with the common type of *Rhizobium*, it is better to inoculate the seed with a known effective strain before planting. If one is not available, a strain known to be effective with a closely related species should be used. Skerman (1977) lists the world sources of both legume seed and *Rhizobium* inoculant and gives instructions for pelleting the inoculum.

The more specialized the host, the less likely it is to accept a wide range of *Rhizobium* species. The shape of the nodule is characteristic of the host not the infecting *Rhizobium*. Work in Australia has shown that these bacteria are not calcium sensitive, and if there is sufficient calcium for the legume host there is sufficient for the bacteria. In acid tropical soils calcium may be deficient or its uptake antagonized by other ions (aluminium, manganese, or rarely potassium) but despite this tropical legumes can often extract adequate calcium from the soil to form nodules on poor acid tropical soils. Adequate phosphate status of the soil is essential for establishing and maintaining legumes in a pasture. A small quantity, grams per hectare, of molybdenum is also essential for the nodules to function.

Skerman (1977) records a case of the fungi which developed on the old roots of cleared scrub producing antibiotics which were toxic to the *Rhizobium* on inoculated legume seed, and the growth of the clover was stunted until the toxins disappeared. This could occur where trees and bush are cleared to plant tree crops with legume cover crops.

Pelleting of seed has been found very useful in Australia but Norris (1966) points out that calcium carbonate is only advisable for the acid-sensitive *Rhizobium* inoculants, to overcome both soil acidity, calcium deficiency and preserve the bacteria. The acid-tolerant species should be pelleted in ground rock phosphate. Norris (1966) gives the division of genera and species as shown in Table 5.5.

Supplementary feeding

Grass with legumes is by far the cheapest food for cattle and it is usually better to spend money on improving pastures, if practicable, than use supplementary feeds. Where grazing is communal or not available or production is greater than the pasture can support, it may be necessary to grow or purchase supplementary feed. Table 5.6 gives an indication of the food value of available supplements which can be used.

As there are few areas of the tropics where grain can be spared for animal feed, supplementary feeding should be based on materials that cannot be eaten by man. This may include cassava reserves that have become woody and are harvested to free the land for other crops. Citrus

Table 5.5 Guide to pelleting legumes

Genera and species with acid-producing ***Rhizobium*****.** Pellet with calcium carbonate	**Genera and species with alkali-producing** ***Rhizobium*****.** Pellet with rock phosphate
Adesmia	*Acacia*
Astragalus	*Alysicarpus*
Cicer	*Arachis*
Hedysarum	*Cajanus*
Lathyrus	*Centrosema*
Lens	*Clitoria*
Leucaena leucocephala syn. *L. glauca*	*Crotalaria*
Lotus corniculatus	*Cyamopsis*
Medicago	*Desmodium*
Melilotus	*Flemingia*
Phaseolus coccineus	*Glycine*
Phaseolus vulgaris	*Indigofera*
Pisum	*Lespedeza*
Psoralea	*Lotononis*
Sesbania	*Lotus uliginosus*
Trifolium	*Lupinus*
Trigonella	*Phaseolus* (other than *P. vulgaris* and *P. coccineus*)
Vicia	*Pueraria*
	Stizolobium
	Vigna

Notes: The genera and species on the right-hand side are the ones likely to provide useful species for the tropics.

Norris (1966) also points out that if legumes are sown at the onset of the rains, the rapid nitrification (see p. 26) may inhibit nodulation.

or banana waste and other by-products may be useful supplements.

Many of the world's oilseeds (cottonseed, groundnut, sim-sim, coconut, oil palm, etc.) are produced in the tropics. In the past, these crops were exported unprocessed and extracted in Europe or America where the oil was utilized and the residue fed to cattle. With the increasing industralization of the tropics, extraction is increasingly carried out in the producing countries. At the moment, the cake is exported but as the demand for animal products increases, it may become economic for cattle owners to feed cake. Starchy foods are also available as maize, rice bran, sweet potatoes and particularly cassava.

With the increasing interest in milk production especially around urban areas, and the more extensive use of exotic cattle, more attention is being paid to supplementary feeding.

For cows producing over (9 litres) of milk each day, a ration based on local by-products was recommended in the Farmers' Diary, Trinidad, as given in Table 5.7.

Table 5.6 Typical values for some cattle feeds available in the tropics

	Total dry matter (%)	**Digestible crude protein**[a]	**Starch equivalent**[a]
Pasture			
Pangola grass			
Wet season	23.4	6.8	54
Dry season	39.3	3.2	56
Forage			
Elephant grass	16	5.5	40
Guatemala grass	23	6	30
Legume cover			
(*Calopogonium, Centrosema*)	25	4	
leaves - dry	87	14	37
Cassava leaves	29	3.3	16.5
Sweet potato tops	19		9.7
Sugar cane tops	30	4.25	15.1
Groundnuts tops - dry	97	5.8	37.5
Maize leaves - dry	97	1.5	37.0
Maize husks	97	0.4	41.8
Roots			
Cassava	30	1.70	88
Sweet potato	28	4.5	89
Oil seed cakes			
Coconut	90	15.5	75
Cotton seed - undecorticated, not delinted	88	15.6	40
Cotton seed - decorticated	90	35	68
Groundnut	90	40	70
Coconut	85	15.5	74
Palm kernel cake	90	17.5	74
Soya bean	89	40	64
Grains			
Maize	87	7.9	78
Millet	87	8.	59
Sorghum	89	7.7	74
Cotton seed	91	13	76
Grain by-products			
Wheat bran	87	11	42
Rice husks	90	0.4	3
Dry cassava	88	1.2	83.5
Urea	100	262	
Molasses			55
Citrus meal		6	60

[a] On a dry matter basis.
See also Topps and Oliver (1978).

Table 5.7 Cattle feed recommendations for Trinidad based on local by-products.

	Nutritional value of ingredient			**Nutritional value in mixture**			
	DCP	**PE**	**SE**	**Pound**	**DCP**	**PE**	**SE**
Coconut meal	15.5	15	75	50	7.75	7.5	37.5
Citrus meal	6	5.5	60	20	1.20	1.1	12
Wheat bran	11.0	10	42	20	2.20	2	8.4
Molasses		—	55	8	—	—	4.4
Bone meal		—	—	1	—	—	—
Cobalt iodized salt		—	—	1	—	—	—
				100	11.15	10.6	62.3
				6	0.669	0.636	3.738

This is recommended at 6 lb (or 3 kg) of mixture for each gallon (5 litres) of milk as below:

Milk produced (litres)	**Concentrate per day** (kg)	
	Poor grass	**Good grass**
6.8	1.4	0
9.1	2.7	1.4
11.4	4.1	2.7
13.7	5.5	4.1
16.0	6.8	5.5
18.0	8.2	6.8

In Uganda, feeding cassava or sweet potatoes, which have a high starch and negligible protein content, has a greater influence on the milk yield of the cattle-grazing indigenous pasture than cottonseed, cotton seed cake, or groundnut cake, the return from these protein feeds being negligible at the average annual level or production of (650-900 litres) plus a calf.

At Kitale in Kenya feeding maize meal at 1.8 kg/day to Jersey cows raised their milk production by an average of 30 per cent. High yielding cows responded much more to supplementary feeding than did low yielding cows (Moberly 1966).

There is an advantage in having a mixture of ingredients in a cattle feed as the mixture will cover the shortfalls of single ingredients. Maize has a low manganese content, 4 to 5 ppm, compared with 50-100 ppm in wheat. Feeding stuffs should be stored in clean dry rat proof buildings as aflatoxin may develop in feeds stored in damp conditions.

Consideration should be given to forage conservation. Haymaking has never been popular in the tropics. Grass drying was being examined in

Malaysia but is unlikely to be economic with the high price of oil. Silage making, using either surplus grass in the wet season or forage maize supplemented with urea, must be both practical and economic on larger farms where machinery is available to move the heavy wet forage.

Urea as a protein source (see Loosli & McDonald 1968)

Urea is a synthetic nitrogenous compound manufactured on a large scale for fertilizer, which is converted into protein by bacterial action in ruminant animals. To be of value it must be fed with fermentable carbohydrates such as grain starches, or more appropriately for the tropics, molasses. On a weight-for-weight basis the energy value of molasses is 70 per cent of that of maize grain (SE 52). Urea has 42 per cent nitrogen and therefore each kilogram of urea is equivalent to 2.62 kg of total or crude protein, or a PE of 131; 1 t of food grade urea with 6 t of cereal grain provide the same amount of protein and nearly as much energy as 7 t of soya beans or cottonseed meal. A small percentage of urea is included in most commercially available cattle feeds. Care has to be taken when feeding urea; it must not be fed to calves (nor to non-ruminant pigs or horses), and a maximum of 2 per cent urea in the feed mixture is recommended for dairy cattle and 3 per cent for beef cattle, or alternatively, up to a quarter of the protein equivalent in growing rations and up to one-third in finishing rations. Readily available carbohydrate must be fed with the urea.

Preston (1972) in Cuba developed a system of fattening beef cattle based on a urea/molasses mixture: 2.5 per cent urea with 0.5 per cent common salt is dissolved in 1.5 per cent water at 50 °C and added to 95.5 per cent blackstrap molasses. Fattening cattle have unrestricted access to this mixture and a mineral mixture. They have a restricted access to fresh forage which has been found to be essential to prevent sickness, and a small quantity of fishmeal as a protein supplement. While the urea could be fed to contribute 50 per cent of the total dietary uptake and molasses 80 per cent of the total dietary energy, far beyond the limits previously considered safe, better results were obtained with the addition of fishmeal and restriction on forage. Molasses is a cheap source of carbohydrate where sugar cane is grown, about 1 t of molasses being available for every 3 t of sugar produced. Molasses has been used as a cattle feed up to 5 per cent or even 10 per cent of the total ration for many years. Molasses toxicity can cause metabolic upsets and deaths. To prevent this it is essential that all the animals have up to 2 kg of dry matter as forage every day. The minimum water must be used to make up the urea/molasses mixture and the troughs in which the mixture is placed should be roofed to prevent dilution by rain. Where it is economically justified, up to 1 kg of ground maize or sorghum is fed daily to each animal to supply glucose.

Other methods of using urea/molasses include mixing with absorbent roughage such as maize straw. To prevent molasses toxicity, in Mexico maize and rice polishings bran, blood meal with grass and guinea grass hay have been fed with about 5 kg per head per day of 2.5 per cent urea in molasses to give liveweight increases of 1 kg/day. Where the urea/molasses mixture is available in the field, devices such as a wooden float on the drum of mixture, are often advised to restrict the intake by the cattle. Urea/molasses must be introduced into the diet gradually to allow the microflora of the rumen to adapt. This is particularly important if the animals have been short of palatable feed.

In South Africa, grazing areas with over 760 mm of rainfall are described as sour veldt as the grasses grow rapidly with the rains, produce good grazing for a few months and then mature to become both unpalatable and indigestible. During the dry season with only this mature herbage, the cattle lose condition. Experimental spraying of this dry vegetation with a mixture of 15 kg urea, 64 litres of molasses, and 155 litres of water at 450 litres/ha of mixture showed that the cattle eagerly grazed the freshly sprayed areas, but 5 to 7 days after spraying it lost its attraction. Instead of losing condition, steers gained 0.1 kg per head per day in the sprayed pasture (Altona 1966). A 10:1 mixture of molasses to urea also improved the value of poor hay. Some cattle died, particularly if the urea got into solution and was drunk in too large quantities. Urea/salt blocks which can only be licked have proved popular. The most satisfactory method of feeding urea has been to use a mixture of 4 parts dried molasses, distillers solubles, urea, with 3 parts of maize meal and 2 parts of salt moistened and made into a hard block in a 200 litre drum. Drums containing the mix are placed in the grazing, protected from rain.

High percentage urea blocks have been shown to be a satisfactory supplement to poor quality grazing in Northern Queensland, for dry cattle and weaners (Burns 1965). It is dangerous to feed urea to weak or hungry cattle who might consume an excess.

With the greater availability of urea throughout the world, urea licks or urea/molasses pasture sprays could become a useful source of protein in areas where the cattle must survive a dry season on mature grazing, or in more intensive areas where elephant grass has grown beyond its high protein stage. It could also lead to the utilization of some of the old grass which is burned each year to make way for the new flush.

Calf rearing

Successful calf rearing is essential to profitable stock keeping in the tropics, particularly where the calves are heifers from quality dams. The milk production development programme in Jamaica was based on the import of batches of non-pedigree Holstein calves 3-4 weeks old by air from Florida. Allen summarized the 10 points for successful calf rearing as follows:

1. If you are starting with bought-in calves make sure you buy from reputable breeders - those you can depend on to have given their calves colostrum.
2. If the calves are from your own farm, make sure they are out of your best producing cows and out of the best bulls available.
3. Be present and help if necessary when the calf is born.
4. Treat the navel cords of the newly born calves with tincture of iodine or other suitable disinfectant.
5. Provide disinfected, dry individual pens for your calves.
6. Be sure your calf gets colostrum (3 pints (1.7 litres) in the first 5 hours of life is vital).
7. Maintain sanitary conditions at all times. (Clean calving pens are stressed in the booklet.)
8. Feed adequate amounts of milk or milk replacer but do not over feed. (Overfeeding can cause persistent scour.)
9. Do not neglect minerals.
10. Be on the lookout for off-conditions and be prompt in treating.

Calf scour

A newly born calf absorbs antibodies very efficiently and very quickly from colostrum, but very quickly the calf's intestine becomes less able to absorb these antibodies. The first 5 hours of the calf's life are very important, after which it can be taken away from the cow. During these 5 hours it needs about 2 litres of colostrum. These antibodies seem to be absorbed better in the presence of the cow.

Five different systems of calf raising are described by Allen:

1. *Nurse cow method* where the cow is milked in the morning and has the calf with her until evening when it goes to a calf pen. This is the simplest, most successful and the most expensive where milk is valuable. Ideal where the calf is very valuable or the milk has no market.
2. *Whole milk* by hand feeding up to 3 months old together with
   (a) up to 2.3 kg/ha (5 lb) of feed mixture, 18 per cent protein, 10-180 days;
   (b) green fodder or Pangola hay from 15-180 days, then high nutritive fodder;
   (c) mineral supplement.
3. *Skimmed milk* where the whole milk is replaced by skim milk with the fat and vitamins A and D replaced by substitutes.

4. *Milk replacement system* where the milk is replaced by a proprietary mixture after about 10 or 14 days.
5. *Limited whole milk* where a limited amount of whole milk is supplied for 4 or 5 weeks and then gradually replaced by a dry calf starter mixture.

Liquid calf feed, whether milk or substitute, should be fed at 32 °C in clean pails. Water should be available in the pens. From about 2 weeks old calves should be able to pick at good hay or fresh grass low in fibre and be supplied with minerals. They should be out at pasture except during the night and the heat of the day. Shade should be available in the pasture. Up to 4 or 5 months calves will make little use of the herbage. Calves should graze the pasture ahead of the adults to avoid worms.

Young calves should ideally have individual pens to stop them sucking each other. As horns are a problem in adult cattle, calves should be dehorned when a few weeks old.

In most parts of Africa where the standard of animal management is not so high as in Jamaica, the cow/calf system with single milking each day is to be advised. A suckling calf takes about 400 litres of milk before it is weaned. Where the milk has no value a free range system will give the best results. Under both these systems, calf deaths from East Coast Fever, which accounts for about half the calf losses in East Africa, should be negligible. The better fed the calf the better the chance of survival from East Coast Fever, and on recovery the animal has resistance to the disease. Adult animals introduced from temperate zones to areas of tick or East Coast Fever often die unless proper veterinary attention is available.

Where calf losses are high, management is bad. Weekly weighing of calves in a sling on a yardarm scale is a useful way of keeping a check on management where the numbers are small. With the cow/calf system, troubles arise when the herdsman is milking twice a day and depriving the calves. This often arises through the herdsman being paid in milk which he extracts from the herd to the serious detriment of the calves.

In the rainy season in Northern Nigeria, newly weaned Zebu calves make poor growth as the herbage is low in protein. A supplement of groundnut cake and bloodmeal is recommended (Lee *et al*. 1959).

Minerals and salt

Cattle require a wide range of elements as explained in Chapter 1. A bullock weighing 340 kg contains about 9 kg of nitrogen, 5.4 kg of calcium, 2.7 kg of phosphate and 0.68 kg of potash. A cow during its lactations has secreted minerals in the milk, and there has been a loss, particularly of potash in the urine. Minerals are also lost in sweating and dribbling saliva.

Most of the elements are available in sufficient quantities in the herbage, and this is understood by nomadic herdsmen who take their cattle to certain areas where the grazing is 'salty'. In other regions the cattle eat certain soils. Burning old pastures makes the minerals in the roughage available in the young grazing. Where one mineral is deficient physiological symptoms usually appear. Mild deficiencies which do not cause visible symptoms or death but cause a loss of production are probably quite common in the tropics. The marked response of crops in parts of West Africa to low rates of phosphate suggests that cattle may be suffering from the same deficiency, though the burning of pastures and the browse plants make phosphate available.

Salt is required by all animals and should be made available mixed with the feed if supplementary feed is given - 0.5 kg of salt, preferably rock salt, is sufficient for 10 animals for a week, distributed daily. This is even more valuable if mixed with ground limestone or chalk and steamed bone flour.

Lack of phosphate is the cause of 'pica' or deprave appetite, where the cows chew bones, often becoming infected with bacterial contaminants of the bones such as anthrax. Lack of phosphate causes emaciation and an unthrifty appearance, stiff jaws and weak bones. Breeding may become irregular and where the deficiency is only mild there may be insufficient phosphate for both milk production and breeding, and the cow may fail to conceive until the lactation ends.

Phosphate deficiency can be corrected by feeding bone meal or dicalcium phosphate, or by adding disodium phosphate to the drinking water. The application of phosphatic fertilizer to the pasture is as efficient a method as any of correcting such a deficiency, as it is accompanied by an improvement in the pasture, particularly if legumes are present. In most tropical areas, however, it is uneconomic at present to apply fertilizers to pastures except intensive dairying areas. High production cows need a high level of calcium and phosphate for the extra milk produced. Where a minor element deficiency occurs application to the pasture is the practical solution.

Young calves need a good supply of minerals. Where supplementary calf feed is used, minerals - particularly calcium and phosphate with other trace elements - may be added to the mixture, otherwise this requirement is met most easily by suspending a proprietary salt lick, compounded into a brick, on the side of the calf pen. This can be held in a simply made wooden holder to avoid soiling. Urea/molasses mineral blocks are useful in the dry season for range cattle.

Better watering facilities

Water is vital to the survival of cattle. The amount of water cattle need increases with growth, increased temperature, lactation and exercise and varies not only between breeds but also between individuals in a breed. In the wet season in the humid tropics the green forage may provide most of their water requirement. But as ecological conditions get more arid and water less available not only does the forage provide less water but the animals' needs increase. In very dry areas the mineral content of the water may rise to unpalatable levels, though cattle can adapt to drinking highly mineralized water, if this continues too long it can upset the animals' metabolism.

Lack of watering facilities in the long dry season limits the cattle numbers in many parts of Africa. FAO estimates suggested that by March 1974 the drought in the Sahelian zone of Africa had caused the death of 3½ million head of cattle. In Mauritania, Mali, Niger, Chad, Upper Volta and Senegal a quarter of the cattle had been wiped out and in some areas, 80 per cent had died.

The Fulani and Masai may have to water the cattle only on alternate days to be able to reach the surviving pasture, perhaps 6 km (nearly 4 miles) from the water, in the peak of the dry season. The construction of dams and watering places improves the condition of the cattle that no longer have the long walk to water and also relieves the overgrazing around old watering places (Fig. 2.6).

The water requirement of cattle is between 5 and 23 litres per head per day. Rollinson *et al.* (1955) found at Entebbe, a cool cattle area, that the average consumption of non-lactating animals was 8.6 litres per head per day, with a range between animals of 5 to 12.7 litres. The water was mainly taken during the day, particularly the hour before midday and the hour before sunset. The cooler the drinking water is the more efficient it is at lowering the body temperature. They quote consumption figures from other work higher and lower, including an estimate from Serere (Uganda) which is in agreement with the results for ranch cattle in Tanzania (Hutchison 1959), where the cattle required 7 per cent of their liveweight daily. Thus a 320 kg animal would require about 22 litres. French (1956) in Central Tanzania found housed cattle required 11 litres a day if it was freely available, but used less if it was only available on alternate days, or every third day. For cattle grazing outside, the requirement is greater (French 1938). Lack of water may reduce the appetite and also affect food uptake, and the provision of adequate drinking water is essential if the productivity of livestock is to be increased.

Disease control

In addition to the cattle diseases of temperate regions the cattle of the tropics are subject to devastating tropical diseases which not only wipe out whole herds, but can seriously upset agricultural operations. Foot and mouth disease is liable to break out in the Punjab in May or June when the cattle are required for threshing. Rinderpest, a highly contagious virus disease of ruminants in the Far East, India and Africa, may eliminate the water buffalo when they are required for the cultivation of the rice paddies.

Dipping

Many of the tropical cattle diseases are tick-borne, and Table 5.8 summarizes their life cycles. With the one host ticks, spraying may eliminate the disease provided that infested animals are not brought in later. With the two and three host ticks, elimination is not possible where the alternative host is free ranging, e.g. birds, hares. Where the alternative host is a domestic animal, these should also be disinfested to break the life cycle.

Ticks cause anaemia and stunting of young stock and a loss of condition in adult stock. Tick wounds are a site for secondary infections. In many major cattle areas, communal dips have been constructed to which the cattle are brought periodically. Dipping tanks are expensive to construct and must be well served with a good water supply, and have an effective system of drainage which will not pollute water supplies. Disposal of 18,000 litres of spent dip is a problem. Such an expenditure must serve a large area and requires close supervision to ensure the concentration of the dip is maintained and the tanks are not misused. The collection of cattle from a large area carries a serious disease hazard.

With the introduction of organic insecticides, it became possible to control ticks adequately by spraying. BHC (HCH) (see Table 4.2) and dieldrin were the original organic chemicals for tick control, but partly due to resistance and partly due to concern that residues of these organo - chlorine compounds might occur in the meat or milk they have been replaced by organo-phosphates such as diazinon or bromophos, or carbamates such as carbaryl (Sevin®). The chlorinated camphene, Toxaphene® is also used. In Uganda (Clifford 1954) a small portable spray pump fitted with two spray lances to spray each side of the animal simultaneously was introduced. Crushes could be constructed from locally available poles and the spraying equipment cost up to £24 in 1954. Special attention was paid to spraying the areas where the ticks collect, such as the ears, under the tail and in the tail brush. Two pints (1 litre) of spray was sufficient for one animal and the residual effect

Plate 5.7 Cattle spray race used for tick control. (Courtesy Cooper, McDongall and Robertson Ltd.)

kept the animals tick free for a number of days. One unit served about 1,000 head of cattle and the cost was under 1s. per head per annum which the cattle owners were quite willing to pay.

Spraying is suitable for large farmers or cattle-owning groups to finance and operate independently. Owners of small herds with 20–25 cattle will often find it worth while to purchase a hand spray pump for their own use. These use about 6–9 litres of spray per animal. A knapsack sprayer is not effective as the jets will not penetrate under the hair. A flat fan jet with pressure will force the insecticide under the hair and to the more inaccessible parts of the skin. A special race has been developed to give a more efficient but less directed cover (Plate 5.7). The surplus spray is collected and recycled. This system uses about 3 litres per animal.

A simple way of treating ticks has been used in South Africa. The cattle, brought together in small groups have been misted with a chemical of low toxicity like diazinon applied with an Ulva or Mini-Ulva sprayer (see p. 337). The cattle must be downwind and move about in the mist. The control is less efficient than dipping or spraying but it is easy, quick and cheap, and better than taking no action when other equipment is unavailable.

Table 5.8 Diseases of livestock in the tropics associated with ticks

Tick	Host	Disease or condition caused
A. ONE HOST TICKS Boophilus annulatus Cattle or Texas Fever Tick.	*Cattle*, horses, sheep, goats, game animals, and dogs.	1. Babesiosis (Piroplasmosis) *B. bigemina* - Texas fever (passes through egg). 2. *Anaplasma marginale* (gall sickness). 3. Spirochaetosis - cattle, sheep, goats, and solipeds. (*Borrelia theileri*).
Boophilus microplus	*Cattle*, horses, sheep, goats, game animals, and dogs.	1. Babesiosis (Piroplasmosis) *B. bigemina* - Texas fever (passes through egg) *B. argentina.* 2. *Anaplasmosis (Anaplasma marginale* - gall sickness *Anaplasma centrale* - benign infection). Spirochaetosis of horses, sheep and goats. 4. Q Fever (*Coxiella burneti*).
Boophilus decoloratus Blue tick (not very appropriate)	*Cattle*, horses, sheep, goats, game animals, and dogs.	Principal carrier of *A.* marginale, *B. bigemina.*
Boophilus calearatus	Cattle, sheep.	1. Babesiosis (*B. bigemina*) 2. *B. berbera.* 3. *Anaplasma marginale.*
B. TWO HOST TICKS Hyalomma aegyptium (*H. sariguzi*).	Larvae and nymphs on small rodents and birds. Adults on cattle, sheep, goats, camels, horses, etc.	Q Fever (*Coxiella burneti*). Tick paralysis.
Hyalomma transiens syn for *H. truncatum.*	Larvae and nymphs on small rodents and birds. Adults on cattle, sheep, goats, camels, horses, etc.	1. Sweating sickness (Toxicosis).

Distribution	Life Cycle	Control
N. America, Australia, Argentina, Brazil, Uruguay, S. Africa, India, Pakistan.	Larva (6 legs), nymph (8 legs) and adult spends whole parasitic life on same animal.	Dipping or spraying with organophosphorus insecticides, Delnav, Diazinon, etc.
S. America as far south as 34th Parallel in Argentina and Uruguay, Asia and N. Australia, Panama.	Larva (6 legs), nymph (8 legs) and adult spends whole parasitic life on same animal.	Dipping or spraying with organophosphorus insecticides. Strains in Australia now resistant to organophosphorus compounds.
Central, East and S. Africa. Prefers relatively moist conditions 0–2,440 m	Larva (6 legs), nymph (8 legs) and adult spends whole parasitic life on same animal. Nearly always attaches to sides of body, shoulders, neck and dewlap.	Strains in most countries have become resistant to arsenic, HCH, Toxaphene, dieldrin, aldrin and DDT. Resistant strains controlled by dipping or spraying in organophosphorus insecticides.
Transcaucasia N. Africa.	Larva (6 legs), nymph (8 legs) and adult spends whole parasitic life on same animal. Nearly always attaches to sides of body, shoulders, neck and dewlap.	Strains in most countries have become resistant to arsenic, HCH, Toxaphene, dieldrin, aldrin and DDT. Resistant strains controlled by dipping or spraying in organophosphorus insecticides.
India Mediterranean	Larvae moult on host, nymphs engorge drop off moult and adults find another host.	By dipping in arsenic, HCH Toxaphene or organophosphorus insecticides.
India Mediterranean	Larvae moult on host, nymphs engorge drop off moult and adults find another host.	By dipping in arsenic, HCH, Toxaphene or organophosphorus insecticides.

Table 5.8 Diseases of livestock in the tropics associated with ticks (continued)

Tick	Host	Disease or condition caused
Hyalomma truncatum (Bent legged tick)	Cattle and goats, other domestic animals, buffalo and large antelopes. Immature stage - hares and birds.	1. Sweating sickness (Toxicosis). 2. Mediterranean fever (*Theileria annulata*). 3. Tick paralysis in man. 4. Q Fever (*Coxiella burneti*).
Rhipicephalus evertsi The Red Legged Tick.	*Cattle*, horses, sheep, goats, game animals,	1. Babesiosis (Piroplasmosis) *B. bigemina.* 2. *Theileria parva* (East Coast Fever). 3. *Nuttallia equi* (Biliary fever) 4. *Borrelia theileri.* 5. Paralysis in lambs.
C. THREE HOST TICKS Rhipicephalus appendiculatus The Brown Ear Tick.	*Cattle*, sheep, goats, horses, mules, donkeys, dogs and wild game.	1. *Babesia bigemina* - passes through egg. 2. *Theileria parva* (East Coast Fever)(main vector) 3. Nairobi Sheep Disease (main vector). 4. Other Theileria (T. mutans).
Rhipicephalus bursa	Sheep, and goats, cattle, horses.	1. *Babesia bigemina* (Gall Sickness). 2. *Babesia ovis.* 3. *Anaplasma marginale* and *ovis* (Gall Sickness). 4. *B. motasi* (Q Fever).
Rhipicephalus pravus East African convex eye brown tick.	Cattle, sheep, goats, dogs, wild antelope and carnivores.	1. East Coast Fever (*T. parva*) 2. Nairobi Sheep Disease.
Rhipicephalus simus The Glossy Tick.	Cattle, sheep.	1. Nairobi Sheep Disease. 2. *Anaplasma marginale.*
Rhipicephalus sanguineus 'Brown Dog Tick' (Kennel Tick).	*Dogs*, cattle, wild carnivores, birds and other animals.	1. *Babesia canis* (Biliary Fever) through egg. 2. *Rickettsia canis* - a fever, through egg. 3. Tick typhus (man). 4. *Anaplasma marginale.*

Distribution	Life Cycle	Control
Africa Mediterranean.	Adults usually attach on udder or scrotum, in the groin, round the anus, in tail tuft and between the hooves. Sometimes 3 hosts.	By dipping in arsenic, HCH Toxaphene or organophosphorus insecticides.
Africa, but cannot stand drought.	Larvae and nymphs are found in the ears or groin region, adults mostly under tail.	Some strains resistant to arsenic and chlorinated hydrocarbons. Can be controlled by dipping or spraying with organophosphorus insecticides. Ticks difficult to control and hand dressing must be resorted to.
Africa. In country between 1,600 and 2,500 m with rainfall of 750 mm per year or more.	Larvae, nymphs and adults on different hosts, predilection site on edges and in ears. head, eyelids, cheeks and tail switch.	By dipping in arsenic or dipping and spraying with BHC, Toxaphene, and organophosphorus insecticides.
Mediterranean basis, N. Africa, Rumania, Russia.	Larvae, nymphs and adults on different hosts, predilection site on edges and in ears, head, eyelids, cheeks and tail switch.	By dipping in arsenic or dipping and spraying with HCH, Toxaphene, and organophosphorus insecticides.
East Africa.	Larvae, nymphs and adults on different hosts, predilection site on edges and in ears, head, eyelids, cheeks tail switch.	By dipping in arsenic or dipping and spraying with HCH, Toxaphene, and organophosphorus insecticides.
Africa - South and East	Larvae, nymphs and adults on different hosts, predilection site on edges and in ears, head, eyelids. cheeks and tail switch.	By dipping in arsenic or dipping and spraying with HCH, Toxaphene, and organophosphorus insecticides.
Very cosmopolitan. India, Pakistan, Africa, practically all countries between 50 °N and 35 °S.	May be found anywhere on body, frequently on ears.	In dogs by bathing in HCH washes. Cattle dipping in arsenic or dipping and spraying with HCH, Toxaphene or organophosphorus insecticides.

Table 5.8 Diseases of livestock in the tropics associated with ticks (continued)

Tick	Host	Disease or condition caused
Amblyomna gemma	Cattle	1. Heartwater
Ambloyomna hebraeum 'The Bont Tick'.	All domestic animals and many wild animals.	1. Heartwater (*Rickertsia ruminantium*). Infection does not pass through egg. 2. Theileriasis.
Amblyomna variegatum 'The Variegated Tick'. 'The Tropical Tick'.	*Cattle*, sheep, goats, horses, wild buffalo and antelopes. Man.	1. Heartwater (*Rickettsia ruminantium*). Principal vector. 2. Nairobi Sheep Disease. 3. Q Fever of man and animals.
Amblyomna americanum 'The Lone Star Tick'.	Wild and domestic animals.	
Amblyomna maculatum 'Gulf Coast Tick'.	Cattle, horses, pigs, and deer in adult stage.	
Ixodes ricinus 'Castor Bean Tick'.	Cattle, sheep, goats and dogs.	1. Babesiosis (*B. bovis*) – through egg of tick. *Anaplasma marginale*. 2. Louping ill Theileriasis. 3. Tick-borne Fever. 4. Paralysis. 5. Q Fever.
Ixodes pilosus 'Paralysis Tick' (Karoo paralysis tick).	Sheep and goats, rarely cattle, and antelopes.	Paralysis.
Ixodes scapularis.	Cattle	1. Anaplasmosis – *A. marginale*.
Dermacentor andersoni.	Horses, cattle, sheep, other animals, and man.	1. Rocky Mountain Spotted Fever in Man. 2. Paralysis (*Anaplasma marginale*) 3. Tularensis (Man) (*Pasteurella tularensis*). Q Fever. 4. Equine Encephalumyelitis.

Distribution	**Life Cycle**	**Control**
Kenya		
South and Central Africa.	*Amblyomna* ticks are generally ornately marked and usually large. They have long hypostomes and palpi and attach firmly	
South and Central Africa. Not in very dry areas 635-760 mm rainfall, 0-2,440 m	causing deep wounds which may become the sites of screw worm attack. 1. *A. hebraeum* and *A. variegatum* is frequently found in the perineal and genital regions. 2. *A. americanum* prefers to attack hairless parts of body. 3. *A. maculatum* larvae and nymphae mainly on	Dipping with arsenic, HCH or Toxaphene.
Coastal area - Gulf of Mexico.	nesting birds. Attaches to inside of ear.	
Europe, including UK, parts of Africa and America, India and Pakistan.		Dipping in organophosphorus insecticides
South Africa (dry areas).	Larvae, nymphs and adults on different hosts.	
USA.		Dipping in arsenic, HCH or Toxaphene.
USA Rocky Mountains and Far East.		Dogs washing and horses spraying with HCH
	Dermacenta eggs usually ornately marked. Short hypostome and palps.	Dipping and spraying with HCH.

Table 5.8 Diseases of livestock in the tropics associated with ticks (continued)

Tick	Host	Disease or condition caused
Dermacentor albipictus 'The Winter Tick'	Cattle, horses, dogs.	1. *Anaplasma marginale, B. equi.* 2. *B. canis.*
Amblyomma cajennense 'The Silver Tick'	Cattle	Q Fever.
Haemaphysalis punctata.	Cattle, sheep and other mammals, adults, larvae and nymphs, also found on reptiles.	1. Babesiosis (*B. bigemina*). 2. *Anaplasma marginale* and *centrale.* 3. (*B. bovis*). 4. *B. molasi* – tick paralysis calves, sheep, goats.

Ticks in the ears are particularly difficult to kill. A Tick-Tag (manufactured by Y-Tek Corporation, Wyoming) has been introduced in the USA. This is an ear identification tag, impregnated with the insecticide Dursban. These are easily attached to the ears and keep them free from ticks.

The intervals between spraying depend upon the life cycle of the tick. *Boophilus*, for example, a one host tick, spends 20 – 24 days on the host. The larvae feed for 3 days, the nymphs for 4, and the adults 7 – 14 days. In this case spraying at weekly intervals will only control the adults. As the larvae can live 4–6 months it is necessary to keep a pasture free from cattle, sheep, goats and horses for 6 months to starve the ticks out. It is not practical to spray the pasture over all unless it is small or special, or the perimeter is adjacent to unsprayed cattle.

Unless the complete farm can be cleared of ticks and reinfestation avoided, dipping or spraying should be carried out to keep the number of ticks down to a level where they are not interfering with the animals. Efficient tick control is more easily achieved where the grazing is fenced and restricted to cattle which are sprayed regularly. Calves reared under tick-free conditions may succumb to tick fever or other tick-borne diseases if transferred to areas not free from ticks. In those areas where East Coast Fever is endemic, it is advisable for young stock to get a mild attack of East Coast Fever, which is not lethal to healthy calves. The same applies to heartwater areas. There has been less trouble in Trinidad with tick-borne diseases in the second generation of animals born in Trinidad from Holstein dams introduced from Canada. The original imports suffered very badly unless well protected by spraying and fencing. Among the herds that are resistant to ticks it has been observed that most of the ticks are carried by a minority of the animals

Distribution	Life Cycle	Control
USA, Europe, Russia, France		Spraying with HCH
Northern and South America and Caribbean, USA.		Spraying with HCH
Europe and N. Africa, Japan.	Eyes absent.	Spraying with HCH.

and by culling these the tick problem is reduced. Jersey cattle appear to have some resistance to ticks (Lee 1979).

Inoculation

Modern veterinary science has done a lot for cattle keepers in the tropics. It is no longer necessary for a cattle owner to watch helplessly while his herd is devastated by rinderpest. Immunization by injection is now possible against virus diseases such as rinderpest and foot and mouth disease and bacterial diseases such as anthrax (which is dangerous to humans) and blackquarter. Modern drugs are available to help exotic cattle introduced into tropical areas recover from tick fever which in the past was too often lethal.

Where anthrax or rinderpest is likely, government inoculation campaigns are carried out. The early opposition of the nomadic herdsmen has been overcome though trouble occasionally flares up.

Natural immunity

In many areas where a profitable dairying industry is developing, pure bred cattle, particularly Holsteins, are being introduced from temperate zones on account of their milk yielding potential. It is generally considered that animals predominantly Holstein or Jersey with about 10 per cent of Zebu blood show a greater resistance than pure bred animals to disease and climate, and hence need less attention and veterinary service, without much loss of production. The Jamaica Hope obtained about 80 per cent of its genes from the Jersey breed and the rest from the Sahiwal (a Zebu) and Holstein breeds, these combine good production (Bodles herd of 139 in milk averaged 4,360 litres per cow from 305 day lactation periods in 1966), with disease resistance. Unfortunately very few Jamaica Hope cattle are available.

Trypanosomiasis (Finelle 1973) (FAO 1979)

It is estimated that nearly 4½ million square miles (11.66 million km^2) of Africa between the 14 °N and 29 °S latitudes, from sea level to 1,800 m, are infested by the tsetse fly (*Glossina* spp.), 22 species of which have been recorded in Africa, each restricted in its distribution to definite ecological zones. Tsetse flies transmit the cattle trypanosomiasis, nagana. Trypanosomiasis is a parasitic disease caused by species of flagellate protozoa of the genus *Trypanosoma*. It is a chronic evolving disease with a variety of symptoms, causing loss of weight and anaemia. Unless appropriate treatment is given it is usually fatal. It can only be diagnosed with certainty by microscopic examination of a blood sample or serological reaction. In a vast area covering about two-thirds of East Africa, over half of West Africa and a quarter of Central Africa, trypanosomiasis limits and in most cases prohibits the keeping of cattle except under continuous drug treatment. Only about 3 per cent of this area has been freed from the tsetse fly in the last 25 years.

Certain forms of trypanosomiasis can be transmitted mechanically by *Stomoxys* and other biting flies. In Africa, *Trypanosoma vivax* is associated with the tsetse fly, but in the northern part of South America, *T. vivax* exists without the tsetse. Apart from the N'dama cattle of West Africa, a small brown humpless animal (Hamitic Longhorn - *Bos taurus*) (Plate 5.8), which under good grazing conditions may reach 320 kg at 3½ years, cattle have little or no resistance to trypanosomiasis. The Guinea and Sierra Leone types of N'Dama are said to be more resistant than the Congo N'dama, and there are individual differences in tolerance between members of one herd. In 1954, the N'dama cattle were introduced into the cotton area of the Ivory Coast for ploughing but resistance to trypanosomiasis broke down when the animals were placed under stress. They were also difficult to train as draught animals. N'dama cattle are an important source of meat for Sierra Leone, Liberia and French Guinea. Occasionally the N'dama are milked but produce little. For a review of tolerance to trypanosomiasis see Murray *et al*. (1979).

If cattle and humans are removed from a tsetse fly area, a reservoir of infection remains among the wild game. Attempts have been made to kill the flies with insecticides either by aerial spraying or selective spraying of likely breeding places with motorized knapsacks. An area can be cleared if it is very heavily settled with farmers and the bush reduced by burning and cultivation. A cleared unsettled area nearly three-quarters of a mile wide is recommended between the settlement and the bush to

Plate 5.8 N'dama cattle. (Courtesy Rose Innes)

prevent invasion, and this leads to administrative difficulties. Land tenure and allocation rights introduce added difficulties to planned settlement schemes.

Several drugs, summarized in Table 5.9, are now available which are very effective against trypanosomiasis (except *Trypanosoma simiae*). They are easy to use provided that the specific instructions are followed.

On a settlement scheme in Western Ethiopia a herd of working oxen, rising to 450 head, were maintained and worked for the 5 years up to 1978, in an area of high tsetse infestation (*Glossina morsitans*) using trypanocidal drugs (Bourn and Scott 1978).

THE IMPROVEMENT OF CATTLE BY SELECTION AND BREEDING*

Natural selection has largely fixed in the native cattle of the tropics, genotypes best suited for the conditions under which they have to exist.

* For a detailed study see Lerner and Donald (1966) or Rice *et al.* (1967).

Table 5.9 Use of trypanocidal drugs.

		Method of treatment			Indications		Toxic effects		
Trypanocide	**Trade name**	**Solution**	**Dosage**	**Injection[1]**	**Highly active on**	**Less active on**	**Good tolerance**	**Possible local reactions**	**Treatment of relapses**
		Percent	*Mg/kg*						
Homidium bromide	Ethidium[(b)]	2 hot water	1	IM	*T. vivax* *T. congolense*		Cattle Sheep Goats	Horses	Diminazene Isometamidium
Homidium chloride	Novidium[(c)]	2 cold water	1	IM	*T. vivax* *T. cougolense*		Cattle		
Diminazene aceturate	Berenil[(d)]	7 cold water	3.5	SC or IM	*T. congolense* *T. vivax*	*T. brucei* *T. evansi*	Cattle Sheep Goats	Horses	Diminazene Isometamidium
Quinapyramine sulphate	Antrycide[(e)] (sulphate)**	10 cold water	5	SC	*T. congolense* *T. vivax* *T. brucei* *T. evansi*		Cattle Sheep Goats Camels	Horses	Isometamidium
Isometamidium chloride	Samorin[(c)] Trypamidium[(f)]	1 or 2 water	0.25 to 1	IM (deep)	*T. vivax* *T. congolense*	*T. brucei*	Cattle* Sheep Goats Horses	Cattle	Diminazene
Suramin sodium		10 cold water	10	IM	*T. evansi* *T. brucei*		Camels Horses		Quinapyramine

[(a)] IM = intramuscular injection; SC = subcutaneous injection. [(b)] Boots Pure Drug Co. (Pharmaceutical) Ltd. [(f)] Specia. Ltd [(c)] May & Baker Ltd. [(d)] Farbwerke Hoechst A.G. [(e)] Imperial Chemical ** Antrycide is no longer available from ICI.

*Possible local reaction

Source: Finelle (1973).

They have resistance to endemic diseases, the ability to utilize the available feed which is often of low quality, and breed. Only limited genetic improvement in desirable features, such as reproduction, growth and production, is possible until nutrition and management are improved to keep pace with the improved genetic potential. Selection for milk production in Zebu herds without adequate nutrition may be reflected in low reproduction rates.

The tissue or somatic cells of cattle have 30 pairs of chromosomes, each of which probably has many hundred genes which determine the hereditary characteristics of the animal. The 30 chromosomes transmitted from each parent can be selected from the 30 pairs in over 1,000 million ways (2^{20}), which explains why a repeated cross gives a wide variation in offspring. The appearance and production of the progeny results from a combination of genetic and environmental factors not readily dissociated. Characters which are genetically controlled may be determined by one or more genes, which if they are on the same chromosome will be genetically linked in their inheritance unless crossing over occurs between chromosome pairs. The gene pairs (alleles) which are located at the same point on each of the pair of chromosomes of the germ cells may be homozygous (identical) or heterozygous.

There are a variety of breeding systems used to improve cattle.

Inbreeding

When considering methods of improving indigenous cattle of the tropics by breeding, the system of close inbreeding and rigorous selection used in the UK by Bakewell, Bates and the Collings brothers, at the end of the eighteenth century, suggests itself.

A policy of inbreeding restricts the herd potential to those genes already present, though they may be recombined in a more desirable and reproducible grouping. Without selection, inbreeding will not change the degree to which a gene is rare or abundant (gene frequency) but in small populations some of the alleles may be lost as inbreeding progresses. As inbreeding progresses the proportion of gene pairs that are homozygous increases and the animals become more uniform genetically but this may well be associated with a loss of hybrid vigour and poorer performance. This should also eliminate lethal recessive such as bulldog calves but care has to be taken to prevent desirable traits being lost at the same time. Investigations however have shown that the degree of heterozygosity in inbred lines is greater than one would anticipate, suggesting that selection favours heterozygotes rather than either homozygote. Since desirable genes are often dominant, good inbred animals will often be prepotent, that is they will stamp their own characteristics on the progeny to a much greater extent than the other parent. Inbreeding is the only known method of increasing prepotency as this depends upon the homozygosity of the dominant genes. As inbreeding usually depresses performance, it is now suggested that it should be used to produce prepotent sires which can be used with predictable results when crossed with commercial herds. This might be applicable in the tropics to produce lines of Zebu cattle well suited for crossing with improved European breeds.

Line Breeding is a form of inbreeding where the parents are not usually so closely related, and its purpose is not to increase homozygosity but to retain a high proportion of the genes of the selected individual, and to maintain the qualities of the herd which would be reduced if a sire was introduced. Line breeding can therefore only be justified in herds with superior sires. At least two sires should be used for line breeding, otherwise there will be too much inbreeding. Line breeding has been used to develop the Santa Gertrudis breed.

Grading is the practice of crossing pure bred sires of a given breed with nondescript scrub or native cows and their heifers, generation after generation. In this way a few pure bred sires can relatively quickly transform a mixed unimproved herd into a group closely resembling the pure breed (F_4 generation) with a performance depending largely on the genetic quality of the bulls used. It is therefore essential to use bulls with the ability to perform well under the conditions which their offspring will have to meet (see also p. 450).

Selection

When improving herds it is essential to be quite clear about the characters to be selected, and to be certain that these are of commercial importance, otherwise the selection pressure which can be applied to the really important characters is reduced. For instance the selection of beef animals for body form and beef characteristics becomes dangerous when the characters selected have no relation to productive value. In all selection, stress must be laid primarily on reproductive efficiency without which selection is very difficult. By the rigorous culling of non-pregnant females, the reproductive rate of a herd can be considerably improved. A slight increase in milk production is a poor substitute for an extra heifer calf.

Williamson and Payne (1978) suggest that each of these approaches has a place in the tropics.

Selection within indigenous breeds

This is applicable where

1. Stress on livestock is very severe and genotype-environment interactions are highly significant.

2. Liveweight gain rather than milk production is required.
3. It is uneconomic to improve the environmental conditions, e.g. semi-arid regions.
4. The indigenous animals possess a specific valuable characteristic, e.g. tolerance to trypanosomiasis.

Williamson and Payne give as specific examples:

1. N'dama and West African Shorthorn breeds in tsetse fly areas.
2. Zebu breeds (e.g. Boran in East Africa and Wadara in West Africa) for meat production in semi-arid areas.
3. Sanga breeds (e.g. Mashona, Tuli and Afrikander) for meat production in Central and South Africa.
4. Breeds such as the Bali (see p. 455) for work and meat production in South–East Asia.

Cross-breeding and/or upgrading of indigenous breeds

Specific examples are:

1. Criollo × Zebu or Criollo × Zebu × Charolais in South and Central America.
2. Possibly the crossing of N'dama, West African Shorthorn and Zebu for tsetse fly areas.
3. In South-East Asia crossing the local cattle such as the Bali with Brahman or European beef breeds.
4. Stabilized *Bos taurus* × *Bos indicus* crosses.

Introduction of exotic breeds

This is likely to be most useful where:

1. Climatic, nutritional, and disease stresses are moderate as on tropical oceanic islands, high altitude tropics, and for dairy cattle in lowland humid areas where grazing is adequate all the year. This is the policy in the Caribbean, South America, South-East Asia.
2. The managerial system is such that the external environment is unimportant, as in the highlands where conditions are more temperate than tropical and in the very hot rich countries where the cattle are kept indoors with air-conditioning.
3. There are no indigenous breeds to exploit the environment such as the lack of Zebu cattle in the semi-arid regions of Australia and the Americas.

Before embarking on a breeding policy it is important to be clear

Table 5.10 Requirements from cattle in different tropical regions.

Purpose	Product	Example
Single	Meat	Ranching.
	Milk[a]	Specialized dairy farms often supplying a milk processing factory as in the Caribbean islands.
Dual	Milk, meat	Nomadic pastoralism in Africa and Asia.
	Meat, work	Subsistence agriculture in S.E. Asia.
Triple	Work, milk, meat	Subsistence agriculture in S.E. Asia.
	Milk, meat, manure	Mixed farming in E. Africa.

[a] The value of the meat from culled dairy cows often means a preference for big Holstein cows over the smaller Jersey which is often better adapted to tropical conditions than the Holstein.

what is required of the animals as this varies widely in different regions according to local demands as outlined by Williamson and Payne (1978) and summarized in Table 5.10.

Current progress

Improvement of Zebu herds by planned breeding has so far been poor, and existing evidence suggests that the methods so far used will only give a poor annual increase in milk production by genetic improvement. In consequence many tropical countries are introducing improved *Bos taurus* animals for crossing with local *Bos indicus* cattle. The National Plan in Cuba, divides the Zebu cattle into two sections, half are graded up for milk production using Holsteins, and the other half for beef using Charolais. Faster growing beef cattle were produced at Diamond estate in Tobago by crossing Sahiwal cows with a Charolais bull. Bullocks from these crosses reached 450 kg liveweight at 2 years of age on Pangola pasture which is much earlier than the progeny of the same cows with a Red Poll bull (Plate 5.9).

By culling and improved management, the average production of many of the Zebu herds in the tropics can be doubled by eliminating the poor breeders and the poor producers, but this only raises milk production to around 1,200 litres of milk per lactation, which is about half that expected from a Zebu × Holstein cross adequately fed.

From time to time pioneers introduced exotic animals or bulls from Europe or North America, only to see them listless and without appetite in the heat, finally losing them to a tropical disease such as tick-borne fever, leaving behind some half-bred animal yielding a little more milk than its dam. The slowness of transport, the cost and the risk of death prevented any large-scale introduction of Friesian, Jersey or Devon

Plate 5.9 Charolais bull used for crossing with Sahiwal cows for early maturing beef animals. The field was originally coconuts severely damaged by a hurricane, now planted to Pangola grass. Diamond Estate, Tobago.

breeds. The introduction of artificial insemination (AI) and the world-wide availability of semen from top class bulls, coupled with more intelligent management such as night paddocking and midday shade, better pastures, adequate drinking water, better drugs and veterinary assistance, has changed all the past thinking, and productivity can now be increased very rapidly where it is economically sound to do this. It is important however to see why the Zebu (*Bos indicus*) cattle are better adapted to tropical conditions than the European cattle (*Bos taurus*) as this will enable a better selection of *B. taurus* animals to be made for the tropics.

ADAPTATION TO TROPICAL CONDITIONS

Causes of adaptation

The daily variation of body temperature in cattle is ½ – 1 °C (1 to 2 °F), around 38 °C (100 – 102 °F). If the body temperature rises from the animal's normal temperature and stays high the animal is usually sick. When exotic cattle are introduced into a tick fever area, their temperatures should be taken night and morning, and a high temperature over 2 or 3 readings calls for veterinary attention. Cattle generate heat by their metabolic processes and their physical activity such as grazing and trekking to water. The skin also absorbs heat from the sun. Heat is dissipated by water evaporation from skin surfaces including those which air passes over when the animal is panting, as well as the skin which sweats.

The better an animal is able to maintain its body temperature the better adapted it is to tropical conditions. This arises from a combination of better heat disposal and the ability to carry out its metabolic and physical processes with a minimum of heat production. *Bos indicus* cattle are able to regulate their body temperatures better in hotter climates than are *Bos taurus* animals.

The range of environmental temperature within which an animal is largely able to maintain its normal body temperature by physical means, such as shivering or sweating, is called the zone of thermoneutrality. The limits of this zone for European breeds (*Bos taurus*) of cattle is roughly 0–22 °C, whereas for those tropical breeds (*Bos indicus*) for which it has been determined, it is in the region of 22–37 °C; and some breeds can graze or trek and maintain their normal body temperature when the shade temperature is over 37 °C (Ferguson 1971). Some forest-adapted breeds, though well adapted to a high ambient temperature, are very susceptible to direct solar radiation.

Rhoad (1944) expressed the ability of an animal to maintain its normal body temperature by a heat tolerance coefficient (HTC) calculated: HTC = 100 - 10 (BT - 101) where BT is the body temperature in °F when exposed to direct sunshine on days with a temperature over 29 °C (85 °F). The lower the temperature rise, the higher the coefficient and the more suited is the animal to tropical temperatures suggests that HTC is more suited for selection for heat tolerance within a breed, and as heat tolerance increases with age, it could be best applied for selection among calves in their first year. This is supported by Payne (1952) who in Fiji found wide differences in HTC within a breed. Webster and Wilson (1980) consider that the index increases up to the second year of life, and remains fairly constant for age and stage of lactation or gestation.

In trials comparing breeds the Zebu always has the highest HTC followed by the Zebu first crosses. Jerseys are generally higher than Friesians (Holsteins) and the Aberdeen Angus is generally low. Dowling (1956) measured the ability of the animal to dispose of heat, by exercising the animal vigorously for an hour or more and the animal is judged by the speed with which its body temperature returns to normal. Kibler and Brody (1950) pointed out that the loose pendulous dewlap and bigger ears gave the Zebu 12 per cent greater surface area than a Jersey of the same weight which aids the dissipation of its body heat. Regular observation of a herd in the tropics will show that some animals appear uncomfortable in the heat and regularly seek shade, while others may graze unconcernedly in the full sun. This is seen with Holsteins imported from Canada to Trinidad, both in the newly calved heifers after their coats are clipped and their calves. Rhoad (1944) found the biggest differences between breeds seeking shade occurred on cloudless windless days, and animals tended to seek shade not at midday but during the late morning and mid-afternoon when they offered a bigger surface to the sun's rays.

Selection on the basis of such observations regularly recorded together with milk records, may develop herds of animals better adapted to local conditions than the original importations. This is no doubt the method which will be used by the practical farmer.

Colour

This has been shown to be very important in the absorption of sunlight.

1. Long waves - (heat and infra-red) and medium waves - (visible light) are reflected by white, yellow or reddish brown hair but not reflected by black hair.
2. Short waves - ultra-violet rays are resisted mostly by black, reddish brown and yellow hides.

A combination of these hair and hide colours suited to the radiation conditions is desirable. For high altitudes with cloudy days, as in the Kenya Highlands, ultra-violet rays are more intense and the black cattle more suitable for these regions than the hot cloudless plains.

Animals without pigment in the hair or hide may be damaged by the ultra-violet light. A pigmented skin is desirable in the tropics.

Coat.

The coat colour is independent of the skin pigment. A white Zebu usually has a pigmented skin. White coats obviously reflect more of the sun's heat than black.

Yeates (1965) suggests that a short, light-coloured coat with a smooth and glossy texture is best for minimizing the effects of solar radiation. This should be combined with a pigmented skin.

The hairs forming the coat of a smooth-coated animal are straight and taper, but in a rough-coated animal, mixed with this type of hair are short curly hairs which mat to the animal's body. A rough coat gives desirable protection in a cold winter but insulates an animal against heat loss under tropical conditions, and is a major factor in the lack of adaptation of temperate cattle to the tropics. The average coat clipped from a rough-coated animal was nearly four times as heavy as that from a smooth-coated animal.

A woolly-coated Afrikander bull which arose as a mutant was used to sire a series of rough-coated progeny. These grew more slowly, had a lower calving percentage and produced less milk than their smooth-coated counterparts (Fig. 5.1). Smooth-coated English breeds lost less weight than the rough-coated animals in a drought (Bonsma 1955b).

If the coat characteristics of a calf are not obvious, a hair sample is taken, moistened and rubbed. The hair from a smooth-coated animal rubs away, whereas the hair of a rough-coated animal felts.

Dowling (1959) has pointed out that the hairs of some cattle may have a central medulla which gives them greater rigidity and the animal a semi-erect coat which allows freer air circulation over the skin with more efficient evaporation of sweat, and hence better cooling of the animal. He found a significant correlation between the presence of these medullated hair fibres and the ability to maintain rectal temperature near the normal.

Hide

Apart from its colour, the thickness of the hide is important. Thick hides are not only more resistant to ticks and heat and after wounding heal better than thin hides, but they are better supplied with blood vessels for heat dissipation. Mobility of the hide and a well-developed panniculus muscle helps to throw off flies.

Sebaceous glands

Domingues (1961) states that the main factors influencing thermal balance of cattle in the tropics are the number and volume of sweat glands, skin pigmentation and thickness of skin and hair. Zebu cattle have about one and a half times as many sweat glands per unit of skin area as European cattle (Dowling 1955) and their sweat glands are two and a half times as large. Jenhinson and Nay (1973) suggested the sweat glands of tropical cattle are smaller but more active than those of temperate cattle.

As the evaporation of sweat is the main source of heat loss in hot environments, the Zebu has a big advantage as it can transpire through its skin much better than European breeds. In selecting Shorthorns in Central Queensland for heat resistance Dowling increased the size of these sweat glands 60 per cent in three generations.

Conformation

Many tropical breeds have a large dewlap and umbilical folds. These increase the surface area of the animal without increasing its volume, and

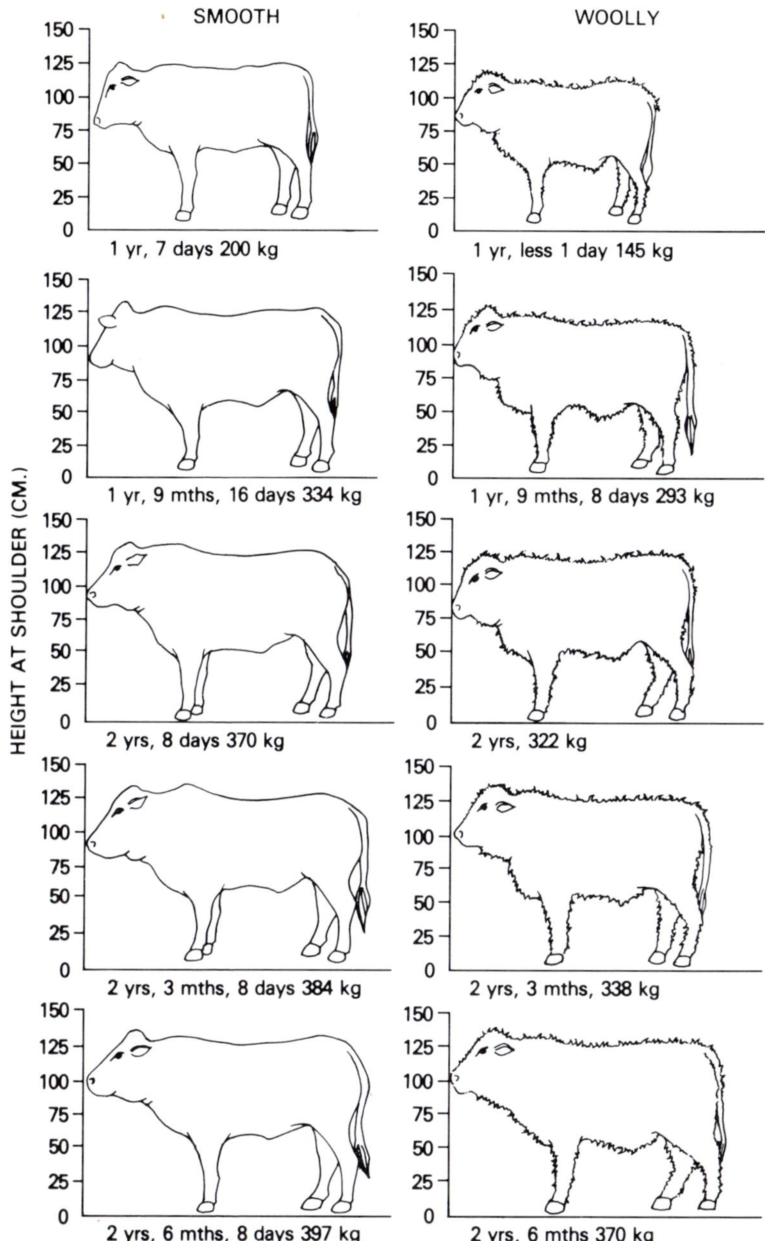

Fig. 5.1 Comparative growth of fed smooth- and whoolly-coated Afrikander cattle. (Bonsma 1955b)

hence increase its ability to dispose of surplus heat.

Bonsma (1955b) suggested that smaller cattle were more adapted to the humid tropics where heat dissipation is more difficult as they present a larger surface area in proportion to volume or body weight than the large animals that are adapted to more arid areas (Fig. 5.2).

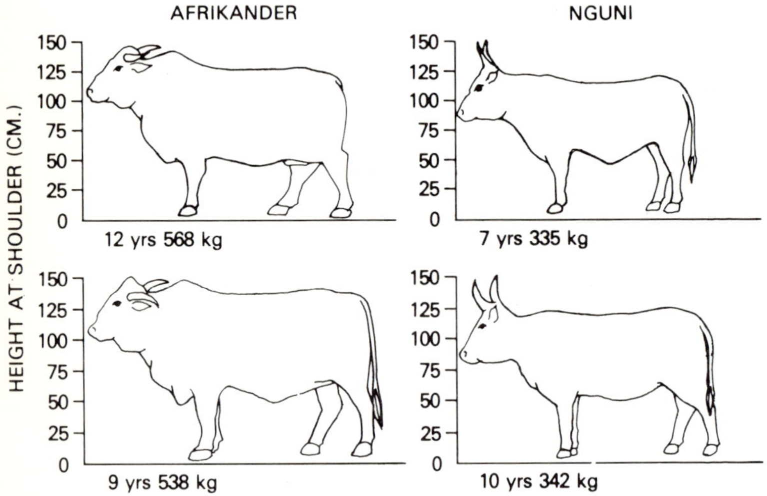

Fig. 5.2 Comparative sizes of mature Afrikander and Nguni (Zulu) cows. (Bonsma 1955b)

Blood

Villares (1940), after studying the native Brazilian cattle, Caracu, and comparing them with introduced breeds, associated their adaptation to tropical climates with a high red-cell count in the blood. The red cell content of European cattle in Europe, USA and New Zealand is between about 6 and 8 millions per cubic centimetre compared with 8 to 13 millions per cubic centimetre for Indian cattle (Mullick and Pal 1944).

Utilization of forage

Cattle adapted to tropical conditions have to utilize poor woody fodder and appear to make better use of such feed particularly the crude protein, than European cattle. One of the first responses to temperatures above the thermo-neutral zone is reduced food intake. This can prejudice normal growth and reproduction, if the food intake cannot be made up by night grazing (Ferguson 1971).

Tolerance to water shortage

The Afrikander cattle when deprived of water showed less weight loss and maintained their appetite better than exotic animals. *Bos indicus* appears to be able to utilize the available drinking water better than *Bos taurus*. This may be associated with the lower water content of the faeces, and higher concentration of the urine (Quarterman *et al*. 1957). The dry season water requirement of Herefords in East Africa is about twice that of Zebu cattle.

Respiration rate

The respiration rate of unadapted cattle increases more rapidly than that of adapted animals as the temperature increases. The rate of respiration increases but not the depth. This increases water vaporization (Rhoad 1940).

Effects of adaptation

1. The adapted animals have better appetites, spend longer grazing and seek shelter less.
2. The adapted animal, feeling the heat less, is freer moving and can travel much further in the heat without showing signs of heat exhaustion. This can be important when grazing is sparse and watering places distant. This is also important in work animals.
3. The weights of calves born to unadapted cows are much smaller when the calf has been carried through a hot period. Air temperature has no effect on the birth weight of calves of adapted animals.
4. The pelvic measurements of an unadapted animal are less than normal when it develops under tropical and sub-tropical conditions.

Results of adaptation

Improvement of cattle

The unadapted cattle suffer from the following disadvantages in comparison with adapted cattle.

Liveweight gain	Slower (see Fig. 5.2)
Fertility	Reduced. Result of slow growth
Calf weights	Below normal

Calving percentage	Lower
Eye cancer	Vulnerable to eye cancer if lacking pigmentation
Conformation	Ewe neck. Deep flat chest. Underdeveloped loin and rump. These are the last parts to mature. Poor tail switch
Disease resistance	Less, due to thinner hide, slower healing, poorer conditions. Lack of acquired immunization
Temperature regulation	Breaks down over 24 °C

Selection

The great difficulty in selecting and breeding cattle is to decide which characters have been inherited genetically and are therefore likely to be passed on to their progeny, and those which are a result of the environment including management. The best breed, for the area and the management system, should be chosen and selection then carried out within that breed.

When the natural increase exceeds the loss from disease, theft, predatory animals, old age, etc., selection becomes possible. At the worst in a buyer's market, this selection is made by retaining those animals which cannot be sold or those of least cash value. A herd of cattle, however, represents a considerable capital asset by most tropical standards. A heifer or a bullock for slaughter will sell for as much, in Uganda, as the average family earn in one year from their standard crops, cotton and coffee. Thus the cattle owners are, in general, the more prosperous farmers and are in a strong position to decide which surplus animals will be disposed of.

A cattle owner with a surplus of stock is able to determine his breeding policy. Until this stage is reached he must concentrate on improving his management, particularly the rearing of calves, to provide a surplus of stock from which a selection can be made, and meanwhile his influence on the future herd is restricted to the choice of bull, or where AI service is available, a choice of semen.

In order to increase his herd, or introduce new blood, he may go outside and purchase from other owners or, usually better, from the government.

It is not possible to choose a good milking cow on its appearance. Beef cattle can be more easily selected by conformation, and their weight estimated by a measuring tape, but this gives no indication of the rate at which the animal has grown.

Memories are often at fault and the basis of any selection must be an accurate system of recording.

Recording

Every cattle owner should have a herd book. Each animal is allocated a separate page, for example in an East African Zebu herd:

No. 37 Meli; Born: 22.7.75; Sire: Omuntu No. 7; Dam: Cheli No. 11

Liveweight	Months:	Birth	1	2	3	4	5	6	7	8	9	10	11
	Weight: (kg)	17	24	36	41	52	54	59	67	74	84	90+	

Calved	*Lactation ended*	*No. of days*	*Total milk produced*
1st Oluli ♀ 15.11.78	19.8.79	299	810 litres
2nd Mwami ♂ 12.10.79	30.7.80	315	920 litres

A note should be made if the animal suffers from a disease to affect the records, or if the calf is lost terminating the lactation prematurely. This recording assumes that a form of milk recording, either daily or weekly, is carried out and the calves are weighed weekly or monthly up to 90 kg. Both these recordings are feasible for the literate peasant farmer with a small herd kraaled near his house. The pedigree of all the animals is readily available in these records. Without a form of recording, gross errors can be made in assessing the animals. A cow producing heavily for a very short time and thus a poor producer for the complete lactation will often without the records be overvalued. Observation on whether the animal regularly seeks shade, collects ticks, or has an unsuitable coat should be added. Any calves born with defects should be noted under both the dam and sire. Irregular breeders should be eliminated.

No owner should part with a heifer until after at least one lactation and preferably two in the herd, but this will depend on facilities available.

Natural increase of pastoralists' herds

When there are no disease epidemics and conditions are favourable, livestock numbers increase fairly rapidly. The cattle of Africa under suitable conditions produce a calf annually for 10 years or more.

The natural increase rate of African herds in Zambia was estimated at 4 per cent in a herd where 45 per cent are breeding cows and heifers, a calf drop of 40 per cent, calf mortality 33 per cent and herd mortality 8 per cent.

For Uganda the gross annual increase was estimated at 16 per cent from 1945-54 and the slaughter rate was 12 per cent.

In Kenya estimates vary from about 20 per cent in Sambaru to about 12 per cent for Masai and Mukogoda, 7 per cent for the Nandi and 3 to 4 per cent in South Baringo.

The estimate for Fulani herds in West Africa ranges from a general figure of 2.2 per cent for Nigeria, 9.2 per cent for certain herds in Adamawa Province, while in a single herd in Bamenda it was between 11 and 17 per cent.

Under favourable conditions without disasters the herds tend to increase from 4 - 10 per cent (Allen 1965).

A well managed small herd with 10 cows and followers, should produce 5 heifer calves per annum, of which 4 should survive. If the cows become unproductive after 10 lactations, one heifer is required annually for replacement. Either the herd size may be increased or three cows or heifers sold annually. There are also four or five bull calves to sell either as calves or bullocks. In the initial years, culling will make room for more heifers, and the daughters of culled cows may be discarded after the first lactation if not promising, but after a few years selection becomes more difficult and calls for an intelligent study of the records, basing selection more on the basis of progeny testing, rather than the individual itself (see p. 449). Selection for milk production on appearance is rarely successful, particularly as the heavy producers are frequently in poor condition.

In a herd which is milked and has not been selected, the distribution of production will be according to the normal curve (Fig. 5.3).

Thus about two-thirds of the milk produced will come from half the herd. A good producing cow may have a poor lactation due to an accident, disease, or the loss of the calf, but on the whole cows are remarkably consistent both in the length of the lactation and the total production. Culling of an unselected recorded herd shows a remarkable increase in average production, often as much as 30 - 40 per cent during the first 2 or 3 years according to the severity of culling, which is turn often depends upon the number of heifers coming forward. Having raised production by improved management and culling, if the feeding standard is as high as economic considerations warrant, further improvements can only come from breeding, which generally means the introduction of exotic blood and this is essentially a longer process than culling.

Where meat rather than milk is the object, the records of milk production and liveweight gain during the first year are useful indicators of early maturity. A calf suckling a low producing cow is hardly likely to gain weight as rapidly as a calf more liberally fed. Weighing of older animals is limited to research stations equipped with a weighbridge. The weight of an adult animal can be fairly accurately estimated by a tape measure as there is a direct correlation of the circumference of the animal with its weight. (For detailed calculations see Williamson and

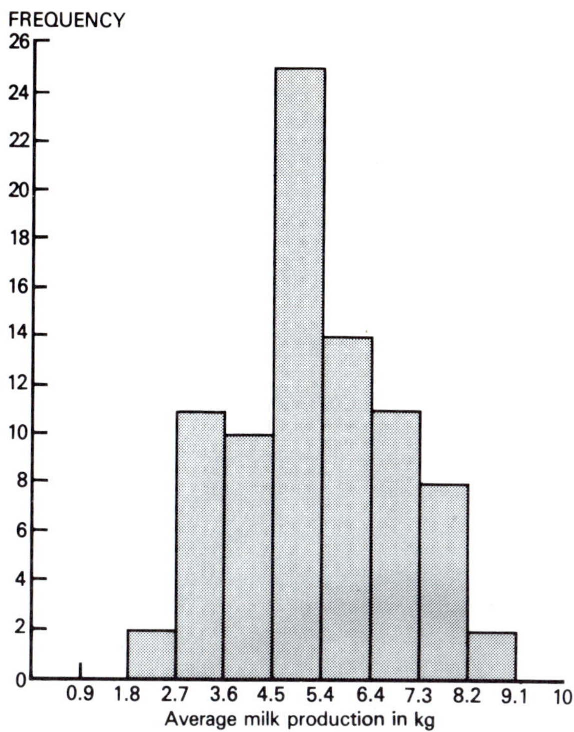

Fig. 5.3 Average milk production of 83 Afrikander cows over a 23-week lactation period. (Bonsma 1955b)

Payne 1978, p. 390). This is sufficiently accurate and easy to be worth while using, particularly when selecting beef animals. Beef animals can also be selected on appearance but where grazing is sparse, care must be taken not to select for size, when a smaller animal would be more productive due to its greater foraging ability. Many of the animals now being used for beef were originally introduced and selected as draught animals which have a different conformation, with flat sides. Selection of beef animals on the basis of conformation should not be on the basis of currently accepted standards for Europe and North America, but to produce the maximum of those beef cuts preferred by the local market.

In India a dual purpose breed is one where the oxen are good workers and the cows good milkers, with the majority of the population vegetarians, the meat is unnecessary. Olver (1938) described the important breeds of cattle in India.

One of the attractions of the Holstein for intensive milk production in the Caribbean is its large size, which provides a valuable meat carcass when its milk producing days are finished.

Characters for selection

Where facilities are available, full use can be made of the heat tolerance test, the increase of respiration rate with rising temperature, and the rate at which the body temperature of an animal returns to normal after being raised by exercise.

The ordinary cattle owner, however, can only apply the outward symptoms of heat tolerance when selecting his animals for breeding.

1. *Fertility*. Animals that fail to breed regularly should be discarded. No cow should be discarded solely on account of its age, but it should be allowed to remain in the herd unless it has other serious disadvantages.
2. *Coat*. All woolly coated animals should be discarded.
3. *Hair colour.* Preferably white, yellow or reddish brown.
4. *Hide colour.* Preferably black, reddish brown, and yellow. The hide should be thick and folded. In the choice of colour, the individual keeper will have his own taste and unless this is an obvious disadvantage as in the case of non-pigmented animals, it is normally satisfactory, as the farmer tends to favour animals that do well.
5. *Tail.* A good switch is a good sign. It helps to keep the flies off.
6. *Conformation.* Full-chested animals with a well-developed dewlap. The scrotum should not be too pendulous as this gives impaired temperature regulation of the testes and infertility. They are also liable to injury. Pendulous udders should be avoided as they are subject to injury. Beef breeds should fatten at the loins and rump.
7. *Heat tolerance.* The animals should move easily even after a period of grazing in the sun and should not seek shade more than average.
8. *Horns.* Unless the horns are needed for defence, they should be bred out. The very long horns of the Ankole-type cattle often damage the other members of the herd and always make the animals difficult to handle. By the use of a polled or stumpy horned bull, a very rapid change can be effected in the appearance of the herd.
9. *Ticks.* Animals that collect many more ticks than the average, should be discarded.

Apart from certain excluding characters, selection should be based on a total score with the characters and production records being attributed points according to their relative importance. Selection for one character at a time is slower and less efficient. In the breeding of the Santa Gertrudis all the 5,000 or 7,000 animals in the 22 herds (8 single sire and 14 multi-sire) concerned were rounded up twice a year and culled on a total score basis. Only 6 per cent of the original heifers reached the top herds with single sires and the culling rate in these top herds was

20 per cent per annum. The sires were selected on the basis of their progeny (Rhoad 1955b).

BREEDING

The yield of indigenous Zebu cattle under good management rarely averages 1,365 litres each lactation in the good herds. Some special Sahiwal herds in India have reached 2,275 litres but in many cases 1,140 litres of milk has been accepted as the limit of production with good feeding, watering and attention. The Serere herd in Uganda achieved a milestone by culling at 1,140 litres, but the herd at Namulonge Research Station in Uganda had two Friesian cows which produced over 4,550 litres, one producing 5,190 litres in 305 days (Passmore 1967). Similar comparisons are available for the West Indies. Where commercial milk production is profitable and the main aim, exotic animals such as Holsteins (Friesians) or Jerseys are chosen and maintained by AI using imported semen. However pedigree animals need regular veterinary attention and are often difficult to get into calf, particularly with AI and exotic animals with about 20 per cent indigenous blood are often equally productive and require less attention. In Jamaica, the risk of tick fever eliminating expensive pedigree animals has been reduced by importing young calves for rearing in Jamaica where they will acquire disease resistance. Ultimately there will be selection within the animals for adaptation to local conditions and a Jamaica Holstein developed which will combine a much higher productivity than the Zebu without the extreme susceptibility to the local conditions of the North American Holstein.

Briefly, the animal much match the quality of management, with the pure Zebu animals for the unimproved conditions which will remain general for many years through much of Africa and Asia; and the improved imported *Bos taurus* animals for the intensive milk production. There remains the question of the average, and regions of poor management which are improving, and herds producing beef more intensively where Zebu × *Bos taurus* crosses are the answer, the amount of *Bos taurus* blood increasing as the management improves.

Progeny testing

Progeny testing is judging an animal by its progeny, rather than its appearance or production. In progeny testing it is essential to consider all the progeny, not just a selection. By this method any undesirable characteristics will become apparent, and animals which can provide a genetic improvement to herds can be selected.

Progeny testing is normally applied to a bull as a large number of calves can be bred in a short time from one bull, particularly if artificial insemination is practised, and the bull or its stored semen can be used after the animal is proven.

The assessment is generally done on a statistical basis from a comparison of all the bull's progeny with similar animals of a different bull or bulls, reared and milked alongside them, or comparing the performance of the progeny with their dam alongside making the necessary corrections for age and lactation differences between mother and daughter. The method of assessing bulls is simply explained in 'Twenty-five million' published by the AI Organization of Great Britain.

In the case of cows, by the time their progeny can be assessed, they are too old, but nevertheless the progeny produced by each cow should be studied in assessing breeding plans. If a cow's daughters have been consistently good producers she should not be discarded if her own milk yield or her total score falls below the standard set.

A breeding programme from 1953-65 involving the progeny testing of small East African Zebu cattle for improved growth rates is described by Stobbs (1966). Prior to 1956, the policy was to line breed within a closed herd to a bull - Kateregá (Plate 5.2), using three full brothers. This was being done before Kateregá was proven. Selection in this programme was limited to growth rate. The selected young bulls were mated to a random choice of 50 cows, and the growth of their offspring compared with the growth rate of the progeny of other bulls in use at the same time. The bulls were loaned out until the test results were known then returned for use on selected cows whose progeny have consistently been superior. These tests showed that one bull (Umorowici) of the 14 compared, produced progeny in 1960 and 1961 that at 2-3 years old were growing 13 per cent faster than the mean of all their contemporaries.

The results also stress that performance testing must be carried out at an environmental level for which the stock are required, as the performance in a favourable environment is dependent upon a different genetic basis than performance in an unfavourable environment, also that the size of the herd is one of the most important factors affecting the speed of genetic advancement.

Grading-up

In the Kenya Highlands, the indigenous cattle have been improved by grading-up. Imported bulls of the Shorthorn or other breeds have been crossed and back-crossed, until after the fifth cross they are 31/32 pure.

The process takes about 30 years but is cheaper than importing pedigree herds and retains some of the natural disease resistance of the indigenous cattle. By the use of AI, grading-up can be done with bulls of much better quality than would be brought to the tropics, and the development of Zebu/Holstein or Zebu/Jersey grade animals is probably ideal for developing areas with a limited management skill and experience. Similarly, grade Zebu beef animals will increase beef production, and here the Charolais may be preferred as it matures rapidly under proper management. Other exotic breeds could be useful for grading-up, for example, the brownish-red South Devon, a dual-purpose breed, the largest of all British breeds. Some South Devon show bulls weigh as much as 1.5 t. The South Devon cows are very good milkers, and the milk has a high percentage of butterfat ideal for ghee manufacture. This breed was very popular in South Africa as it is well suited to the climate and produced good trek oxen.

It is generally accepted that under adverse tropical conditions it is desirable to retain about 10 per cent Zebu blood. After the third top cross the progeny are 1/8 or 12.5 per cent Zebu and the problem arises with the choice of the next sire. Another top cross reduces the Zebu blood to a level that valuable resistance might be lost. This course would probably be chosen unless there were signs of resistance and heat tolerance having reached a minimum. To use a Zebu bull would be a regressive step from a production point of view, and to use a grade animal could result in too much segregation. Up-grading has been carried too far in Venezuela by crossing the local Criollo cattle, descendants of cattle imported from Spain over 400 years ago, with Zebus, so that the productivity of the Criollo cattle is lost. There is concern in adjacent Trinidad that the reverse may happen with Holstein imports and the disease resistance of the local cattle lost. Experience in Trinidad suggests such crosses can produce high yielding cows much easier to manage than pure Holsteins. There must be few grade animals which have been progeny tested or used for AI service. There is an obvious requirement for the organizations offering world-wide supplies of semen to have some of their animals progeny tested under tropical conditions, on both Zebu and European cattle.

Artificial insemination

In most cattle countries an AI service is now available provided by the government. Some progressive large herds, even in the tropics, have staff to carry out AI themselves as it is very important to have the inseminator available at short notice. The heat period of Zebu cattle is often short and not easily detected. Most of the failures result from the long delay between the cow being observed to be on heat and the specialist arriving. Where communications are bad, as in many tropical areas, it is often difficult to get cows in calf by AI.

Marples and Trail (1967) gave the following results for Lunyo farm in Uganda, 1.6 km (1 mile) away from the AI Service Centre.

	No.	**%**
In calf to first insemination	64	43.2
In calf to second insemination	34	23.0
In calf to third insemination	16	10.8
In calf to subsequent insemination	20	13.5
In calf but wrong calving date for recorded insemination	9	6.1
Not conceived (3,5,8,10,11 attempts)	5	3.4
	148	100

At the Kendal dairy in Trinidad, situated some 32 km (20 miles) from the AI Service Centre, only 31 per cent of the Holstein/Zebu grade animals conceived at the first insemination.

There are a number of advantages for AI which make it well worth while if available, and make it essential for any country wishing to develop its livestock industry.

1. The bulls available are far better than the individual could afford to purchase or would risk introducing into a tropical climate. These bulls are usually progeny tested for various attributes such as increasing butter fat, or total yield.
2. A wide choice of bulls is available, including bulls from many foreign countries.
3. It is not necessary to have a bull on the farm. This eliminates a source of trouble on the farm and enables a small farmer to keep an extra cow. A bull is usually necessary in a large herd for holidays and animals difficult to get into calf.
4. Infertility diseases, many of which are spread venereally, are controlled. Disposable plastic pipettes are used only once.
5. Bull infertility which may not be detected for 2 or 3 months is eliminated.
6. Calving can be planned to coincide with the higher milk price or better grazing conditions. In exotic herds the difficulty of getting the animals into calf may override planned calving.

Artificial insemination in brief offers the economic mass improvement of better managed herds by the use of proven superior bulls.

Semen is stored mainly in liquid nitrogen at around −230 °C in special containers which can be airfreighted around the world. An inseminator can be trained in a 6 week intensive course. The cost of the

Plate 5.10 Windsor Lad. Jamaica Hope bull. Jamaica (Courtesy Ministry of Agriculture and Lands, Jamaica)

semen depends upon the chosen sire but in many cases is cheaper than keeping a bull.

Progeny testing - AI bulls

In the UK from the date the first semen is taken from a young bull it takes about 5 years for his progeny record to come through. This is not the final judgement on merit as further information continues to accumulate on which the bull can be constantly reappraised.

New breeds

Where conditions are not so favourable to European stock, a cross with a lower proportion of European *Bos taurus* and a higher proportion of *Bos indicus* is required. In Trinidad, white Zebus of the Nellore type were being crossed with Friesians and the 3/4 and 5/8 crosses were very satisfactory, but it was not considered possible, to stabilize that cross. On the other hand, the Jamaica Hope (Plate 5.10) is an example of a

Plate 5.11 Camote, 21-month-old Santa Gertrudis show bull weighing 698 kg. Texas. (Courtesy King Ranch, Texas; photograph by John Cypher Jnr.)

stabilized cross, 80 per cent Jersey with some Zebu (Sahiwal) and Holstein blood. It took over half a century to breed, and was favoured by the Jersey animals being adapted to local conditions. Unfortunately the number available is inadequate. In 1972 there were only about 7,000 registered Jamaica Hope cattle of which less than 3,000 were purebreds. The bulls could be useful animals for grading-up. Similarly, the Santa Gertrudis (Plate 5.11) breed developed at the King Ranch, Texas, is 3/8 Brahman imported from India, and 5/8 Shorthorn. After over 30 years of breeding, the production and quality of the Shorthorn was combined with the heat tolerance, fly resistance and the ability of the Brahman breed to utilize sparse herbage. At the outset of this breeding programme 5,000 - 7,000 animals were selected. Few governments could offer such resources. The Brahman has also been combined with the French breed, Charolais, to give the Charbray cattle, and the Hereford and Shorthorn to give the Beefmaster. The Red Poll has been used with the Zebu in Jamaica to give the Jamaica Red, and in Senegal to produce the Nellthropp. The Jersey would appear to be a very suitable breed to use for grading-up or for creating a new breed with the Zebu as it has better heat tolerance than the other European breeds, is easy to handle, and breeds regularly. The Jersey has been used in North India to cross with the Red Sindhi to give the Jersind which calves earlier than the Red Sindhi and yields 2,275 litres in its first lactation. Unfortunately the Jersey does not give such a good meat carcass as the Holstein, it has a lower potential for milk production than the Holstein, and the number of animals and proven sires available throughout the world is much smaller than in the case of the Holstein.

BALI CATTLE (Bos banting synonym Bos sondaicus)

These cattle appear to have been domesticated since prehistoric times in South-East Asia and have not been crossed with other cattle although they have the same number of chromosomes as *Bos taurus* and *Bos indicus* and will cross. There are about a million of these animals, practically all in Indonesia where they form 15 per cent of the total cattle population. Bali cattle are easily trained and well suited to working in small fields. Two animals can carry out all the cultivations on 3 ha of land. They are generally hardy, appear resistant to ticks and thrive on available forage and stubbles without any concentrates. Normally they are not milked as they are poor producers. They are very fertile, and have a high killing out percentage at 375 to 400 kg, around 57 per cent or more. The carcasses have a low fat content, and the meat, dark red in colour, is very popular in Hong Kong and other Far Eastern markets. These animals are ideal for peasant farmers in South-East Asia and their usefulness could be spread throughout that area (Payne & Rollinson 1973).

WATER BUFFALOES (Bubalus bubalis)*

There are some 100 million domesticated buffaloes in 38 different countries in the world, about half in India, and a quarter of them in China.

The buffalo, which was probably domesticated some 20 centuries B.C., is an essential part of the farming system of the East. Utilizing pasture, poor even for a Zebu, the buffalo provides a reasonable milk yield which has about 7 per cent butterfat, and meat. It pulls carts, hauls timber and turns mills, but its supreme function is in paddy cultivation, where it has been described as the 'living tractor'.

* (Cockrill 1974)

No other animal can haul the primitive cultivation implements, hock deep through waterlogged land to puddle the soil in the small rice paddies. After planting it raises irrigation water, and finally hauls away the crop and drives the threshing equipment.

The buffalo is well adapted for working in hot, marshy or waterlogged lands as occur in continental India and Asia, though they are better when worked in the cooler parts of the day. Wallowing in rivers or mudholes assists in maintaining their body temperature and also appears to give them pleasure.

Despite the considerable importance of the buffalo in tropical agriculture, and particularly rice production, little scientific investigation has been carried out on the buffalo. In India, successful selection for milk yield has been achieved, and they make a useful contribution to the milk supply of many towns such as Bombay where the highly successful Kaira Cooperative, one of the largest in the world, also supplies 'tone' milk, a blend of dried skim milk, water and water buffalo milk with 9-10 per cent solids-not-fat, for the lower income groups. In Pakistan they have virtually replaced cows for urban milk supply. The buffalo milk is also used for making ghee (clarified butter) which is very popular in India, Pakistan and Arab countries.

MIXED FARMING

Traditional cattle-owning tribes are mainly in areas not particularly suited to crops and enjoy their nomadic existence, though certain of them, such as the Fulani of West Africa occasionally take up a more settled form of life and cultivate land.

In certain areas where cattle thrive and rainfall is adequate for cropping, there is often a case for integrating these two historically independent forms of husbandry, cattle and crops, into a system of mixed farming.

This was encouraged in Northern Nigeria from the late 1920s, the numbers of successful farmers increased steadily.

1932 - about 100
1936 - about 1,000
1953 - about 10,000
1965 - 36,000

In the 10 years following the Second World War an average of 1,900 joined the scheme and 567 failed each year. The success was due to it being an adaptation of a traditional system as the farmers already knew the value of manure. Fulani herdsmen have long been encouraged to graze their herds on land after cereals have been harvested.

It has been estimated by Samaru that in a fully integrated mixed farm, where animal power is maintained on a grass/legume sown pasture, a minimum area of 10 ha is needed and this is not possible in heavily populated intensely cultivated areas. In parts of East Africa, where conditions are suitable mixed farming has also been encouraged.

The more agricultural areas, where many farmers own cattle, are the most suitable for the development of mixed farming. The big step is to persuade the farmer to herd his cattle individually rather than communally, or to purchase some cattle. The purchase of cattle requires a relatively large captial outlay which was provided in Northern Nigeria by short-term loans from the local government and in Uganda by the Land Bank. Loans were subject to the security offered by the applicant and an assessment of the possibility of the borrower achieving success. Further capital is required for housing and fencing though locally available materials are largely used. As the standard of management provided by individual herding should reduce calf losses to a minimum, the capital investment in cattle should therefore show a return of about 20 per cent per annum, and the loan repaid in about 10 years. The expenditure of extra capital and labour on integrating the crops and stock is justified by the following benefits.

Improved cattle

The condition of cattle herded in small groups, receiving close personal attention from the owner, regular watering and the benefit of supplementary food from crop residues, is very markedly superior to the communally herded cattle.

Maintenance of soil fertility

Once the land is protected from indiscriminate grazing, the pasture can be improved by seeding legumes such as Stylo into the grass. Even better but more costly the original cover can be replaced with a selected grass/legume mixture if seed is available. Ultimately these leys can be cultivated and cropped, and the cropping area planted to a ley. In each of these treatments the legume is bringing nitrogen into the system.

Reduced stock losses

The close attention enables veterinary assistance to be given to sick animals to reduce avoidable deaths. Calf losses in particular are reduced. Loss by theft is also virtually eliminated.

Improved nutrition

The availability of milk at the homestead makes a marked improvement in the diet of the owner and his family.

Additional source of income

The cattle owner has an extra source of income from the sale of stock and surplus milk. One animal is probably worth more than the produce of half a hectare of cultivation.

Assistance for cultivation

The short grass areas are suitable for cultivation by ox-drawn implements, particularly the plough and weeder and this can take some of the drudgery out of hoe-based cultivation. Two heifers, two working oxen and an ox-plough were considered the basic equipment of the mixed farming system of Northern Nigeria. Many farmers also have a cart. It is estimated that there are 50,000 teams of bullocks, mainly white Fulani but also Adamawa Gudali and Shuwa, in Northern Nigeria. They are trained at 4 years old and work till about 10 years old. With the Emcot ridger, the main implement, 0.5 - 1.5 ha of land can be ridged per day (Laurent 1969). In the wetter areas, the cultivated areas are often cleared from forest and contain too many tree stumps for ox-drawn implements. The ox-drawn cart is a vital means of transport in Sri Lanka and India. Between 1951 and 1958, 2,780 ox-carts were bought by Malawi farmers. These were subsidized by the Agricultural Production and Marketing Board. Ox-carts have declined in importance in Africa though they are still of use to the small farmer for carrying manure and crops. Oxen may also serve as pack animals for carrying produce to market, lift water for irrigation, and help in threshing. A number of ox-drawn implements, suitable for West Africa, are manufactured in Senegal.

Provision of cattle manure

Cattle manure is available to improve the fertility of the land. As about 80 per cent of the nitrogen and a large part of the potash and phosphate ingested by cattle is excreted in the urine and faeces, this is very valuable. A cart is useful for haulage.

In Zaria Province of Northern Nigeria, grain stalk and coarse grass, unsuitable for feeding, are used for bedding and converted into manure. Two bullocks produce 5 t of manure a year (King 1939). This is sufficient to treat the cultivated area of the average farmer with 2 t per annum. Such a dressing may double the yield of Guinea corn in this area, where crops, particularly cereals and groundnuts, show a very marked response to manure (Hartley & Greenwood 1933; Hartley 1937) due to the phosphate content, though the organic matter becomes important in overcropped land.

Farmers growing maize, coffee and bananas can obtain marked improvement in yields from the application of cattle manure, and the same probably applies to cocoa. Apart from the added plant nutrients, the improved water retention of the soil has a visible effect on yield in areas marginal for rainfall or with a marked dry season.

Long-term trials at Serere (Uganda) showed a marked effect from infrequent applications of cattle manure, i.e. 5 t every 5 years. The effect was cumulative and cotton, sorghum, *Eleusine* and groundnuts responded. Even 5 years after the last application, the crops showed a response (Annual Report Dept. of Agriculture, Uganda 1958).

Where night grazing is carried out the year before ploughing, an increase of 110–220 kg of the following cotton crop was recorded at Serere.

In short, mixed farming provides a better fed farmer with superior cattle, utilizing crop waste to produce more milk, meat and manure. The manure in turn increases the production from the cultivated area; a further contribution to the improved nutrition and income of the farmer.

There is considerable scope for increasing production from cattle as illustrated (Table 5.11) by the rough comparisons of productivity in the different regions of the world.

Table 5.11 Production of beef and milk per head of cattle population, 1966/67

Region	Kilograms per head per year	
	Beef	**Milk**
North America	87	520
Europe inc. USSR	59	1016
Oceanic	47	512
Latin America	27	89
Africa	14	76
Asia inc. mainland China	11	80

Source: The state of food and agriculture 1970. FAO.

References

Abercrombie, K. C. (1972) Agricultural mechanization and employment in Latin America, *Int. Labour Rev.*, **106**, 11.

Adams, S. N. and **McKelvie, A. D.** (1955) Environmental requirements of cocoa in the Gold Coast. *Cocoa Conference, London*, 1955. The Cocoa, Chocolate and Confectionary Alliance, 22-7.

Ahn, P. A. (1970) West African Agriculture (3rd edn.) Vol. 1 *West African Soils*. Oxford University Press. 332 pp.

Ainsworth, G. C. (1966) A General purpose classification of fungi. *Bibliography* of *Systematic Mycology* (1966), 1-4. Commonwealth Mycological Institute, Kew, Surrey.

Aiyer, R. S. *et al.* (1972) Long term algalization field trial with high-yielding varieties of rice (*Oryza sativa* L.), *Indian J. Agric. Sci.*, **42** (5), 380-3.

Akehurst, B. C. (1981) *Tobacco* (2nd. edn.) Longman, London. 388 pp.

Alderfer, R. B. and **Merkle, F. G.** (1943) The comparative effects of surface application vs. incorporation of various mulching materials on structure, permeability, run-off and other soil properties, *Soil Sci. Soc. Amer. Proc.*, **8**, 79-86.

Allan, W. (1965) *The African Husbandman.* Oliver and Boyd, London. 505 pp.

Allard, R. W. (1960) *Principles of Plant Breeding*. New York, John Wiley.

Allen, C. J. *Calf Rearing*. 16 pp. Booklet. Ministry of Agriculture and Lands, Jamaica.

Allen, C. J. (1967) A programme to increase milk production in Jamaica, Seminar on Development and Progressing of the Agricultural Sector, May 22-26, 1967. Santo Domingo, Dominican Republic.

Allen, O. N. and **Baldwin, I. L.** (1954) Rhizobia - legume relationships, *Soil Sci.*, **78**, 415-27.

Allison, F. E. and **Morris, H. J.** (1930) Nitrogen-fixation by soil algae. *Proc. 2nd Int. Congr. Soil Sci.*, Leningrad, **3**, 25-8.

Altona, R. E. (1966) Urea and biuret as protein supplements for range cattle and sheep in Africa, *Outlook on Agriculture*, **5** (1), 22-7.

Andrews, D. J. (1970) Breeding and testing dwarf sorghums in Nigeria, *Expl. Agric.*, **6**, 41-50.

Andrews, D. J. (1972a) Intercropping with sorghum in Nigeria, *Expl. Agric.*, **8**, 139-150.

Andrews, D. J. (1972b) Intercropping with guineacorn - a biological co-operative, *Samaru Agric. Newsletter*, **14**, 20 and 40.

Annual Report (1958) *Annual Report, Tea Research Institute of East Africa*.

Annual Report (1959) *Annual Report of the Department of Agriculture*, 1958, Uganda Protectorate.

Annual Report (1959) *Annual Report Tocklai Experimental Station*, Indian Tea Association.

Annual Report (1959b) *Annual Report, Coffee Research Station, Ruiru, and Coffee Research Science, Kenya*, for the year 1957/58.

Annual Reports of the West Africa Cocoa Research Institute: (1959) for the year 1957/58; (1960) for the year 1958/59.

Annual Report (1961, 1962) *Annual Report of the West African Rice Research Institute*. Rokupr, Sierre Leone.

Annual Report (1965) *Annual Report of the International Rice Research Institute*, Los Banos, Laguna, Philippines.

Anon. (1936) Use of leguminous plants in tropical countries as green manure, as cover, and as shade, *Rome. Internat. Inst. Agric*.

Anon. (1959) Water and crop growth: Record of research, *Annual Report, 1959, EAAFRO*, 7-21.

Anon. (January 1960) RRIM 600 series clones, *Planters' Bull. Rubb. Res. Inst. Malaya*, **46**, 8-13.

Anon. (1963) Review of the Planters' Conference (cover crops), *Planters' Bull. Rubb. Res. Inst. Malaya*, **68**, 111-16.

Anon. (1965) Gamma-B.H.C. effectively controls stem borers. *IRRI Reporter*, Jan. 1965, 1.

Anon. (1966a) Pangola grass. Impressive response to nitrogen in coastal Queensland, *Rural Research in CSIRO*, **57** (Dec. 1966), 16-18.

Anon. (1966b) R.R.I.M. clonal seedling trials, *Planters' Bull. Rubb. Res. Inst. Malaya*, **82**, 10-15.

Anon. (1966c) Resistance to stem borer, *IRRI Reporter*, **2** (3), 3-4.

Anon. (1966d) Selection of clones, *Tea Estate Practice, 1966*, **64**. Tea Boards of Kenya, Tanganyika and Uganda.

Anon. Editorial (1966) Towards a million tons of rubber, *J. Rubb. Res. Inst. Malaya*, **82**, 1-3.

Anon. Editorial (1967) *Planters' Bull. Rubb. Res. Inst. Malaya*, **88**.

Anon. Editorial (1977) *Planters' Bulletin* No. 153 (Nov.)

Anon. (1973) Introducing Caribbean Stylo, *Rural Research* (CSIRO, Australia) **82**, 7.

Anon. (1974a) *Mulches: effect on plants and soil.* (1974-66 indexed), Commonwealth Bureau of Soils, Harpenden, UK.

Anon. (1974b) *Irrigation Water (1966-74)* Annotated bibliography, Commonwealth Bureau of Soils, Harpenden, UK.

Anon. (1979) *A Handbook of Rice Diseases in the Tropics*. IRRI, Philippines.

Anon. (1976) Performance of clones in commercial practice - Tenth report, *Planters' Bulletin*, No. 144 (May), p. 60.

Anon. (1977) Enviromax planting recommendations 1977-79, *Planters' Bulletin*, No. 153 (Nov.), pp. 163-82.

Anstead, R. D. (1915) *Coffee; its Cultivation and Manuring in S. India, Bangalore*. 3 pp.

Anstead, R. D. (1921) The variability of yield of individual coffee bushes, *Tropical Agriculturist (Ceylon)*, **56** (6), 338-9.

Anthony, K. R. M. and **Jones, A. J.** (1966) A cotton research programme in Thailand, *Emp. Cott. Gr. Rev.*, **43**, 257-62.

Anthony, K. R. M. and **Willimott, S. G.** (1957) Cotton interplanting experiments in the S.W. Sudan, *Emp. J. Exp. Agric.*, **25**, 29-36.

Archibald, J. F. (1955) The propagation of cacao by cuttings. West African Cocoa Research Institute, *Technical Bulletin*, No. 3, 8 pp.

Arnold, R. M. (1955) Growth of the dairy industry in Jamaica (1940-53), *Trop. Agric. Trin.*, **32**, 38-44.

Ashby, D. G. and **Pfeiffer, R. K.** (1956) Weeds a limiting factor in tropical agriculture, *World Crops*, **8**, 227-9.

Ashby, S. F. (1929) Strains and taxonomy of *Phytophthora palmivora*. Butler. *Trans. Brit. Mycol. Soc.*, **14**, 18-38.

Atger, P. and **Jacquemard, Ph.** (1965) Maladies bacteriennes de *Diparopsis watersi* Rotsch. (*Lepidoptera Noctuidae*). II, Isolement d'un bacille pathogene, *Coton et Fibr. Trop.*, **20**, 287-8.

Athwal, D. S. (1972) I.R.R.I.'s current research programme in rice, science and man. International Rice Research Institute.

Atwal, A. (1976) *Agricultural Pests of India and S.E. Asia*. Kalyani, Delhi. 502 pp.

Aubert, G. and **Fournier, F.** (1957) Sols Africain, **III**, 90.

Baker, R. E. D. and **Simmonds, N. W.** (1951, 1952) Bananas in East Africa, *Emp. J. Exp. Agric.*, **19**, 283-90; **20**, 66-76.

Balls, W. L. (1916) The influence of natural environmental factors upon the yield of Egyptian cotton. *Phil. Trans. Roy. Soc., London, B*, **208**, 157-223.

Balls, W. L. (1951) Below soil level, *Emp. Cott. Gr. Rev.*, **28**, 81-96.

Balls, W. L. (1953) *The Yields of a Crop*. E. & F. N. Spon, Lond. 144 pp.

Bals, E. J. (1970) Ultra low volume and ultra low dosage spraying, *Cott. Gr. Rev.*, **47**, 217.

Bals, E. J. (1971) Some thoughts on the concept of ULD (ultra-low-dosage) spraying,. *Report on the International Conference on Low Volume and Ultra Low Volume Applications, Belgrade*, 1970, p. 27.

Bals, E. J. (1973) Some observations on the basic principles involved in ultra-low volume spray applications, *PANS*, **19**, 193.

Barnes, J. M. (1959) Toxicity of pesticides, *Bull. Hyg.*, **34**, 1205 - 1219.

Bartley, B. G. D. (1957) Trinitario - Scavina hybrids - new prospects for cocoa improvement, *Cocoa Conference 1957*. Cocoa, Chocolate and Confectionary Alliance, London, 36 - 40.

Beauchamp, R. S. A. (1953) Sulphates in African inland waters, *Nature Lond.*, **171**, 769 - 71.

Beckinsale, R. P. (1957) The nature of tropical rainfall, *Trop. Agric. Trin.*, **34**, 76 - 98.

Beirnaert, A. and **Vanderweyen, R.** (1941) Contribution a l'étude génétique et biométrique des variétés d'*Elaeis guineensis* (Jacquin,) 1941, *INEAC, Série Scientifique*, No. 27.

Ben-Dyke, R., Sanderson, D. M. and **Noakes, D. N.** (1970) Acute toxicity data for pesticides (1970), *World Rev. Pest Control*, **9**, 119.

Bernstein, L. (1966) Soil salinity and crop productivity, *Span*, **9**, 76 - 9.

Bigger, M. (1966) Giant looper - its up to you, *Tanganyika Coffee News*, 6, 4, 177, 179, 181.

Birch, H. F. (1958) Pattern of humus decomposition in East African soils, *Nature Lond.*, **181**, 788.

Bishop, A. H. (1979) *World Animal Review*. No. 29

Blackburn, F. H. B., Hanschell, D. M. and **Clarke, M.** (1952) Some aspects of weed control in Trinidad, Conference of the B.W.I. Sugar Technologists Association in British Guiana (Reprinted *Trop. Agric. Trin.*, **29** (1 - 3) January - March 1952).

Blair Rains, A. (1963) Grassland research in Northern Nigeria. Samaru Miscellaneous Paper No. 1.

Boa, W. (1966) Equipment and methods for tied ridge cultivation. *Inform. Working Bull. Agric. Engineering*, FAO, **28**, 54 pp.

Bock, K. R. and **Guthrie, E. J.** (1978) African mosaic disease in Kenya, *Proceedings Cassava Protection Workshop*, CIAT, Cali, Colombia, 1977, p. 41 - 4.

Boerma, A. H. (1973) The world food and agricultural situation, *Phil. Trans. R. Soc. Lond. B*, **267**, 5 - 12.

Bogdan, A. V. (1977) *Tropical Pasture and Fodder Plants (Grasses and Legumes)*. Longman, London. 490 pp.

Bonsma, Jan. C. (1949) Breeding of cattle for increased adaptability to tropical and subtropical environments, *J. Agric. Sci.*, **39**, 204 - 221.

Bonsma, Jan. C. (1955a) Degeneration of the British beef breeds in the tropics and subtropics, in *Breeding of Cattle for Unfavourable Environments*. Rhoad (1955), pp. 17 - 20.

Bonsma, Jan. C. (1955b) The improvement of indigenous breeds in subtropical environments in *Breeding of Cattle for Unfavourable Environments*. Rhoad (1955), pp. 170 - 86.

Bourn, D. and **Scott, M.** (1978) The successful use of work oxen in agricultural development of tsetse infested land in Ethiopia, *Tropical Animal Health and Production* UK. 10 (4) 191 - 203.

Bouyoucos, G. J. (1949) Nylon electrical resistance for continuous measurement of soil moisture in the field, *Soil Sci.*, **67**, 319 - 30.

Bouyocos, G. J. and **Mick, A. M.** (1940) An electrical resistance method for the continuous measurement of soil moisture under field conditions, *Mich. Agric. Exp. Sta. Tech. Bull.*, 172.

Boyd, J. (1980) *Tools for Agriculture: A Buyers Guide to Low-Cost Agricultural Implements*. 2nd edn. Intermediate Technology Publications, London. 172 pp.

Bredón, R. M., Torell, D. T. and **Marshall, B.** (1967) Measurement of selective grazing of tropical pastures using oesophageal fistulated steers, *J. Range Management*, **20**, 317.

Bridges, E. M. (1970) *World Soils*. Cambridge University Press. 89 pp.

Brook, R. H. (1979) Sulphur in agriculture. *Abstracts on Tropical Agriculture*, **5** (9), 9.

Brooks, F. T. (1953) *Plant Diseases* (2nd edn.) Oxford University Press. 457 pp.

Brooks, C. H. (1953) *Egyptian Cotton*. Leonard Hill, London.

Brown, G. E. (1972) *Pruning of trees, Shrubs and Conifers*. Faber and Faber, London.

Brown, K. J. (1963) Rainfall, tie ridging and crop yields in Sukumaland, Tanganyika. *Emp. Cott. Gr. Rev.*, **XL**, 34-40.

Brown, K. J. (1965) Progress reports from experiment stations. Season 1964-65. Northern Nigeria, *ECGC Crop Physiology Section*, 11-13.

Brown, K. J. (1972) Nyanza Province. Kibos Research Station, *Cotton Research Reports, Kenya*, 1970-71.

Bunting, A. H. (1959) Agricultural research in the groundnut scheme, 1947-51 *Nature Lond.*, **168**, 804-6.

Bunting, A. H. and **Curtis, D. L.** (1969) Local adaptation of sorghum varieties in Northern Nigeria. UNESCO Natural Resource Studies VII. Reprinted 1970 as *Samaru Research Bulletin* No. 106.

Buntjer, B. J. (1971) Aspects of the Hausa system of cultivation around Zaria, *Samaru Agricultural Newsletter*, **13**, 18.

Burges, H. D. (1964) Control of insects with bacteria, *World Crops*, **16** (3), 70-6.

Buringh, P. (1979) *Introduction to the study of soils in tropical and subtropical regions*. Agric. Univ. Wageningen, Netherlands.

Burns, M. A. (1965) Urea block licks, *Queensland Agric. J.*, **91**(1), 12-15.

Burns, W. (1938) *The Progress of Agricultural Science in India during the Past Twenty-Five Years*. Indian Science Congress Association Calcutta, 1937, 133-86.

Burt, E. O. (1938) Agricultural and animal husbandry in India, *Indian Science Congress Silver Jubilee Sessions*, 1938. Ind. Sci. Cong. Ass. Calcutta, 150.

Butler, Sir Edwin J. and **Jones, S. G.** (1949) *Plant Pathology*. Macmillan, London, 979 pp.

Butterworth, M. H. (1961) Studies on Pangola grass. II. The digestibility of Pangola grass at various stages of growth, *Trop. Agric. Trin.*, **39**, 189-93.

Buzacott, J. H. (1962) The defects for which the majority of seedlings are discarded during selection, *Proc. Int. Soc. Sug. Cane Tech. Mauritius*, **11**, 410-13.

C.B.C. and **D.M.B.** (1973) The screw worm strikes back, *Nature Lond.*, **242**, 493.

Cabato, F. H. (1970) Cover cropping in coconut plantations, in *Coconut Production*, Ed. R. G. Emata, p. 92.

Camp, A. F. and **Walker, M. N.** (1927) Soil temperature studies with cotton, *Florida Agric. Exp. Sta. Bull.*, 189.

Capot, J. (1966) La production de boutures de clones sélectionés de caféièrs canephora, *Café, Cacao, Thé*, **10** (3), 219-27.

Capot, J. (1973) L'amélioration du caféier par hybridation interspécifique, *Bull. Séanc. Acad, Royale Sci. Outre-Mer*, **2**, 280.

Carpenter, P. H. (1938) The application of science to modern tea culture, *Emp. J. Exp. Agric.*, **6**, 1-10.

Carvalho, A., Monaco, L. C. and **Fazuoli, L. C.** (1978) Coffee breeding. Studies of S_2 and S_3 Mundo Novo and Burbom Amarelo progenies and Fl hybrids between these cultivars (in Portuguese), *Bragantia (Brazil)*, **37** (15), 129-38.

Chan, E. (1979) Growth and early yield performance of Malayan Dwarf × Tall coconut hybrids on the coast clays of Peninsular Malaysia (in French), *Oléagineux (France)*, **34** (2), 65-70.

Chantran, P. and **Grimal, R.** (1971) La formation des forgerons appui de la culture cottonière attelée au Mali, *Promotion Rural*, **41**, 9-13.

Chapas, L. C. and **Bull, R. A.** (1956) Effects of soil applications of nitrogen, phosphorus, potassium and calcium on yields and deficiency symptoms in mature oil palms at

Umudike, *J.W. Afric. Inst. Oil Palm Res.*, **2**, 74-84.

Chapman, H. D. (1964) Foliar sampling for determining the nutrient status of crops, *World Crops*, **16** (Sept), 36-46.

Charreau, C. (1972) Problemes poses par l'utilization agricole des sols tropicaux par des cultures annuelles, *Tropical Soil Research Symposium*, IITA, Ibadan, May 1972.

Charter, C. F. (1949) The characteristics of the principal cocoa soils, *Report of the Cocoa Conference, London*, August, 1949. Cocoa, Chocolate and Confectionery Alliance, London, 105-12.

Chenery, E. M. (1954) Acid sulphate soils in Central Africa, *Fifth Int. Congress of Soil Science, Leopoldville*, 16-21 Aug. 1954, **15**, 195-8.

Chenery, E. M. (1960) An introduction to the soils of the Uganda Protectorate, Memoirs of Res. Div.; Series 1. Soils, No. 1, Department of Agriculture, Uganda.

Chenery, E. M. (1966) Factors limiting crop production: 4 Tea, *Span*, **9** (1), 45-8.

Chew, P. S. and **Khoo, J.** (1976) Growth and yield of intercropped oil palms on a coastal clay soil in Malaysia, *Proc. Malaysia Int. Agric. Oil Palm Conference*, Kuala Lumpur, 1976, pp. 541-53.

Cheyne, O. B. M. (1952) Seed coconuts 'block-nuts' *v*. 'mother palm nuts', *Ceylon Coconut Quart.*, **3**, 123-6.

Child, R. (1959) *Report of the Director*, Tea Research Institute E. Afr. A. R., 1958, 12-17.

CIMMYT Review (1977) Centro Internacional de Mejoramiento de Maizy Trigo, Mexico, D.F., 99 pp.

Clements, H. F. (1965) Effects of silicate on the growth and leaf freckle of sugar cane in Hawaii, *Proc. Int. Soc. Sug. Cane Tech. Puerto Rico*, **12**, 197-215.

Clifford, H. C. (1954) Spraying of cattle as a method of control of cattle ticks in Uganda, *Trop. Agric. Trin.*, **31**, 19-26.

Coaker, T. H. and **Passmore, R. G.** (1958) *Stomoxys* sp. on cattle in Uganda, *Nature Lond.*, **182**, 606-7.

Cockrill, W. Ross (1966) The buffalo: a physiological phenomenon, *Span*, **9** (2), 83-5.

Cockrill, W. Ross, ed. (1974) The husbandry and health of the domestic buffalo, FAO, Rome.

Coffee, R. A. (1979) Electrodynamic energy - A new approach to pesticide application. *Proc. 10th. British Crop Protection Conf. Pests and Diseases*.

Collins, G. A. and **Coward, L. D. G.** (1971) The improvement of hand-operated groundnut decorticating machines, Trop. Prod. Inst. London, Publication G. 68.

Colonial Research (1958-59). HMSO, London.

Constable, D. C. and **Hodnett, G. E.** (1953) Manuring of *Hevea brasiliensis* at Dartonfield, Ceylon, *Emp. J. Exp. Agric.*, **21**, 131-6.

Constantinesco, I. (1976) Soil conservation for developing countries, *Soils Bulletin (FAO)*.

Cope, F. and **Trickett, E. S.** (1965) Measuring soil moisture, *Soils and Fertilizers*, **28** (3), 201-8.

Copeland, E. B. (1924) *Rice*. Macmillan, London. 352 pp.

Cormack, J. M. (1948) The construction of small earth dams, *Rhod. Agric. J.*, **46**, 355-62.

Coster, C. L. (1942) The work of the West Java Research Institute Buitenzorg, 1938-41, *Emp. J. Exp. Agric.*, **10**, 22-30.

Coulter, J. K. (1950) Peat formation in Malaya, *Malay agric. J.*, **33**, 63-81.

Cowgill, W. H. (1958) The sun hedge system of coffee growing, *Coffee and Tea Ind.*, **81**, 87-90.

Craufurd, R. Q. (1964) The relationship between sowing date, latitude and duration for rice (*Oryza sativa* L.), *Trop. Agric. Trin.*, **41** (3) 213-24.

Crowther, F. (1941) Form and rate of nitrogenous manuring of cotton in the Sudan Gezira, *Emp. J. Exp. Agric.*, **9**, 125-36.

Crowther, F. (1944) Report of Plant Physiology Section, Research Division, Dept. of Agric., Sudan.

Cunningham, R. K. (1960) Effect of major nutrients on cocoa, in *Annual Report, West African Cocoa Research Institute*, 1958-59, 51.
Cunningham, R. K. (1963) What shade and fertilizers are needed for good cocoa production? *Cocoa Growers Bulletin*, **1**, 11-16.
Cunningham, R. K. and **Lamb, J.** (1958) Cocoa shade in a manurial experiment in Ghana, *Nature Lond.*, **182**, 119.
Dagg, M. and **Blackie, J. R.** (1965) Studies on the effects of changes in land use on the hydrological cycle in East Africa by means of experimental catchment areas, *Bull. IASH*, **10**, 63-75.
Daniels, J. (1959) The inbreeding and close breeding of sugar cane, *Proc. Int. Soc. Sug. Cane Tech. Hawaii*, **10**.
Danthanarayana, W. (1966) Shot-hole borer control, *Tea Quart. Ceylon*, **37** (3), 100-105.
De, P. K. (1939) The role of blue-green algae in nitrogen fixation in rice fields. *Proc. Roy. Soc. B*, **127**, 121-39
De Bach, P. (1964) *Biological Control of Insect Pests and Weeds*. Chapman and Hall, London.
Deepak Lal (1972) Wells and welfare, *Development Centre Studies. Case Study No. 1.* Development Centre of the Organization for Economic Cooperation and Development.
de Gee, J. C. (1950) Preliminary oxidation potential determinations in a 'Sawah' profile near Bogor (Java), *Trans. 4th Int. Congr. Soil Sci.*, **1**, 300-3.
de Geus, J. G. (1954) *Means of Increasing Rice Production*. Centre d'Etude de l'Azote, Geneva. 143 pp.
Dennison, E. B. (1961) The value of farmyard manure in maintaining fertility in Northern Nigeria, *Emp. J. Exp. Agric.*, **29**, 330-6.
Dent, J. M. (1947) Some problems of empoldered rice lands in Sierra Leone, *Emp. J. Exp. Agric.*, **15**, 206-12.
Deuse, J. and **Lavabre, E. M.** (1979) Le désherbage des cultures sous les tropiques, G.-P. Maisonneuve et Larose, Paris. 312 pp.
Deveria, V. (1978) T.R.F. super clones, *Quarterly News Letter Tea Research Foundation of Central Africa (Malawi)*, No. 49.
Devuyst, A. (1953) Selection of the oil palm (*Elaeis guineensis*) in Africa, *Nature Lond.*, **172**, 685-6.
De Weille, G. A. (1957) Possibilities of forecasting epidemics caused by fungi using sunshine records (Dutch), *Ber. alg. Proefst. AVROS Medan*, No. 8, 1 (mimeo).
Dias, G. R. W. (1967) Eradication of water weeds in Ceylon, *World Crops*, **19**, 64-8.
Dijkman, M. G. (1951) *Hevea. Thirty Years of Research in the Far East.* University of Miami Press, Florida.
Dillewijn, C. van (1946). *Sugar Cane Plant Breeding*. Int. Inst. of Agric. Rome.
Djokoto, R. K. and **Stephens, D.** (1961) Thirty long term fertilizer trials under continuous cropping in Ghana, *Emp. J. Exp. Agric.*, **29**, 181-95. 245-57.
Doll, J. D. (1978) *Proceedings Cassava Protection Workshop, CIAT*, Cali, Colombia. Nov. 1977. p. 65.
Domingues, O. O. (1961) *O gado nos tropicos* (Cattle in the Tropics). Instituto de Zootecnia, Ser Monografias, Rio de Janeiro (Brazil).
Dominguez, R. P. F. (1971) Avance del trabajo sobre selecciòn de plantas de cacao (*Theobroma cacao* L.) por resistecia al Longo *Ceratocystis fimbriata, Rev. Fac. Agron. Venezuela*, **6**, 5.
Doughty, L. R. (1953) the value of fertilizers in African agriculture: field experiments in East Africa, 1947-51, *E. Afri. Agric. J.*, **19**, 30-1.
Douglas, L. A. (1965) Some aspects of coconut agronomy in Papua and New Guinea, *Papua and New Guinea Agric. J.*, **17** (2), 87-91.
Dowling D. F. (1956) An experimental study of heat tolerance in cattle, *Aust. J. Agric. Res.*, **7**, 469-81.

Dowling, D. F. (1959) Medullation in heat tolerance of cattle, *Aust. J. Agric. Res.*, **10**, 736-43.

du Bois, H. (1957) Types d'assolement en culture extensive de la zone cotonière Nord, *Bull. d'Information de l'INEAC*, **6**, 227-41.

Dudal, R. (1958) Paddy soils, *Newsletter, International Rice Commission, FAO, Bangkok*, **7** (2), 19-27.

Dudal, R. (1963) Dark clay soils of tropical and sub-tropical regions, *Soil Science*, **95** (4), 264-70.

Dudal, R. (1965) Dark clay soils of tropical and sub-tropical regions, R. Dudal (Ed.). Rome, *FAO Agricultural Development paper 83*. 161 pp.

Dudal, R. (1966) Soil resources for rice production, Mechanization and the World's Rice Conference, Massey-Ferguson with FAO.

Dudal, R. and **Moormann, F. R.** (1964) Major soils of South East Asia, their characteristics, use and agricultural potential, *J. Trop. Geogr.*, **18**, 54-80.

Edgar, A. T. (1960) *Manual of Rubber Planting (Malaya), 1960*. Incorp. Society of Planters, Kuala Lumpur, 705 pp.

Efferson, J. N. (1952) *The Production and Marketing of Rice*. Rice Journal, New Orleans, USA, 534 pp.

Ellis, R. T. (1967) The prospects for irrigation of tea in Central Africa, *Investors' Guardian*, **209** (6479), 889-91.

Ene, L. S. O. (1977) Control of cassava bacteria blight (CBB), *Tropical Root and Tuber Crops Newsletter (USA)*, No. 10, 30-1.

Evans, H. (1936) The root system of the sugar cane, Pt. II, *Emp. J. Exp. Agric.*, **4**, 208-20.

Evans, H. (1955a) Factors affecting cocoa yields - discussion, *Report of the Cocoa Conference, London*, September 1955. Cocoa, Chocolate and Confectionery Alliance, London, p. 28.

Evans, H. (1955b) Nutritional requirements of cocoa - discussion, *Report of the Cocoa Conference, London*, September 1955. Cocoa, Chocolate and Confectionery Alliance, London, p. 50.

Evans, H. (1961) Personal communication, 24 November 1961, commenting on the section on shade in the 1st edition of *Tropical Agriculture*.

Evans, H. C. (1973) New developments in black pod epidemiology, *Cocoa Growers Bulletin*, **20**, 10.

FAO (1966) *Fertilizers and their Use*. A pocket guide for extension officers.

FAO (1970) *Provisional Indicative World Plan for Agricultural Development*. 672 pp.

FAO (1978a) Research on the control of the coconut palm rhinoceros beetle. Phase II, Fiji, Tonga, Western Samoa, *Technical Report*, FAO, Rome.

FAO (1978b) Ruminant nutrition. Selected articles from the World Animal Review, FAO Animal production and health, Paper 12.

FAO (1979) The African trypanosomiasis, FAO Animal Health Paper No. 14.

Farbrother, H. G. and **Manning, H. L.** (1952) *Climatology*. Progress Reports from Experiment Stations, 1951-52, Namulonge, Uganda. Emp. Cott. Growing Corp., Lond., 3-6.

Feakin, S. D. (1970) (Ed.) Pest control in rice, *PANS Manual No. 3*, Min. Overseas Development, London. 270 pp.

Fielden, G. St. Clair (1940) (Compiler) Vegetative propagation of tropical and subtropical plantation crops, Imperial Bureau of Horticulture and Plantation Crops - *Tech. Comm.*, **13**, 99 pp.

Fennah, R. G. and **Murray, D. B.** (1957) The cocoa tree in relation to its environment, *Report of the Cocoa Conference, London*, September 1957. Cocoa, Chocolate and Confectionery Alliance, London, pp. 222-7.

Ferguson, W. (1971) Adaptive behaviour of cattle to tropical environments, *Trop. Sci.*, **13**, 113.

Fernie, L. M. (1965) The behaviour of arabica coffee clones on different pruning systems,

Report Coffee Res. Station, Lyamungu, Tanganyika, 60-7.

Ferwerda, F. P. (1959) The supply of better planting material. 2. Canephoras (Robusta), *Advances in Coffee Production Technology*. B. Sachs and Pierre E. Sylvain (Eds.). Coffee and Tea Industries, NY, 36-40.

Finelle, P. (1973) African animal trypanosomiasis, *World Animal Review*, **7**, 1. Part 1, Disease and chemotherapy.

Fleming, S. D. (1958) Perennial grasses as a limiting factor in tropical agriculture and methods for their control, *African Weed Control Conference*, July 1958 - Southern Rhodesia. 179-197.

Freeman, W. E. (1967) Monthly general meeting of the Agricultural Society of Trinidad and Tobago at La Deseada Estate, 23 February 1967, *J. Agric. Soc. Trin. Tob.*, **67**, 111-23.

French, E. W. and **Gay, W. B.** (1963) Weed control in rice fields. *World Crops*, **15**, 3-11.

French, M. H. (1938) *Ann Rep. Dept. Vet. Sci. Anim. Ind. Tanganyika*, Pt. II, 80.

French, M. H. (1956) The effect of infrequent water intake on the consumption and digestibility of hay by Zebu cattle, *Emp. J. Exp. Agric.*, **24**, 128-36.

Gadd, C. H. (1935) Drought conditions in relation to tea culture, *Tea Quart.*, **8**, 20.

Gadd, C. H. (1949) Monographs on tea production in Ceylon issued by the TRI Ceylon, No. 2. *The Commoner Diseases of Tea*.

Garmany, H. F. M. (1956) Grass mulching of seedbeds: saves time and labour, *Rhod. Tobacco*, **13**, 6.

Garrard, N. M. (1966) Paddy rice production, Mechanization and the World's Rice Conference, Massey-Ferguson with FAO Stoneleigh, England.

Gasser, J. K. R. (1964) Fertilizer urea, *World Crops*, **16**, 25-32.

George, E. F. (1962) Annual Report Mauritius Sugar Research Institute.

Geortay, G. (1956) Données de base pour le gestion de paysannats de cultures vivriere en région équatoriale forestière, *Bulletin information de l'INEAC (Congo Belge)*, August 1956, 227-9.

Giglioli, E. G. (1956) *Red Rice Investigations 1951-56*. British Guiana Rice Development Co. Ltd. and British Guiana Department of Agriculture, 68 pp.

Giglioli, E. G. (1966) The farming world, No. 398, BBC broadcast, Overseas Service.

Gilbert, S. M. (1945) The mulching of *Coffea arabica*, *E. Afr. Agric. J.*, **11**, 75-9.

Gokhale, N. G. (1955) Estimating the decrease in yield on ceasing to manure unshaded tea, *Emp. J. Exp. Agric.*, **23**, 96-100.

Goldsworthy, P. R. (1970) The growth and yield of tall and short sorghums in Nigeria, *J. Agric. Sci.*, **75**, 109.

Goldsworthy, P. R. and **Heathcote, R.** (1963) Fertilizer trials with groundnuts in Northern Nigeria, *Emp. J. Exp. Agric.*, **31** (124), 351-66.

Goodchild, N. A. (1959) Note on the relative effectiveness of sulphate of ammonia and urea as a fertilizer for tea, *Tea Quart., TRI East Afr.*, **2**, 62-3.

Grassl, C. O. (1962) Problems and potentialities of intergeneric hydridization in a sugar cane breeding programme. *Proc. Int. Soc. Sugar Cane Tech. Congr., Mauritius* (1962), **11**, 447-455.

Gray, B. S. (1965) *The Oil Palm* (Discussion, p. 11) Tropical Products Institute Conference, London.

Green, M. J. (1964) *Vegetative Propagation of Tea*. Pamphlet No. 20. Tea Research Institute of East Africa. 21 pp.

Greene, H. (1948) *Using Salty Land*. FAO Agricultural Studies, No. 3.

Greene, H. (1960) Paddy soils and rice production, *Nature, Lond.*, **186**, 511-13.

Greene, H. (1961) Some recent work on soils of the humid tropics, Soils and Fertilizers, **24**, 325-27.

Greenland, D. J. and **Lal, R.** (1977) (Eds.) *Soil Conservation and Management in the Humid Tropics*. Wiley, Chichester. 283 pp.

Greenland, D. J. and **Nye, P. H.** (1959) Increase in the carbon and nitrogen contents of

tropical soils under natural fallows. *J. Soil Sci.*, **9**, 284-99.
Greenwood, M. (1951) Fertilizer trials with groundnuts in Northern Nigeria, *Emp. J. Exp. Agric.*, **19**, 225-41.
Griffith, G. ap. (1951) Factors influencing nitrate accumulation in Uganda soil, *Emp. J. Exp. Agric.*, **19**, 1-12.
Griffith, G. ap. and **Mills, W. R.** (1952) *Dept. of Agric. Uganda, Record of Investigations No. 2.* 1949-50, Govt. Printer, Entebbe, 25.
Grimes, R. C. and **Clarke, R. T.** (1962) Continuous arable cropping with the use of manure and fertilizers, *E. Afr. Agric. J.*, **28**, 74-80.
Grist, D. H. (1975) *Rice*, 5th edn. Longman, London, 602 pp.
Hagenzieker, F. (1956) Studies on subsoil placement of fertilizer at Urambo, Tanganyika Territory, *Emp. J. Exp. Agric.*, **24**, 109-20.
Haines, W. B. and **Crowther, E. M.** (1940) Manuring of *Hevea* III. Results on young buddings in British Malaya, *Emp. J. Exp. Agric.*, **8**, 169-84.
Hale, J. B. (1947) Mineral composition of leaflets in relation to the chlorosis and bronzing of oil palms in West Africa. *J. Agric. Sci.*, **37**, 236-44.
Hall, D. W. (1970) Handling and storage of food grains in tropical and subtropical areas, *FAO Agricultural Development Paper No. 90*. 350 pp.
Handog, A. S. and **Bartolome, R.** (1966) The effect of spacing on the yield of arabica coffee, *Coffee Cacao J.*, **9** (1), 10-18.
Hanna, A. D., Judenko, E. and **Heatherington, W.** (1955) Systemic insecticides for the control of insects transmitting swollen shoot virus disease of cocao in the Gold Coast, *Bull. Ent. Res.*, **46**, 669-710.
Harding, W. A. T. and **Cameron, D. G.** (1972) New pasture legumes for the wet tropics, *Queensland Agric. J.*, **98**, 394.
Hardy, F. (1933) Cultivation properties of tropical red soils, *Emp. J. Exp. Agric.*, **1**, 103-12.
Hardy, F. (1936) Some aspects of tropical soils, *Trans. 3rd Int. Congr. Soil Sci.*, **2**, 150-63.
Hardy, F. (1946) Seasonal fluctuations of soil moisture and nitrate in a humid tropical climate, *Trop. Agric. Trin.*, **23**, 40-9.
Hardy, F. (1960) *Cacao Manual*. Inter-American Institute of Agricultural Sciences, Turrialba, Costa Rica, 395 pp.
Hardy, F. (1970a) *Suelos Tropicales.* Herrero Hermanos, Sucesores, S.A., Mexico. 334 pp.
Hardy, F. (1970b) *Edafologia Tropical.* Herrero Hermanos, Sucesores, S.A., Mexico. 416 pp.
Hardy, F. and **Derraugh, L. F.** (1947) The water and air relations of some Trinidad sugar cane soils. Part I, *Trop. Agric. Trin.*, **24**, 76-87.
Harler, C. R. (1956) *The Culture and Marketing of Tea*. Oxford.
Harrison, W. H. and **Aiyer, S. P. A.** (1916) The gases of swamp rice soils, *Mem. Dept. Agr. India*. (*Pusa*), **IV**, 1-18, 135-49.
Harrison Church, R. J. (1974) *West Africa*. 7th edn. Longman, London.
Hartley, C. W. S. (1958) Advances in oil palm research in Nigeria in the last 25 years, *Emp. J. Exp. Agric.*, **26**, 136-51.
Hartley, C. W. S. (1977) *The Oil Palm*. 2nd edn. Longman, London. 824 pp.
Hartley, K. T. (1937) An explanation of the effect of farmyard manure in Northern Nigeria, *Emp. J. Exp. Agric.*, **5**, 254-63.
Hartley, K. T. and **Greenwood, M. G.** (1933) The effect of small applications of farmyard manure on the yields of cereals in Nigeria. *Emp. J. Exp. Agric.*, **1**, 113-21.
Haynes, D. W. M. (1960) Agricultural engineering development in N. Nigeria, *Agric. Mechaniz. Bull.*, **1** (2), 13-24.
Haynes, D. W. M. (1966) The development of agricultural implements in Northern Nigeria, *Samaru Res. Bull.*, 65.
Hayward, J. A. (1972) Relationship between pest infestation and applied nitrogen on cotton in Nigeria, *Cott. Gr. Rev.*, **49**, 224.

Heathcote, R. G. (1970) Soil fertility under continuous cultivation in Northern Nigeria. 1. The role of organic manures, *Exp. Agric.*, **6**, 229.

Henzell, E. F. and **Norris, D. O.** (1962) Processes by which nitrogen is added to the soil/plant system. From *A Review of Nitrogen in the Tropics with particular reference to Pastures*. A Symposium. Bulletin 46. Comm. Bur. Past. Fld. Crops.

Herklots, G. A. C. (1972) *Vegetables in South-East Asia.* Allen and Unwin, London. 525 pp.

Hernandez, C. C. (1956) The establishment of pasture land in the Philippines, *J. Soil Sci. Soc. Philipp.*, **8** (19), 150-2.

Hill, D. S. (1975) *Agricultural Insect Pests of the Tropics and their Control.* Cambridge University Press. 516 pp.

Holm, L. G. (1966) The role of weed control in agricultural development, *Eighth British Weed Control Conference*, 689-701.

Holm, L. G., Plucknett, D. L., Pancho, J. V. and **Herberger, J. P.** (1977) *The World's Worst Weeds - Distribution and Biology*. University Press of Hawaii, Honolulu.

Horrel, C. R. (1958) Herbage plants at Serere Experiment Station, Uganda, 1954-57. II: Legumes, *E. Afr. Agric. J.*, **24**, 133-8.

Howard, A. and **Howard, G. L. C.** (1910) The fertilizing influence of sunlight, *Nature Lond.*, **82**, 456-7.

Howard, C. R. (1980) The draft ox. Management and uses, *Zimbabwe-Rhodesia Agric. J.*, **77**, 19-34.

Hubble, G. D. and **Martin, A. E.** (1960) Nitrogen in tropical agriculture. Symposium Brisbane, *Nature Lond.*, **186**, 941-2.

Hudson, J. C. (1965) Agronomic use of a soil survey in Barbados. *Exp. Agric.*, **1** (3), 215-24.

Hurov, H. R. (1961) Green bud strip budding of two- to eight-month-old rubber seedlings. *Proc. Nat. Rubber Res. Conf. Kuala Lumpur*, 1960, 419-28.

Hutchinson, G. E. (1944) *Amer. Scient.*, **32**, 178.

Hutchinson, Sir Joseph (1970) High Cereal Yields. Paper read to the Commonwealth Section of the Society.

Hutchinson, Sir Joseph (1971) High cereal yields, *Journal of the Royal Society of Arts*, No. 5174, Vol. CXIX, p. 104. Jan. 1971.

Hutchinson, Sir Joseph (1973) Closing remarks. *Agricultural Productivity in the 1980's*. Meeting of the Royal Society, London.

Hutchinson, J. B. and **Panse, V. G.** (1937) Studies in plant breeding technique. II. The design of field tests of plant breeding material, *Ind. J. Agric. Sci.*, **7**, 531-64.

Hutchinson, H. G. (1959) Variation in liveweight of cattle on farm and ranch in Tanganyika, *E. Afr. Agric. J.*, **24**, 279-85.

Huxley, P. A. (1963) *Solar Radiation Levels Throughout the Year for some Localities in Africa and elsewhere*. Technical Bulletin No. 2. Makerere University College, Kampala, Uganda.

Huxley, P. A. (1965) Climate and agricultural production in Uganda, *Exp. Agric.*, **1**, 81-97.

IITA (1976) *Annual Report of the International Institute of Tropical Agriculture, Ibadan, Nigeria* (for 1975). 228 pp.

Imms, A. D. (1937) *Recent Advances in Entomology*. Churchill, London.

Imms, A. D. (1977) *A General Textbook of Entomology*. 9th edn. Methuen, London. Revised by Richards and Davies, 886 pp.

Ingram, J. S. and **Humphries, J. R. O.** (1972) Cassava storage - a review. *Trop. Sci.*, **14**, 131.

IRRI (1966) *IRRI Reporter* March 1966, **2**, 2.

IRRI (1973) *The IRRI Reporter* 1/73. International Rice Research Institute, Manila.

IRRI (1975) *The IRRI Reporter* 3/75. Improved rice varieties needed for world's deep water regions. International Rice Research Institute, Manila.

IRRI (1975a) *The IRRI Reporter* 4/75. International Rice Research Institute, Manila.

IRRI (1977) *The IRRI Reporter* 3/77. New IRRI machines for small farmers. International Rice Research Institute, Manila.

Ishag, H. M. (1965) The effect of sowing date on growth and yield of groundnuts in the Gezira, *Sols Afr.*, **10** (2-3), 509-20.

Jack, H. W. and **Sands, W. N.** (1929) Observations on the dwarf coconut palms in Malaya, *Malay Agric. J.*, **17**, 140-65.

Jadin, P. and **Jacquemart, J. P.** (1978) Effet de l'irrigation sur la précocité des jeunes cacaoyers, *Café, Cacao, Thé (France)*, **22** (1), 31-6.

Jayaraman, V. and **de Jong, P.** (1955) Some aspects of nutrition of tea in Southern India, *Trop. Agric., Trin.*, **32**, 58-65.

Jenkinson, D. Mc. E., and **Nay, T.** (1973) The sweat glands and hair follicles of Asian, African and South American cattle, *Aust. J. biol. Sci.*, **26**, 259-75.

Jennings, P. R. (1966) The evolution of plant type in *Oryza sativa. Econ. Botany*, **20** (4), 396-402.

Jennings, P. R. and **Johnson, L.** (1966) Breeding for improved rice production, Mechanization and the World's Rice Conference, Massey-Ferguson with FAO Stoneleigh, England.

Jenny, H. (1930) Consistency of organic matter in soils as dictated by the zonality principle, *Missouri Agr. Expt. Sta. Research Bull.*, **152**.

Jenny, H. (1950) Causes of high nitrogen and organic matter content of certain tropical forest soils, *Soil Sci.*, **69**, 63-9.

Jenny, H., Gessel, S. P. and **Bingham, F. T.** (1949) Comparative study of decomposition rates of organic matter in temperate and tropical regions, *Soil Sci.*, **68**, 419-32.

Jeppson, L. R. Keifer, H. H. and **Baker, E. W.** (1975) *Mites Injurious to Economic Plants*. Univ. of Calif. Press, Berkeley. 614 pp.

Jewitt, T. N. (1942) Loss of ammonia from ammonium sulphate applied to alkaline soils, *Soil Sci.*, **54**, 401-10.

Jewitt, T. N. (1945) Nitrification in Sudan Gezira soils, *J. Agric. Sci.*, **35**, 264-71.

Johnston, Bruce F. (1958) *The Staple Food Economies of Western Tropical Africa*. Stanford University Press, Stanford, California.

Johnstone, D. R. (1972) Micrometeorological and operational factors affecting ULV application of insecticides onto cotton and other crops. Mimeo Report, Porton.

Jolly, A. L. (1956) Clonal cuttings and seedling cocoa, *Trop. Agric.*, **33**, 233-7.

Jones, G. H. G. (1942) The effect of a leguminous cover crop in building up soil fertility, *E. Afr. Agric. J.*, **8**, 48.

Jones, K. H., Sanderson, D. M. and **Noakes, D. N.** (1968) Acute toxicity data for pesticides. *World Review of Pest Control*, **7**, 135-43.

Jones, L. H. (1974) Propagation of clonal oil palms by tissue culture, *Oil Palm News*, **17**, 1-9.

Jones, L. H. (1977) Tissue culture for tropical crops, *Spectrum, Br. Sci. News*, **147**, 13-16.

Jones, R. J. (1972) The place of legumes in tropical pastures, *Tech. Bull., ASPAC Food and Fertilizer Technol. Center*, **9**, 1.

Jones, T. A. and **Maliphant, G. K.** (1958) Yield variations in tree crop experiments with specific reference to cacao, *Nature Lond.* **182**, 1613-14.

Kasasian, L. (1971) *Weed Control in the Tropics*. Leonard Hill, London. 307 pp.

Kasasian, L., Cunningham, R. K., Smith, R. W. and **Brind, D. W.** (1978) British Overseas Aid Agricultural Research (Crop and Soil Sciences) 1968-73. Overseas Research Publication No. 25, HMSO London.

Kawaguchi, K. and **Matsuo, Y.** (1956) Movement of active oxide in dry paddy soil profiles, *Trans. 6th Int. Congr. Soil Sci. Paris*, Vol. C, 533-7.

Kay, D. E. (1979) *Crop and Product Digest No. 3 - Food Legumes*. Tropical Products Institute, London.

Kehl, F. H. (1950) Vegetative propagation of tea by nodal cuttings, *Tea Quart.*, **21**, 3-17.

Kelley, R. B. (1959) *Native and Adapted Cattle*. Angus and Robertson, Sydney.

Kendrew, W. G. (1953) *The Climates of the Continents.* (4th edn) Oxford University Press, 607 pp.

Kibler, H. H. and **Brody, S.** (1950) Effects of temperature 50 ° to 105 °F and 50 ° to 9 °F on heat production and cardiorespiratory activities in Brahman, Jersey, and Holstein cows, *Res. Bull.*, No. 464. Mo. Agric. Exp. Sta.

King, H. E. (1957) Cotton yields and weather in Northern Nigeria, *Emp. Cott. Gr. Rev.*, **34**, 153-4.

King, H. E. (1960) *Survey of Progress Reports for the Season 1958-59.* Emp. Cott. Growing Corp., Lond., 10 pp.

King, H. E. (1967) *Progress Reports from Experiment Stations*, Season 1964-65. Survey of Reports, Cotton Research Corporation, London.

King, J. M. (1939) Mixed farming in Northern Nigeria, *Emp. J. Exp. Agric.*, **7**, 271-98.

Klingebiel, A. A. and **Montgomery, P. H.** (1961) Land Capability Classification. US Dept. Agric. Handbook 210.

Knight, R. L. (1957) Blackarm disease of cotton and its control, *Plant Protection Conference* 1956. Fernhurst, Butterworths, London, 53-9.

Knipling, E. F. (1963) Alternate methods of pest control, 23-38, in *New Developments and Problems in the Use of Pesticides*, Nov. 1962, N.A. Sci., Nat. Res. Counc., USA, Pub. No. 1082.

Knipling, E. F. (1964) A new era in pest control: the sterility principle, *Agric. Sci. Rev.*, **1** (1), 2-12.

Koenigs, F. F. R. (1950) A 'Sawah' profile near Bogor, Java, *Trans. 4th Int. Congr. of Soil Sci.*, pp. 297-300.

Kranz, J., Schmutterer, H. and **Koch, W.** (1977) *Diseases, Pests and Weeds in Tropical Crops.* Verlag Paul Parey, Hamburg and Wiley. 666 pp.

Krug, C. A. (1959) The Supply of Better Planting Material. 1. Arabicas. Advances in Coffee Production Technology - Coffee and Tea Industries. Reprinted from *Coffee and Tea Industries and the Flavour Field*, **81**, 52.

Kung, P. (1966) Desirable techniques and procedures on water management of rice culture. Mechanization and the World's Rice Conference, Massey-Ferguson with FAO., Stoneleigh, England.

Kverno, N. B. and **Mitchell, G. C.** (1976) Vampire bats and their effect on cattle production in Latin America, *World Animal Review*, **17**, 1.

Lal, R. (1976) Soil erosion investigations on an alfisol in southern Nigeria, *IITA, Monograph* No. 1, IITA, Ibadan.

Lambert, A. R. and **Crowther, F.** (1935) Further experiments on the interrelation of factors controlling the production of cotton under irrigation in the Sudan, *Emp. J. exp. Agric.*, **3**, 276-94.

Lamusse, M. J. M. (1965) The effect of weed competition on the sugar content and yield of sugar cane, *Trop. Agric., Trin.*, **42** (1), 31-7.

Laurent, C. K. (1969) The use of bullocks for power on farms in Northern Nigeria, *Nigerian Inst. Soc. Econ. Res.*, Re. 61.

Lawes, D. A. (1962) Rainfall conservation and the yield of cotton in Northern Nigeria, *Emp. Cott. Growing Corporation, Research Memoirs*, No. 44.

Lawes, D. A. (1963) A new cultivation technique in tropical Africa, *Nature Lond.*, **198** (4887), 1328.

Lawes, D. A. (1966) Rainfall conservation and the yields of sorghum and groundnuts in Northern Nigeria, *Exp. Agric.*, **2** (2), 139-46.

Lawton, R. M. (1978) A study of the dynamic ecology of Zambian vegetation, *J. Ecology*, **66**, 175.

Leach, J. R., Shepherd, R. and **Turner, P. D.** (1971) Underplanting coconuts with cocoa in West Malaysia, *Cocoa Growers' Bull.*, **17**, 21; **18**, 5.

Leach, R. (1937) Observations on the parasitism and control of *Armillaria mellea, Proc. Roy. Soc. B*, **121**, 561-73.

Leather, R. I. (1972) Coconut research in Fiji, *Fiji Agr. J.*, **34** (1), 3.

Lee, B. (1979) Resistant cattle for tick control, *Rural Research*, **105** (Dec.), p. 4.

Lee, R. P. *et al.* (1959) Field investigations into gastro-enteritis and poor post-weaning weight gains in Nigerian Zebu cattle, *Bull. Epizootic Diseases, Africa*, **7**, 349-54.

Legge, J. B. B. (1960) *Some Notes on the Nature and Control of Rosette and Bushy Top Virus Diseases*. Leaflet No. 1. Tobacco Research Board of Rhodesia and Nyasaland.

Lenton, C. M. (1978) Owls as rat controllers - a preliminary report, *Planter (Malaysia)*, **54**, 72-83.

Leon, J. (1971) Germplasm collections in Central America, *Plant Genetic Resource Newsletter*, **26**, 10.

Leong, Y. S. and **Mayakrishnan** (1978) Planting material used in Peninsular Malaysia 1975, *Planters Bulletin*, No. 154 (March 1978), 15-20.

Lerner, I. M. and **Donald, P.** (1966) *Modern Developments in Animal Breeding*. Academic Press.

Levandowsky, D. W. (1959) Propagation of clonal *Hevea brasiliensis* by cuttings, *Trop. Agric. Trin.*, **36**, 247-57.

Little, E. C. S. (1967) Progress report on transpiration of some tropical water weeds, *PANS (C)*, **13** (2), 127-32.

Littlehales, J. C. G. (1960) Irrigation in coffee, *Kenya Coffee*, **25**, 97.

Loomis, R. S. and **Williams, W. A.** (1963) Maximum crop productivity: an estimate, *Crop Sci.*, **3**, 67.

Loosli, J. K. and **McDonald, I. W.** (1968) Non protein nitrogen in the nutrition of ruminants, *FAO Agric. Stud.*, **75**, 94 pp.

Loué, A. (1951 and 1953) Etude de la nutrition du cafeier par la methode du diagnostic foliare. Bingeville, Cote d'Ivoire, *Bull. Trim.*, No. 3, 10-35, (1951); *Bull. Trim.*, No. 8, (1953), 97-104 and 113-56.

McCulloch, J. S. G., Pereira, H. C., Kerfoot, O. and **Goodchild, N. A.** (1966) Shade tree effects in tea gardens, *World Crops*, **18** (3), 26-7.

Macmillan, H. F. (1956) *Tropical Planting and Gardening*. Macmillan, London. 560 pp.

Mainstone, B. J. (1960) Effects of ground cover type and continuity of nitrogenous fertilizer treatment upon the growth to tappable maturity of *Hevea brasiliensis, Proc. Natur. Rubber Res. Conf., Kuala Lumpur*, 1960.

Mainstone, B. J. (1963) Manuring of *Hevea*. Some long-term manuring effects with special reference to phosphorus, in one of the Dunlop (Malaya) experiments, *Emp. J. Exp. Agric.*, **31** (122), 175-85.

Mainstone, B. J. (1976) Cocoa on inland soils in peninsular Malaysia, *Planter (Malaysia)*, **52** (598), 16-24 January.

Manning, H. L. (1949) Planting date and cotton production in the Buganda Province of Uganda, *Emp. J. Exp. Agric.*, **17**, 245-58.

Manning, H. L. (1956) The statistical assessment of rainfall probability and its application in Uganda agriculture, *Proc. Roy. Soc. B*, **144**, 460-80.

Manning, H. L. and **Kibukamusoke, D. E. B.** (1960) The cotton crop, *Progress Reports from Experimental Stations*, 1958-59, Uganda. Empire Cotton Growing Corporation, London, 6-8.

Marlatt, C. L. (1900) Biological control of cottony cushion scale. *Year book US Dept. Agric.*, 247.

Marples, H. J. S. and **Trail, J. C. M.** (1967) An analysis of a commercial herd of dairy cattle in Uganda, *Trop. Agric., Trin.*, **44**, 1, 69-75.

Marrs, G. J. and **Middleton, M. R.** (1973) The formulation of pesticides for convenience and safety, *Outlook on Agriculture*, **7**, 231.

Marsh, R. W. (1977) (Ed.) *Systemic Fungicides*. 2nd edn. Longman, London, 321 pp.

Marshall, C. E. (1949) *The Colloid Chemistry of the Silicate Minerals. Agronomy*. Monograph Series, Vol. 1. New York Academic Press, 195 pp.

Martin, H. (1973) *The Scientific Principles of Crop Protection*. 6th edn. Edward Arnold, London, 1973.

Mason, T. G. (1938) Note on the technique of cotton breeding, *Emp. Cott. Gr. Rev.*, **15**, 113-17.

Mason, T. G. and **Lewin, C. J.** (1925) Growth and correlation in the oil palm (*Elaeis-guineensis*), *Ann. Appl. Biol.*, **12**, 410-21.

Matthews, G. A. (1972a) Ultra-low volume spray application on cotton in Malawi, *14th Int. Congr. of Entomology*, 1972, Canberra. Sec. 12. Reprinted *PANS*, **19**, 48.

Mathews, G. A. (1972b) Effect of nitrogen, sulphur, phosphorus and boron on cotton in Malawi, *Exp. Agric.*, **8**, 219.

Matthews, G. A. (1979) *Pesticide Application Methods.* Longman, London. 334 pp.

Matsushima, S. (1966) Some experiments and investigation on rice plants in relation to water in Malaysia, *Symposium Service Centre for South East Asian Studies*, Kyoto University, **3**, 115-23.

May, P. J. (1971) Grain sorghum in the Ord valley - three crops a year? *J. Agric. W. Australia*, **12**, 113.

Mayne, W. W. (1932) Physiological specialization of *Hemileia vastatrix, Nature Lond.*, **129**, 510.

Medcalf, J. C. (1956) Preliminary study on mulching young coffee in Brazil, *IBEC Res. Inst. Bull.*, No. 12.

Mehlich, A. (1966) Production of maize for grain and mulching materials, *Kenya Coffee*, **31** (303), 105-9.

Menendez, T. and **Shepherd, K.** (1975) Breeding new bananas, *World Crops*, **27**, 104-12.

Mercado, B. L. and **Talatala, R. L.** (1977) Competitive ability of *Echinochloa colonum* L. against direct-seeded lowland rice. *Proc. Asian Pacific weed Sci. Soc. Conf.* (Indonesia), 6. 161-165.

Meredith, C. H. (1944) Antagonism of soil organisms to *Fusarium oxysporum cubense, Phytopathology*, **34**, 426-9.

Mertz, E. J., Bates, L. S. and **Nelson, O. E.** (1964) Mutant gene that changes protein composition and increases lysine content of maize endosperm, *Science*, **145**, 279.

Metcalfe, J. R. (1959) A preliminary reassessment of *Diatraea saccharalis* (F) in Barbados, West Indies, *Trop. Agric., Trin.*, **36**, 199-209.

Meyer, D. R. and **Anderson, A. J.** (1959) *Nature Lond.*, **183**, 61.

Mills, W. R. (1953) Nitrate accumulation in Uganda soils, *E. Afr. Agric. J.*, **19**, 53-4.

Milne, G. (1935) Composite units for the mapping of complex soil associations, *Trans. 3rd Int. Congr. Soil Sci.*, **1**, 345-7.

MJG (1958) Clonal selection schemes, *Indian Tea Association Scientific Department Tea Encyclopedia*, Tocklai, Serial No. 129, B.4, November 1958.

Moberly, P. K. (1966) Maize meal as an energy supplement for the Jersey cow, *E. Afr. Agric. J.*, **32**, (2), 155-8.

Mohr, E. C. J., Van Baren, F. A. and **Von Schuyleborgh**, J. (1972) *Tropical Soils*, 3rd edn. Mouton-Ichtiar Baru-Van Hoeve. The Hague.

Moomaw, J. C. and **Vergara, B. S.** (1964) The environment of tropical rice production. The mineral nutrition of the rice plant, 3-14, *Symposium at the Int. Rice Res. Inst.*, Feb. 1964.

Mulder, D. and **De Silva, R. L.** (1960) A forecasting system for blister blight control based on sunshine records, *Tea Quart.*, **31**, 56.

Mullick, D. M. and **Pal, A. K.** (1944) Studies on the composition of the blood of farm animals in India. 1. A study of some haematological and chemical constituents in the blood of normal cattle, *Indian J. Vet. Sci.*, **13**, 146-9.

Murray, D. B. (1958) Response of cacao to fertilizers, *Nature Lond.*, **182**, 1613.

Murray, D. B. (1966) Cocoa - prospects for the future, *J. Agric. Soc. Trin. Tob.*, **66** (2) 163-70.

Murray, M., Morrison, W. I., Murray, P. K., Clifford, D. J. and **Trail, J. C. M.** (1979) Trypanotolerance - a review, *World Animal Review*, **31**, 2.

Nagarajah, S. and **Pethiyagoda, V.** (1965) The influence of 'lungs' on carbohydrate reserves and growth of shoots, *Tea Quart. Ceylon*, **36** (3), 88-102.

Narayana, G. V. and **John, C. M.** (1949) Varieties and forms of the coconut, *Madras Agric. J.*, **36**, 349-66.

Neal, M. E. B. (1966) Upland rice production, Mechanization and the World's Rice Conference, Massey-Ferguson with FAO, Stoneleigh, England.

Newsam, A. (1963) Covers and root diseases, *RRI Plant. Bull.*, **68**, 177-81.

Newsam, A. (1964) Conference review. Root diseases of *Hevea, RRI Plant. Bull.*, **75**, 207-9.

Normand, C. W. B. (1938) The weather of India - Indian Science Congress. Silver Jubilee Session, 1938. Ind. Sci. Cong. Ass. Calcutta.

Norris, D. O. and **Henzell, E. F.** (1960) Nitrogen in tropical agriculture - Symposium Brisbane. *Nature Lond.*, **186**, 941-2.

Norris, D. O. (1966) *The Legumes and their Associated* Rhizobium *in Tropical Pastures*. W. Davies and C. L. Skidmore, 89-105. Faber & Faber. 215 pp.

Nutman, F. J. (1933) The root system of *Coffea arabica, Emp. J. Exp. Agric.*, **1**, 271-96.

Nutman, F. J. (1959) Evidence that the spores of coffee leaf-rust are not dispersed by wind, *Kenya Coffee*, **24**, 451-3.

Nye, P. H. (1953) A survey of the value of fertilizers to the food farming areas of the Gold Coast. Part II, *Emp. J. Exp. Agric.*, **21**, 262-74.

Nye, P. H. (1954) A survey of the value of fertilizers to the food farming areas of the Gold Coast. Part III, *Emp. J. Exp. Agric.*, **22**, 42-54.

Nye, P. H. (1965) Discussion, *Symposium on Soil Resources of Tropical Africa, London*, Sept. 1965. African Studies Association of the United Kingdom.

Nye, P. H. and **Greenland, D. J.** (1960) The soil under shifting cultivation, *Comm. Bur. Soils Tech. Comm*. No. 51.

Obeng, H. B. and **Smith, J. K.** (1963) Land capability classification of the soils of Ghana, *Ghana J. Sci.*, **3** (1), 52-65.

Obi, J. K. (1967) The influence of preceding crops on subsequent crops following bush fallow in Umudike, Eastern Nigeria, *Trop. Abstracts*, **22** (q), 1636.

Ogborn, J. E. A. (1970) Methods of controlling *Striga hermontheca* for West African farmers, *Samaru Agric. Newsletter*, **12** (6), 90.

Ollagnier, M. and **Gascon, J. P.** (1965) La sélection du palmier à huile a IRHO, *The Oil Palm*. Tropical Products Institute Conference, London, May 1965.

Ollagnier, M. the **Ochs, R.** (1972) Les déficiences en soufre du palmier à huile et du cocotier, *Oleágineux*, **27** (4), 193.

Olsen, F. J. (1972) Effect of large applications of nitrogen fertilizer on the productivity protein content of four tropical grasses in Uganda, *Trop. Agric., Trin.*, **49**, 251-60.

Olver, A. (1938) A brief survey of some of the important breeds of cattle in India, *Misc. Bull*. **17**, Imp. Counc. Agric. Res. India.

Onwueme, I. C. (1978) *The Tropical Tuber Crops*. Wiley, Chichester. 234 pp.

Ooi Cheng Bin (1977) Propagation and planting materials in Hevea, RRIM refresher course on rubber planting and nursery techniques, 8-13 Aug. 1977. RRIM, Kuala Lumpur, Malaysia.

Ou, S. H. (1975) *A Handbook of Rice Diseases in the Tropics*. IRRI, Philippines. 76 pp.

Owen, H. (1951) Cocoa pod diseases in West Africa, *Ann. Appl. Biol.*, **38**, 715-18.

Padmanabhan, S. Y. (1967) Blast disease of rice, *PANS, Sec. B*, **13** (1), 62-9.

Padmanabhan, S. Y. and **Jain, S. S.** (1966) Effect of chlorination of water on control of bacterial leaf blight of rice caused by *Xanthomonas oryzae*. (Uyeda and Ishiyama) Dowson, *Current Sci.*, **35**, 24.

Padwick, G. W. (1956) *Losses caused by Plant Diseases in the Colonies*. Commonwealth Mycological Inst. Kew, 60 pp.

Pant, C. P. and **Gratz, N. G.** (1979) Malaria and agricultural development, *Outlook on Agric.*, **10** (3), 111-15.

Parish, D. H. and **Feillafe, S. M.** (1960) A comparison of urea with ammonium sulphate as a nitrogen source for sugar cane, *Trop. Agric., Trin.*, **37**, 223-5.

Passmore, R. G. (1967) *Progress Reports from Experiment Stations, Season 1965-66*. Uganda. p. 13. Cotton Research Corporation, London.

Patil, N. P. (1963) Economics of drill sowing as against broadcasting in Ragi, *Agric. Situation India*, **18** (6), 407-9.

Payne, W. J. A. (1952) Breeding studies, *Agric. J. Fiji*, **23**, 9-13.

Payne, W. J. A. and **Rollinson, D. H. L.** (1973) Bali cattle, *World Animal Review*, **7** 13.

Peachey, J. E. (1969) (Ed.) Nematodes of tropical crops, *Comm. Agric. Bur. Tech. Comm.* No. 40, 355 pp.

Pearsall, W. H. (1950) The investigation of wet soils and its agricultural implication, *Emp. J. Exp. Agric.*, **18**, 289-98.

Peat, J. E. and **Brown, K. J.** (1962) The yield response of rain grown cotton at Ukiriguru in the Lake Province of Tanganyika, *Emp. J. Exp. Agric.*, **30**, Part I,. 215-31; Part II, 305-14.

Penman, H. L. (1948) Natural evaporation from open water, bare soil and grass. *Proc. Roy. Soc. London A*, **193**, 120-45.

Penman, H. L. (1950) *Quart. J. R. Met. Soc.*, **76**, 372.

Penman, H. L. (1963) Vegetation and hydrology, *Comm. Bur. Soils*, Tech. Comm. No. 53.

Percy, H. C. (1975) In *ULV Spraying for Cotton Pest Control.* N. Morton (Ed.). Cotton Research Corporation, London, p. 14.

Peregrine, D. J. (1973) Toxic baits for the control of pest animals, *PANS*, **19**, 523.

Pereira, H. C., Chenery, E. M. and **Mills, W. R.** (1954) The transient effect of grasses on the structure of tropical soils, *Emp. J. Exp. Agric.*, **22**, 148-60.

Pereira, H. C. and **Jones, P. A.** (1954a) Field responses by Kenya coffee to fertilizers, manures and mulches, *Emp. J. Exp. Agric.*, **22**, 23-36.

Pereira, H. C. and **Jones, P. A.** (1954b) Tillage study in Kenya coffee, *Emp. J. Exp. Agric.*, **22**, 231-40; 323-31.

Pereira, H. C., Wood, R. A., Brzostowski, H. W. and **Hosegood, P. H.** (1958) Water conservation by fallowing in semi-arid tropical East Africa, *Emp. J. Exp. Agric.*, **26**, 213-28.

Phillips, L. L. (1976) Cotton, **Chap. 56**, pp. 196-200, in *Evolution of Crop Plants*. N.W. Simmonds (Ed.). Longman, London and New York.

Pieterse, A. H. (1978) The water hyacinth (*Eichhornia crassipes*) - a review, *Abstracts Trop. Agric.*, **4** (2), 9-42. February 1978. Amsterdam.

Pieterse, A. H. (1979) The broom rapes (Orobanchaceae) - a review, *Abstracts Trop. Agric.*, **5** (3), March 1979. 9-35. Amsterdam.

Pillsbury, A. F. and **Degan, A.** (1968) Sprinkler irrigation, *FAO Agricultural Development Paper*, No. 88, 179 pp.

Portsmouth, G. B. (1949) Report of the plant physiologist for the year 1949, *TRI Ceylon. Bull.*, **31**, AR 32-37.

Pound, F. J. (1934) The progress of selection, *Fourth Annual Report of Cocoa Research*, 1934. ICTA, Trinidad.

Pound, F. J. (1938) *Cacao and Witch Broom Disease of South America*. Report, Dept. of Agric., Trin. and Tob., 58.

Prendergast, A. G. (1957) Observations on the epidemiology of vascular wilt disease of the oil palm (*Elaeis guineensis* Jacq.), *J. W. Afr. Inst. for Oil Palm Res.*, **2**, 148-75.

Prentice, A. N. (1957) Some notes on the couch grass, *Digitaria scalarum*. First East Africa Herbicide Conference, *E. Afr. Agric. J.*, **23**, 11.

Preston, T. R. (1972) Fattening beef cattle on molasses in the tropics, *World Animal Review*, **1**, 24.

Prevett, F. F. (1973) (Ed.) Tropical stored products information. No. 25. Special Issue. Ibadan Grain Storage Seminar.

Purseglove, J. W. (1968) *Tropical Crops: Dicotyledons*. Longman, London and New York. 719 pp.

Purseglove, J. W. (1972) *Tropical Crops: Monocotyledons*. Longman, London and New York. 607 pp.

Py, C. and **Tisseau, M.-A.** (1965) *L'Ananas*. G.-P. Maisonneuve et Larose, Paris, 298 pp.

Quarterman, J., Phillips, G. D. and **Lampkin, G. H.** (1957) A difference in the physiology of the large intestine between European and indigenous cattle in the tropics, *Nature Lond.*, **180**, 552-3.

Quinn, J. P. (1978) India seeks top spot in tea. *Tea and Coffee Trade Journal (USA)*, **150** (8), 28.

Rai, B. K. (1973) The red rice problem in Guyana, *PANS*, **19** (4), 557.

Ramdas, L. A. and **Dravid, R. K.** (1936) Soil temperatures in relation to other factors controlling the disposal of solar radiation at the earth's surface. *Proc. Nat. Inst. Sci. India*, **2**, 131-43.

Ramly, A. W. H. (1966) A simple groundnut stripper, *World Crops*, **18** (1), 34-5.

Rayner, R. W. (1958) *Ann. Rep. Coffee Res. Station, Ruiru, Kenya, 1957-58*, 97.

Reese, W. E. (1966) Guide to sixty soil and water conservation practices, *Soils Bull. FAO*, **4**.

Rhoad, A. O. (1940) A method of assaying genetic differences in the adaptability of cattle to tropical and subtropical climates, *Emp. J. Exp. Agric.*, **8**, 190-8.

Rhoad, A. O. (1944) The Iberia heat tolerance test for cattle, *Trop. Agric. Trin.*, **21**, 162-4.

Rhoad, A. O. (1955a) *Breeding Beef Cattle for Unfavourable Environments*. A symposium presented at the King Ranch Centennial Conference. University of Texas, Austin Press.

Rhoad, A. O. (1955b) *Procedures used in Developing the Santa Gertrudis Breed*, p. 203 in Rhoad (1955a).

Rhodes, E. R. and **Nangju, D.** (1979) Effects of pelleting cowpea and soyabean seed with fertilizer dusts, *Exp. Agric.*, **15**, 27-32.

Rice, V. A., Andrews, F. W., Warrick, E. J. and **Legates, J. E.** (1957) *Breeding and Improvement of Farm Animals*, McGraw-Hill, New York.

Richards, A. V. (1964) Progress in planting of clonal tea in Ceylon, *Tea Quart. Ceylon*, **35** (3), 176-7.

Richards, A. V. (1966) The breeding, selection and propagation of tea, *Tea Quart. Ceylon*, **37** (3) 154-60.

Richards, O. W. and **Davies, R. G.** (1977) *Imms' General Textbook of Entomology*. 10th edn. Vols. I & II, 1354 pp. Chapman & Hall, London.

Richardson, E. F. (1965) Cover crop recommendations, *Kenya Sisal Board Bull.*, **53**, pp. 13, 15, 17.

Richardson, H. L. (1965) The use of fertilizers, *Symposium on Soil Resources of Tropical Africa*, London, September 1965.

Richardson, H. L. (1966) The Freedom from Hunger Campaign - five years of the FAO fertilizer programme, *Outlook on Agric.*, **5** (1), 3-16.

Robinson, J. B. D. (1951) A brief review of sugar cane manuring in Barbados, *Proc. B.W.I. Sug. Tech.*, 1951, 73-7.

Robinson, J. B. D. and **Chenery, E. M.** (1958) Magnesium deficiency in coffee with special reference to mulching. *Emp. J. Exp. Agric.*, **26**, 259-73.

Robinson, J. B. D. and **Hosegood, P. H.** (1965) Effects of organic mulch on fertility of a latosolic coffee soil in Kenya, *Exp. Agric.*, **1** (1), 67-80.

Roelofsen, P. A. (1941) *Natuurwet. Tijdschr. Ned.-Ind.*, **101**, 179.

Rollinson, D. H. L., Harker, K. W. and **Taylor, J. I.** (1955) Studies on the habits of Zebu cattle. III. Water consumption of Zebu cattle, *J. Agric. Sci.*, **46**, 123-9.

Rombouts, J. E. (1953) Micro-organisms in the rhizosphere of banana plants in relation to susceptibility of resistance to Panama disease, *Plan and Soil*, **4**, 276-88.

Rose, C. W. (1966) *Agricultural Physics*. Pergamon Press, London. 226 pp.

Rose Innes, R. (1977) *A Manual of Ghana Grasses*. Ministry of Overseas Development, Land Resources Division, Surbiton, England. 263 pp.

Rosher, P. H. (1957) Means of increasing rice production in Trinidad, *Agric. Soc. Trin. Tob. J.*, **57**.

Ruskin, F. R. (1977) (Ed.) *Leucaena. Promising forage and tree crop for the tropics*. National Academy of Sciences, Washington, D.C.

Russell, E. W. (1973) *Soil Conditions and Plant Growth*. 10th edn. Longman, London. 849 pp.

Russell, T. A. (1953) The spacing of Nigerian cocoa, *Emp. J. Exp. Agric.*, **21**, 145-53.

Ruston, D. F. (1962) Effects of delay in sowing cotton, *Cott. Grow. Rev.*, **39**, 10.

Ruthenberg, Hans (1976) *Farming Systems in the Tropics*. 2nd edn. Clarendon Press. 366 pp.

Saint, S. J. (1930) *Manurial Trials, Sugar Cane*. Rep. Dept. Agric., Barbados, 1929-30, 82-93.

Sanchez, Pedro A. (1976) *Properties and Management of Soils in the Tropics*. Wiley-Interscience. 618 pp.

Sanders, R. N. (1966) Animal selection for the tropical environment, p. 115-28, in *Tropical Pastures*. W. Davies and C. C. Skidmore, Faber & Faber. 215 pp.

Sanyasi Raju, M. (1952) The role of organic manures and inorganic fertilizers in soil fertility, *Madras Agric. J.*, **39**, 130-47.

Schlippe, P. de (1956) *Shifting Cultivation in Africa. The Zande System of Agriculture*. Routledge & Kegan Paul, London.

Schofield, R. K. (1935) *Trans. 3rd Int. Congr. Soil Sci. Oxford*, **2**, 37.

Schwabe, W. W. (1973) The long slow road to better coconut palms, *Spectrum*, 9.

Scott Russell, R. (1977) *Plant Root Systems. Their function and interaction with the soil*. McGraw-Hill, London. 298 pp.

Sen, A. N. (1958) Nitrogen economy of soil under rahar (*Cajanus cajan*), *J. Indian Soc. Soil Sci.*, **6**, 171-6.

Shalo, P. L. and **Hansen, K. K.** (1973) Maziwa Lala - fermented milk, *World Animal Review*, **5**, 33.

Shankaracharya, N. B. and **Mehta, B. V.** (1971) Note on the losses of nitrogen by volatilization of ammonia from loamy-sand soil of Anand treated with different nitrogen carriers under field conditions, *Indian J. Agric. Sci.*, **41**, 131.

Sharma, P. C. (1968) Establishment and maintenance of shade on North Bank, Assam, *Two and a Bud*, **15** (3), 96-7.

Shephard, C. Y. (1937) *The Cacao Industry of Trinidad*. Imperial College of Tropical Agriculture Trinidad.

Simmonds, N. W. (1956) A banana collecting expedition to South-East Asia and the Pacific, *Trop. Agric., Trin.*, **33**, 251-71.

Simmonds, N. W. (1966) *Bananas*. 2nd edn. Longman, London. 512 pp.

Simmonds, N. W. (1969) Genetical bases of plant breeding, *J. Rubb. Res. Inst. Malaya*, **21**, 1.

Simmonds, N. W. (1976) (Ed.) *Evolution of Crop Plants*. Longman, London. 339 pp.

Simmonds, N. W. (1979) *Principles of Crop Improvement*. Longman, London. 408 pp.

Siregar, H. (1954) The influence of different dates of sowing upon the yield and other agronomic characters of a photosensitive variety, *FAO Working Party on Rice Breeding, Tokyo, Japan*. Paper No. IRC/BP/1954/19.

Skerman, P. J. (1977) *Tropical Forage Legumes*. FAO, Rome. 609 pp.

Smith, A. N. (1959) Chemical Department Report. Tea Research Institute of East Africa, *Annual Report 1958*, 34.

Smith, A. N. (1963) The chemical and physical characteristics of tea soils, *E. Afr. Agric. Forestry J.*, **28** *(3), 123-5.*

Smith, K. M. (1957) *A Textbook of Plant Virus Diseases*. 2nd edn. Churchill, London, 652 pp.

Sparnaaij, L. D. (1965) Variations de la production annuelle du palmier à huile. *Oleagineux*, **20** (11), 655-9.

Sprague, G. F. (1960) Inbreeding compared with recurrent selection in corn improvement. *Proc. of 10th Congress of Int. Soc. of Sugar Cane Technologists*, Hawaii, 1959, Elsevier, 653-61 pp.

Sprague, G. F. and **Miller, P.** (1952) The influence of visual selection during inbreeding on combining ability in corn, *Agron. J.*, **44**, 258-62.

Srinivasan, C. S. and **Vishveshwara, S.** (1979) Cultivation of San Ramon (dwarf) hybrid coffee, *Indian Coffee*, **43** (3), 43.

Steiner, L. F, Harris, D. J., Mitchell, W. C., Fujimoto, M. S. and **Christenson, L. D.** (1965a) Melon fly eradication by overflooding with sterile flies, *J. Econ. Entomol.*, **58** (3), 519-22.

Steiner, L. F., Mitchell, W. C., Harris, D. J., Kozuma, P. T., and **Fujimoto, M. S.** (1965b) Oriental fruit fly eradication by male annihilation, *J. Econ. Entomol.*, **58** (5), 961-4.

Stephen, R. C. (1957) The influence of planting date on tobacco growth and on the influence of *Cercospora* leaf spot disease in Southern Rhodesia, *Emp. J. Exp. Agric.*, **25**, 291-300.

Stephens, D. (1966) Two experiments on the effects of heavy application of triple superphosphate on maize and cotton in Buganda clay loam soil, *E. Afr. Agric. Forestry J.*, **31** (3), 283-90.

Stessels, L. and **Fridmann, M.** (1972) Utilisation de l'énergie solaire pour le conservation du café en région tropicale humide, *Café Cacao Thé*, **16**, 135.

Stevenson, G. C. (1960) Inbreeding with sugar cane in Barbados, *Proc. of 10th Cong. Int. Soc. Sugar Cane Technologist*, Hawaii, 1959, Elsevier, 670-82.

Stevenson, G. C. (1965) *Genetics and Breeding of Sugar Cane*. Longmans Green, London, 284 pp.

Stobbs, T. H. (1966) The improvement of small East African Zebu cattle, *Exp. Agric.*, **2** (4), 287-93.

Stockinger, K. R. (1971) Nutrient availability under irrigation, *Samaru Agricultural Newsletter*, August, **13**, 75.

Storey, H. H. and **Leach, R.** (1933) A sulphur deficiency disease of the tea bush, *Ann. App. Biol.*, **2**, 23-56.

Stout, B. A. (1966) *Equipment for Rice Production*. FAO Agricultural Development Paper No. 84, 169 pp.

Swamy Rao, A. A. (1964) A report on the preliminary investigations, design, and development, testing and economic analysis of the new single and double bullock harness at the Development Centre from May 1962 to April 1964. Allahabad, Agricultural Institute, 92 pp.

Teoh, C. H., Adham Abdullah and Reid, W. M. (1979) Critical aspects of legume establishment and maintenance, *Proc. Rubb. Res. Inst. Malaysia Planters' Conf.* 252-271.

Theron, J. J. (1951) The influence of plants on the mineralization of nitrogen and the maintenance of organic matter in the soil, *J. Agric. Sci.*, **41**, 289-96.

Theron, J. J. and **Haylett, D. G.** (1953) The regeneration of soil humus under a grass ley, *Emp. J. Exp. Agric.*, **21**, 86-98.

Thomas, A. S. (1940) *Robusta Coffee in Agriculture in Uganda*. J. D. Tothill (Ed.). Oxford, London, pp. 289-311.

Thomas, A. S. (1947) The cultivation and selection of Robusta coffee in Uganda, *Emp. J. Exp. Agric.*, **15**, 65-81.

Thompson, C. G. (1958) A polyhedrosis virus for control of the Great Basin tent caterpillar. *Malacosoma fragile, Trans. I. Int. Conf. Insect. Pathology and Biol. Control*, Prague, 1958, 201-4.

Tinsley, T. W. (1977) Viruses and the biological control of insect pests, *Bio. Science*, **27** (10), 659.

Togun, S. and **Ajibike, B.** (1964) Preliminary evaluation of green budding of rubber in Nigeria, *Proc. Agric. Soc. Nigeria*, **3**, 17-20.

Toovey, F. W. and **Broekmans, A. F. M.** (1955) The Deli palm in West Africa, *J.W. African Inst. for Oil Palm Res.*, **1** (3), 9-50.

Topper, B. F. (1957) New method of vegetative propagation for cocoa, *World Crops*, **9** (1), 38-9.

Topps, J. H. and **Oliver, J.** (1978) *Animal Foods of Central Africa - Technical Handbook No. 2.* Rhodesia Agricultural Journal, Zimbabwe.

Tothill, J. D. (1940) (Ed.) *Agriculture in Uganda.* Oxford Univ. Press.

Tothill, J. D. (1948) (Ed.) *Agriculture in the Sudan.* Oxford Univ. Press.

Tothill, J. C. and **Peart, J. R.** (1972) The case for more efficient use of native pastures in an extensive beef enterprise. I. By agronomic methods. II. By supplements, *Trop. Grasslands*, **6**, 240.

Trapnell, C. G. (1958) Climatic types of vegetation in Southern Kenya, *Annual Report 1958. Record of Research*, 42-47. East Afr. High Commission.

Trinick, M. J. (1973) Symbiosis between *Rhizobium* and the non-legume, *Trema aspera, Nature* (quoted *Rural Research, CSIRO, Australia*, **81**, 16).

Tunstall, J. P. (1958) Biology of the Sudan bollworm *Diparopsis watersi* in the Gash Delta, Sudan, *Bull. Ent. Res.*, **49**, 1-23.

Tunstall, J. P., Mathews, G. A. and **McKinley, D. J.** (1971) The introduction of cotton insect control in Malawi, *Proc. 6th Br. Insectic. and Fungic. Conf.*, Brighton, 785.

Turner, P. D. (1965) The incidence of *Ganoderma* disease of oil palms of Malaya and its relation to previous crop, *Ann. App. Biol.*, **55**, 417-23.

Underwood, E. J. (1966) *The Mineral Nutrition of Livestock*. Commonwealth Agricultural Bureaux with FAO.

Vageler, P. (1933) *An Introduction to Tropical Soils* (Transl. H. Greene). Macmillan, London.

Van Der Burg, B. (1969) Cacao budding - a neglected technique, *World Crops*, **21**, 105.

Van der Plank, J. E. (1963) *Plant Diseases: Epidemics and Control*. Academic Press, London and New York.

Van der Plank, J. E. (1968) *Disease Resistance in Plants*. Academic Press.

van Dierendonck, F. J. E. (1959) *The Manuring of Coffee, Cocoa, Tea and Tobacco*. Centre d'Etude de l'Azote, Geneva. 205 pp.

Verliere, G. (1966) Valeur fertilisante de deux plantes utilisèes dans les essais de paillage du caféiere: *Tithonia diversifolia* et *Flemingia congesta, Cafe, Cacao, The*, **10** (3), 228-36.

Vernon, A. J. (1967) New developments in cocoa shade studies in Ghana, *J. Sci. Fd. Agric.*, **18**, 44-8.

Vicente-Chandler, J., Caro-Costas, R., Pearson, A. W., Abruna, F., Figarella, J. and **Silva, S.** (1964) *The Intensive Management of Tropical Forages in Puerto Rico.* University of Puerto Rico Agric. Exp. Station, Rio Pedras. Bulletin 187, Dec. 1964.

Villares, J. B. (1940) *Rev. Indust. Anima.*, **3**, 7-33.

Vine, H. (1953) Experiments on the maintenance of soil fertility at Ibadan, Nigeria, *Emp. J. Exp. Agric.*, **21**, 65-85.

Visser, T. (1958) The validity of assessing tea yields on a basis of intermittent plucking and test plucking, *Tea Quart.*, **29**, Pt. 1, 21-9.

Visser, T., Shanmunganathan, N. and **Sabunayagam, J. V.** (1961) The influence of sunshine and rain on tea blister blight, *Exobasidium vexans* Massee, in Ceylon, *Ann. App. Biol.*, **49**, 306-15.

Wahid, A. (1973) Pakistani buffaloes, *World Animal Review*, **7**, 22.

Walker, D. I. T. and **Simmonds, N. W.** (1981) Varieties 2: Breeding, selection and trials, in *Sugar Cane*. Blackburn, Longman.

Warner, John N. and **Grassl, C. O.** (1958) The 1957 sugar cane expedition to Melanesia, *Hawaiian Planters Record*, **55**, 3.

Watson, G. A. (1960) Cover plants and soil nutrient cycle in *Hevea* cultivation, *Proceedings of the Natural Rubber Research Conference, Kuala Lumpur*, 1960.

Watson, G. A., Wong, P. W. and **Narayan, R.** (1963) Effect of cover plants on growth of *Hevea*. IV. Leguminous cover crops compared with grasses, *Mikania scandens* and mixed indigenous covers, *J. Rubb. Res. Inst. Malaya*. 18. 123-4.

Webb, L. J. (1956) Note on the studies on rain forest vegetation in Australia. Study of tropical vegetation. *Proc. Kandy Symposium*, UNESCO, 1956, 171-3.

Webster, C. C. and **Wilson, P. N.** (1980) *Agriculture in the Tropics*. Longman, London. 640 pp.

Wellman, F. L. (1961) *Coffee, Botany, Cultivation and Utilization*. Leonard Hill, London, 448 pp (pp. 129-51).

Whitby, S. (1919) Variation in *Hevea brasiliensis, Ann. Bot.*, **33**, 313-21.

Whitehead, R. A. (1966) Some notes on dwarf coconut palms in Jamaica, *Trop. Agric., Trin.*, **43** (4) 277-93.

Whitehead, R. A. and **Thompson, B. E.** (1966) Introduction and exchange of coconut planting material. *Nature Lond.*, 209, (5023), 634-5.

Whittlesey, D. (1936) Major agricultural regions of the earth, *Ann. Assoc. Amer. Geogr.*, **26**, 199-242.

Wight, W. (1958) The shade tradition in tea gardens of North India, *1958 Report Ind. Tea Assn. Sci. Dept. (Tocklai)*, 75-122.

Wight, W. and **Barua, P. K.** (1954) Morphological basis of quality in tea, *Nature Lond.*, **173**, 630-1.

Wijewardene, R. (1978) Appropriate technology in tropical farming systems, *World Crops*, **30** (3), 128.

Wild, A. (1972) Mineralization of soil nitrogen in Nigeria, *Exp. Agric.*, **8**, 91.

Wilkinson, G. E. (1970) The infiltration of water into Samaru soils, *Samaru Agric. Newsletter*, **12** (5), 81.

Williams, C. N. and **Joseph, K. T.** (1970) *Climate, Soil and Crop Production in the Humid Tropics*. Oxford University Press, Singapore. 177 pp.

Williamson, G. and **Payne, W. J. A.** (1978) *Animal Husbandry in the Tropics*. 3rd edn. Longman, London. 755 pp.

Willatt, S. T. (1971) A comparative study of the development of young tea under irrigation. II. Continued growth in the field, *Trop. Agric. Trin.*, **48** (3), 271-7.

Willimott, S. G. and **Anthony, K. R. M.** (1958) The response of *Eleusine* to different forms of nitrogen fertilizer, *Emp. J. Exp. Agric.*, **26**, 373-8.

Wilson, P. N. and **Osbourn, D. F.** (1963) Experimental work on Pangola grass (*Digitaria decumbens Stent.*) at the Imperial College of Tropical Agriculture, Trinidad. Bulletin No. 82. Agricultural Experiment Station, Paramambo, Surinam.

Winchester, J. A. (1966) Biological control of plant nematodes. The effect of higher plants, *Monogr. Centro Cooper, Cient. Unesco Amer. Latina*, **1**, 195-9.

Withers, B. and **Vipond, S.** (1974) *Irrigation Design and Practice*. Batsford, London. 306 pp.

Wood, B. J. (1968) Studies on the effect of ground vegetation on infestations of *Oryctes rhinoceros* (L.) (Col., Dynastidae) in young oil palm replantings in Malaysia, *Bull. Ent. Res.*, **59**, 85.

Wood, G. A. R. (1975) *Cocoa*. 3rd Edn. Longman. 292 pp.

Worthing, C. R. (1979) *The Pesticide Manual - A World Compendium*. 6th edn. British Crop Protection Council. 655 pp.

Wright, N. C. (1954) 'The Ecology of Domesticated Animals.' Chap. 5 in Vol. I of Hammond, J. (Ed.) Progress in the Physiology of Farm Animals. Bulterworths: London.

Wrigley, G. (1966) Modern herbicides in rice cultivation, Mechanization and the World's Rice Conference, Massey - Ferguson with FAO, Stoneleigh, England.

Wrigley, G. (1973) Mineral oils as carriers for ultra-low-volume (ULV) spraying, *PANS*, **19**, 54.

Wycherley, P. R. (1969) Breeding of *Hevea, J. Rubb. Res. Inst. Malaya*, **21**, 38.

Wycherley, P. R. (1976) Rubber, in *Evolution of Crop Plants*. N. W. Simmonds (Ed.), pp. 77-80.

Wyniger, R. (1962) *Pests of Crops in Warm Climates and their Control*. Verlag fur Recht und Gesellschaft Ag. Basel. Acta Tropica. 555 pp.

Yeates, N. T. M. (1965) *Modern Aspects of Animal Production*. Butterworth, London.

Young, A. (1976) *Tropical Soils and Soils Survey*. Cambridge University Press, Cambridge. 476 pp.

Index